Lecture Notes in Computer Science 9471

Commenced Publication in 1973
Founding and Former Series Editors:
Gerhard Goos, Juris Hartmanis, and Jan van Leeuwen

More information about this series at http://www.springer.com/series/7409

Tie-Yan Liu · Christie Napa Scollon
Wenwu Zhu (Eds.)

Social Informatics

7th International Conference, SocInfo 2015
Beijing, China, December 9–12, 2015
Proceedings

 Springer

Editors
Tie-Yan Liu
Microsoft Research Asia
Beijing
China

Wenwu Zhu
Tsinghua University
Beijing
China

Christie Napa Scollon
Singapore Management University
Singapore
Singapore

ISSN 0302-9743   ISSN 1611-3349 (electronic)
Lecture Notes in Computer Science
ISBN 978-3-319-27432-4   ISBN 978-3-319-27433-1   (eBook)
DOI 10.1007/978-3-319-27433-1

Library of Congress Control Number: 2015955907

LNCS Sublibrary: SL3 – Information Systems and Applications, incl. Internet/Web, and HCI

Printed on acid-free paper

This Springer imprint is published by SpringerNature
The registered company is Springer International Publishing AG Switzerland

Preface

This book constitutes the proceedings of the 7th International Conference on Social Informatics (SocInfo 2015), held in Beijing, China, in December 2015. The 19 full papers presented in this volume were carefully reviewed and selected from 42 submissions. The papers are organized in topical sections such as user modeling, opinion mining, user behaviors, crowd sourcing, etc.

SocInfo is an interdisciplinary venue for researchers from computer science, informatics, social sciences, and management sciences to share ideas and opinions and present original research work on studying the interplay between socially centric platforms and social phenomena. The ultimate goal of social informatics is to create a better understanding of socially centric platforms not just as a technology, but also as a set of social phenomena. To that end, we have invited interdisciplinary papers on applying information technology in the study of social phenomena, on applying social concepts in the design of information systems, on applying methods from the social sciences in the study of social computing and information systems, on applying computational algorithms to facilitate the study of social systems and human social dynamics, and on designing information and communication technologies that consider the social context.

In addition to the presentations of the papers at the conference, we were very pleased to have six keynotes from leading experts in our community:

- Peter A. Gloor, "The Stupidity of the Crowd and the Wisdom of the Swarm"
- Jaideep Srivastava, "Interpersonal Trust Dynamics in Online Systems – Models and Applications"
- Sam Gosling, "The Psychology of Social Behavior in Online and Offline Spaces"
- Feiyue Wang, "Intelligence 5.0: From Social Informatics to Social Intelligence via Social Computing"
- Xing Xie, "Building User Knowledge Graph Based on Human Behavioral Data"
- Jie Tang, "Modeling Social influence in Large Social Networks"

The conference could not have been a success without the contribution of many people. First, we gratefully acknowledge our financial sponsors: Academy of Mathematics and Systems Science, Chinese Academy of Sciences; National Center for Mathematics and Interdisciplinary Sciences, Chinese Academy of Sciences; East China Normal University; and Yahoo Research Barcelona. Second, SocInfo Steering Committee member Ee-Peng Lim provided valuable advice and support during the whole process. The general chairs, Xiaoguang Yang and Aoying Zhou, took care of organizing issues and many other things to make sure the event ran smoothly. The publicity chair, Jinhu Lv, made great efforts in promoting the conference and editing the conference proceedings. Our special thanks go to the local co-chairs, Peng Cui and Xijin Tang, and their teams, whose contributions were indispensable to making the event run. Last but not least, a big thank you to all participants of SocInfo 2015, who made it such a great event!

December 2015

Tie-Yan Liu
Christie Napa Scollon
Wenwu Zhu

.

Organization

Organizing Committee

General Chairs

Xiaoguang Yang	Academy of Mathematics and Systems Science, Chinese Academy of Sciences, China
Aoying Zhou	East China Normal University, China

Program Co-chairs

Tie-Yan Liu	Microsoft Research Asia, China
Christie Napa Scollon	Singapore Management University, Singapore
Wenwu Zhu	Tsinghua University, China

Publicity Chairs

Jinhu Lv	CAS Academy of Mathematics and Systems Science, China
Hao Shen	Communication University of China

Steering Committee Liaison

Ee-Peng Lim	Singapore Management University, Singapore

Workshop Chairs

Jinhu Lv	CAS Academy of Mathematics and Systems Science, China
Weining Qian	East China Normal University, China

Local Organizing Chairs

Peng Cui	Tsinghua University, China
Xijin Tang	CAS Academy of Mathematics and Systems Science, China

Program Committee

Co-chairs

Tie-Yan Liu	Microsoft Research Asia, China
Christie Napa Scollon	Singapore Management University, Singapore
Wenwu Zhu	Tsinghua University, China

Senior Members

Wei Chen	Microsoft Research Asia, China
Xujin Chen	CAS Academy of Mathematics and Systems Science, China
Munmun De Choudhury	Georgia Institute of Technology, USA
Bruno Gonçalves	Aix-Marseille Université, France
Adam Jatowt	Kyoto University, Japan
Haewoon Kwak	Qatar Computing Research Institute, Qatar
Mounia Lalmas	Yahoo Labs London, UK
James She	Hong Kong University of Science and Technology, SAR China
Huawei Shen	CAS Institute of Computing Technology, China
Jie Tang	Tsinghua University, China

Members

Palakorn Achananuparp	Singapore Management University, Singapore
Yong-Yeol Ahn	Indiana University Bloomington, USA
Fred Amblard	IRIT - University of Toulouse 1 Capitole, France
Jisun An	Qatar Computing Research Institute, Qatar
Ching Man Au Yeung	Axon Labs Ltd., Hong Kong, China
Nick Beauchamp	Northeastern University, USA
Piotr Bródka	Wroclaw University of Technology, Poland
Matthias R. Brust	Singapore University of Technology and Design, Singapore
James Caverlee	Texas A&M University, USA
Fabio Celli	University of Trento, Italy
Freddy Chong Tat Chua	HPLabs, USA
Michele Coscia	Harvard Kennedy School, USA
Victor M. Eguiluz	IFISC, CSIC-UIB, Spain
Emilio Ferrara	Indiana University Bloomington, USA
Vanessa Frias-Martinez	University of Maryland, USA
Wai-Tat Fu	University of Illinois, USA
Manuel Garcia-Herranz	Universidad Autonoma de Madrid, Spain
Andreea Gorbatai	Haas School of Business, UC Berkeley, USA
Przemyslaw Grabowicz	Max Planck Institute for Software Systems, MPI-SWS, Germany
Christophe Guéret	Data Archiving and Networked Services, The Netherlands
Alexander Hanna	University of Wisconsin-Madison, USA
Stephan Humer	Universität der Künste, Berlin, Germany
Andreas Kaltenbrunner	Barcelona Media, Spain
Kazuhiro Kazama	Wakayama University, Japan
Przemysław Kazienko	Wroclaw University of Technology, Poland
Brian Keegan	Northeastern University, USA

Konstantinos Konstantinidis	Information Technologies Institute, ITI, Greece
Farshad Kooti	USC Information Sciences Institute, USA
Nicolas Kourtellis	Telefonica Research, Spain
Matteo Magnani	Uppsala University, Sweden
Winter Mason	Facebook, USA
Yelena Mejova	Qatar Computing Research Institute, Qatar
Stasa Milojevic	Indiana University, USA
Mikolaj Morzy	Poznan University of Technology, Poland
Tsuyoshi Murata	Tokyo Institute of Technology, Japan
Keiichi Nakata	University of Reading, UK
Natalie Pang	Nanyang Technological University, Singapore
André Panisson	Data Science Lab, ISI Foundation, Italy
Mario Paolucci	Institute of Cognitive Science and Technology, CNR, Italy
Symeon Papadopoulos	Information Technologies Institute, ITI, Greece
Paolo Parigi	Stanford University, USA
Ruggero G. Pensa	University of Turin, Italy
Michal Ptaszynski	Kitami Institute of Technology, Japan
Hemant Purohit	Ohio Center of Excellence in Knowledge-enabled Computing, USA
Jose J. Ramasco	IFISC, CSIC-UIB, Spain
Georgios Rizos	CERTH-ITI, Greece
Luca Rossi	University of Copenhagen, Denmark
Giancarlo Ruffo	Università di Torino, Italy
Jitao Sang	CAS Institute of Automation, China
Claudio Schifanella	RAI Research Center of Turin, Italy
Rossano Schifanella	University of Turin, Italy
Frank Schweitzer	ETH Zurich, Switzerland
Xiaolin Shi	Microsoft Research, USA
Emma Spiro	University of Washington, USA
Araz Taeihagh	Singapore Management University, Singapore
Bart Thomee	Yahoo Labs, USA
Michele Trevisiol	Universitat Pompeu Fabra and Yahoo Research, USA
Carmen Vaca	Yahoo Research, Spain
Elad Yom-Tov	Microsoft Research, Israel
Emilio Zagheni	University of Washington, USA

Sponsors

Academy of Mathematics and Systems Science, Chinese Academy of Sciences
National Central for Mathematics and Interdisciplinary Sciences, Chinese Academy of Sciences
East China Normal University
Yahoo
Springer

Academy of Mathematics and Systems Science
Chinese Academy of Sciences

EAST CHINA NORMAL UNIVERSITY

the language of science

Contents

Culture, Imagined Audience, and Language Choices of Multilingual Chinese and Korean Students on Facebook

Ha Kyung Kong(✉), Y. Wayne Wu, Brian Bailey, and Karrie Karahalios

Computer Science Department, University of Illinois Urbana Champaign,
Champaign, IL, USA
hkong6@illinois.edu

Abstract. Multilingual users of social networking sites (SNSs) write in different languages for various reasons. In this paper, we explore the language choice of multilingual Chinese and Korean students studying in the United States on Facebook. We survey the effects of collectivist culture, imagined audience, and language proficiency on their language choice. Results show that multilingual users use language for dividing and filtering their imagined audience. Culture played two contrasting roles; users wanted to share their culture in English but share their emotions in their native language. Through this work, we hope to portray language choice not as a tool for exclusion but of consideration for the potential audience and adherence to one's culture.

1 Introduction

Context collapse is a phenomenon that occurs on social networking sites (SNSs) where multiple audiences are collapsed into one [19]. This merging of audiences can be problematic as a person presents oneself differently based on the audience. The modified presentation of oneself does not only affect the presenter but also the audience through the new topics and languages, to which they are exposed on SNSs. In face-to-face conversations, the audience is typically visible and non-dynamic, so multilingual speakers often choose a language that accommodates the audience [10]. They are socially expected to choose a mutually shared language based on the audience that is present, and it can be considered rude to speak in a language that obviously excludes a person from the conversation.

However, the situation is different for online communication where the audience is dynamic and invisible. The context collapse creates a new environment where people are exposed to many more languages that they do not understand. Although some people enjoy the global atmosphere, others may feel left out when major conversations are held in a foreign language. The social obligation of language choice becomes more subjective as people deal with context collapse. The following thread of comments on a picture posted on Facebook is just one of many examples of conflicting standards of appropriate language choice online:

© Springer International Publishing Switzerland 2015
T.-Y. Liu et al. (Eds.): SocInfo 2015, LNCS 9471, pp. 1–16, 2015.
DOI: 10.1007/978-3-319-27433-1_1

Fig. 1. Two users show different opinions on the appropriateness of writing in a language that the potential audience does not understand. Pseudonyms were used for privacy.

David: (in Korean) Jay Kim, you are becoming good looking. What are you taking? Lol

Jay: (in Korean) Saudi sandstorm lol Anyhow, this is someone else's post so if we speak Korean it's awkward [for the non-Korean poster]

David: don't care :)

(Original thread shown in Fig. 1)

Amy is a non-Korean speaker who tagged Jay on a picture. Jay and David are Korean speakers. Because Amy does not understand Korean, Jay thinks it is inappropriate to use Korean in the comments for that picture while David thinks it does not matter.

This uncertainly of appropriate language choice on SNSs led us to investigate factors that affect the language choice of multilingual users online. Specifically, our qualitative research studies the language choice of multilingual Chinese and Korean students in the United States. This population represents a large sector of the multilingual international students in the United States. In the 2013-2014 academic year, the number of international students in the U.S. reached a record high of 886,000 students: China was the first country of origin making up 31% of international students, India the second with 12%, and South Korea the third with 8% [13]. We excluded Indian students from our study as English is often a dominant language in many regions within India. From here on in this paper, Chinese and Korean students will be referred to as "multilingual students". We interviewed multilingual students about the influence of the content of the post and their imagined audience on their language choice for a Facebook post [2,4,17]. We also briefly surveyed the participants' perceptions of received foreign language posts on Facebook. More specifically, we investigate the following research questions:

RQ1. How does the topic and the content of the post affect the language choice of multilingual international students on Facebook? What role does culture play in the decision?

RQ2. How does the imagined audience affect the language choice of multilingual international students on Facebook?

2 Related Work

In order to understand Chinese and Korean students' usage of Facebook, this section will briefly describe how culture affects the characteristics of a SNS and its usage. We then define language choice and the imagined audience to better explain the roles they play in multilingual international students' language choice on Facebook.

2.1 Collectivistic Culture and SNS Usage

SNSs have been growing in importance, as they have become a global meeting place where people from various cultures and backgrounds can share information. As of 2015, Facebook has 1.44 billion monthly active users worldwide [25]. Despite the global nature of SNSs, they often exhibit characteristics of their local culture or origin. Studies have shown that characteristics of Eastern cultures and Western cultures are reflected through the usage and features of Cyworld and RenRen, respectively the Korean and Chinese local sites, and Facebook, an American site. Qiu et al. examined the cultural differences between Facebook and RenRen, and found that people perceived RenRen as more collectivistic and Facebook as more individualistic [22]. This is consistent with research suggesting a prominent collectivistic culture in China and Korea, and a more individualistic culture in North America [11,12]. Collectivism is marked by "the subordination of individual goals to the goals of a collective, and a pursuit of harmony and interdependence within the group" [24]. These cultures are typically high-context cultures as well, where communication heavily relies on the context and subtle cues to emphasize a shared experience and to avoid offending others. On the other hand, individualistic cultures value independence and direct communication [11].

As [22] found that RenRen was perceived as more collective than Facebook, results from [5] and [14] revealed that Cyworld displays a collectivistic nature compared to Facebook. The local SNS was also found to reflect the high-context culture of Korea. Perhaps more interestingly, they found that despite the individualistic style of Facebook, Koreans favored the collectivistic and indirect ways of relating with others on Facebook as they had previously interacted on Cyworld. In our study, we similarly examine how the collectivistic culture of Chinese and Korean students in the U.S. play a factor in their language choice.

2.2 Language Choice

In this paper, "language choice" refers to a multilingual student's choice between his or her first language (L1) and English for a Facebook post. There are many studies on code switching [3,6,18,21], a specific type of language choice that involves the switching of languages within a thread of conversation. Since the language choice is made in the context of a conversation, the language choice in code switching is heavily dependent on the language choices that were made earlier. However, this work explores language choice for posts that are independent of previous posts (ex. status updates). Despite our focus on language choice of independent posts, we utilized Appel and Muysken's six functions for code switching to frame the multiple choice options to the question "What were your reasons for choosing this language?" in Part 2 of our study since the functions were generalizable to language choice in general [3]. The multiple choice options offered for the question were: More appropriate for expressing the content, To address a specific audience, To quote something, Habitual expressions (ex. greetings), Didn't know the terms/appropriate expression in the other language, and Other. We chose Appel and Muysken's functions over other categories such as Malik's ten reasons for code switching based on a previous study that it was more appropriate for online written discourse [6,18].

Previous studies have focused on different influences on language choice on SNS such as the function of the SNS [16], a specific demographic of posters [22], and the difference between the user's offline and online language choices [7]. Our study focuses on the impact of the imagined audience,the content of the post, and the culture of the poster on language choice. Comparably, [23] addressed language choice and addressivity, matching one's speaking or writing style to the audience, between Thai-English speakers on Facebook. They found that bilingual users presented their local identity through Thai as expected, and the use of English marked global orientation and their localised experience as the participants resided in the United Kingdom. So, rather than using languages to distinguish international audience from Thai audiences, both English and Thai were used "to establish a translocal community operating in an online, semi-public space" [23]. This differs from the results of our study that the multiple students used L1 to distinguish their imagined audience. The difference in the results may be due to the difference in the nature of the two populations that were studied. A new area that we will explore is how the language choice is influenced by culture, which is addressed in the Discussions section.

2.3   The Imagined Audience

Audience and addressivity on Facebook differ from those in other types of communication in various ways. Previous work has explored how multilingual speakers maintain their identity and self-presentation in different social networks and workplace settings through language choice [15,27]. Unlike in face-to-face communication, posters on Facebook do not know the actual audience who will read a post, so they imagine an audience [2,4,9]. There are four categories of audience on SNSs: addressee, active friends, wider friends, and the internet as a whole [26]. Addressee is the main target audience, and as expected, they were the most influential in language choice of a post. However, active friends and wider friends also play a part in language choice as will be shown in the Discussions section (5.2).

Overall, our research combines the effect of audience on language choice with previously found reasons for codeswitching to examine the language choice of multilingual international students in the United States. Although the composition of imagined audience largely determined a user's dominant language on

Facebook, we found that language choice was based on a combination of many factors including the topic/content of the post, the culture of the poster, and the respect for the imagined audience.

3   Methodology

3.1   Participants

For the purpose of the study, we recruited nineteen subjects from the local community of a Midwestern university. Qualified participants were Chinese/Korean-English bilingual speakers who had published at least fifty status updates, comments, and shares in the last two years, and had been in the United States for at most four years as of the time of the study. Each participant was entered into a raffle for a $100 Amazon gift card. We recruited participants who had resided in the U.S. for fewer than five years because this is roughly the time span required to complete a college or graduate level degree. We chose this population because we wanted participants who had sufficient exposure to their native culture and language prior to coming to the U.S.

We interviewed 19 participants (Chinese: 4 males, 4 females; Korean: 7 males, 4 females) for the study. Their ages ranged from 20 to 40 (mean = 25.53, s.d. = 4.31). Six participants were from a computer science background, five from other engineering majors, five from sciences, and three from other majors. The average length of their stay in the States was 2.32 years with a standard deviation of 1.29.

3.2   Experiment Flow

The experiment consisted of four parts; participants completed Part 1 at home and the remaining three parts in a lab setting.

Part 1. Online Scenario-Based Open-Ended Exercise. Before the participants attended the on-site study, they were asked to fill out a scenario-based survey online. The scenario-based survey was conducted online to avoid on-site priming from the presence and language of the researchers. Having the survey online also allowed the participants to take it at home on their own computers, providing an environment where they might write actual Facebook posts.

The survey started with a sample scenario described in both English and L1. The participants were then asked to compose a post, as they would compose a status update on Facebook, based on the sample scenario. After viewing a sample scenario and a sample post to understand the basic mechanics for the survey, the participants were asked to compose a post for each of the eight scenarios. The scenarios were selected based on related works on language choice and existing Facebook posts in Chinese, Korean, and English.

The first two scenarios were closely linked to the participants' own ethnic backgrounds. We hypothesized that the participants would choose to write these posts in L1 since their main audience was likely to be other L1 speakers. Scenarios

3 and 4 were related to class work and a Western holiday, where the imagined audience was likely to be non-L1 speakers. For Scenarios 5 and 6, we chose two scenarios that people would likely broadcast to both Korean/Chinese and American friends. We hypothesized that people would write in both languages to reach a wider audience. Scenarios 7 and 8 included topics where the target audience was ambiguous (see Appendix A for details).

These scenarios were presented in a randomized order to investigate how people formed their imagined audience based on the topic of the posts. Although we conducted the survey online to avoid on-site priming and presented each scenario in both English and L1, weak priming could not be avoided since some participants read the instructions solely in L1. After the participants finished the online survey, they were asked to visit our research lab for a forty-minute on-site session.

Part 2. Two Surveys: Demographics and Facebook Posts. The on-site session started with a survey that addressed participants' basic background information, English education history, and English proficiency. While the participants were filling out the survey, we gathered their Facebook posts (status updates, comments, and shares) through Facebook Query Language (FQL) with the participants' consents[1]. After the participants finished the survey, they were presented a set of six of their Facebook posts that were randomly chosen to include two posts in L1, two posts in English, and two posts containing both languages. The posts were weighted according to their language independence during the selection process where status updates were most heavily weighted and comments the least. Comments were considered last since the language choice for a comment is highly dependent on the language of a previous comment or the original post, and shares considered after status updates since language choice for shares is influenced by the language and contents from the source [10].

For each post, the participants were presented with a series of multiple choice questions about the intended audience for the post, whether they cared if people other than the intended audience saw the post, and their obligation level for choosing that language. In addition, they were asked to select the reasons for writing the post in that language from options mentioned in Section 2.2.

Part 3. Exploratory Visualization of Language Usage on Facebook. Next, we proceeded with an exploratory visualization of the participants' Facebook language usage. The visualization consists of two components: a yearly view on the left and a monthly view on the right (see Figure 2). The yearly dot-line graph displays the general trend of the participant's language usage over a year with months on

---

[1] We originally designed the study to include a Facebook app that collected the necessary information for the study. However, after multiple requests for Facebook API application permission, we were not granted the permission to access the participants' data without explanation. Due to this technical constraint, we focus on language choice of independent posts since we were not able to collect the original posts on which the participants commented.

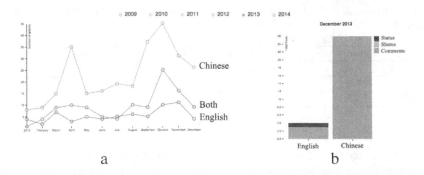

Fig. 2. Yearly (a) and Monthly (b) distribution of languages for the participant's posts.

the x-axis and the number of posts in each language on the y-axis (Figure 2a). The three lines represent Facebook post counts in L1, English, and both languages. When a participant clicks on a point on the dot-line graph, the corresponding month is displayed in the monthly bar chart on the right. The monthly bar chart includes the breakdown of the participant's posts into three types of Facebook posting activities: sharing, commenting and posting (Figure 2b).

Through this visualization, the participants were able to view their Facebook usage patterns over their entire Facebook history, and the visualization acted as a prompt during the interview process. The participants reported enjoying this part of the study and often stopped to examine periods of rapid language trend changes. Their interest in the visualization and exploration of unexpected or forgotten language patterns led to their active participation in the interview.

Part 4. Semi-Structured Interview. The study ended with a 20-minute semi-structured interview that delved deeper into participants' language choice motivations. After browsing the visualization, participants were asked to describe patterns such as active/inactive periods and the language choice trends, and whether it matched their expectations of their language usage on Facebook. The participants were then asked to describe their perceptions of English and L1 use on Facebook and the language norms amongst their friends. Finally, the interviewer selected some of the participants' Facebook posts and scenario-based posts (from Part 1 and 2) that related to the research questions to probe further about their target audience, language choice, and language usage obligations.

3.3 Data Analysis

We used qualitative analysis through a mixed methods approach to answer our research questions based on the answers from the online open-ended exercise (Part 1), the participants' actual Facebook posts gathered through FQL (Part 2), and the interviews (Part 4). To analyze the interviews, one researcher iteratively created topic categories from the interview transcriptions via open coding.

Table 1. Responses for Interview Q4. What were your reasons for choosing this language?

	appropriate	audience	quote	habitual	knowledge	other
Eng	22	16	2	6	2	3
L1	24	12	1	4	3	6
Both	23	8	3	4	4	2

After the primary categories and subcategories were established, two researchers used NVivo [1], an annotation tool, to classify sections of the transcriptions into one of the categories. After the categorization phase, we used axial coding to finalize the relationships between the categories and derived two main themes that were consistent throughout the interviews. The two main themes were the effects of culture and content on language choice and the division of imagined audience. We discuss these themes in more detail in the Discussion section.

4   Results

To find out the participants' motivation for language choice, the Facebook survey (Part 2) included a question regarding the specific reasons for using a particular language for a post [3]. The results are presented in Table 1. Participants most often chose the *appropriateness* of a language as the reason for writing in that particular language, and the interviews revealed that the appropriateness is determined by the content of the post as well as the cultural norms. This was also shown through the scenario-based exercise (Part 1).

4.1   Scenarios

The results for six of the eight scenarios aligned with our hypothesis as the majority of the posts were in the predicted language (see Table 2). The two scenarios that did not conform to our hypothesis were the Both-predicted scenarios. We hypothesized that people would write in both languages for scenarios requiring a broader audience. Surprisingly, the majority of the participants only wrote in L1 or English for the car sale scenario where a broader audience might be beneficial. People who chose English did so because even the Chinese/Koreans

Table 2. Language choice results for different scenarios

	New Year	Food	Meeting	Christmas	Car	Birthday	Professor	NYC Trip
Eng	4	6	12	12	9	7	6	5
L1	10	13	7	5	6	7	13	13
Both	6	0	1	3	4	6	0	2
Hypo.	L1	L1	English	English	Both	Both	Either	Either

Table 3. Responses for Interview Q3. Who was your target audience for this post? We used Min et. al's categories for classifying social relationships [20].

	family	work	friends	hobby	neighbor	religious	everyone	other
Eng	5	6	19	0	2	3	8	1
L1	3	5	23	1	2	2	9	1
Both	5	10	31	3	4	5	9	0

who might buy the car, live in the U.S. and speak English, so they deemed it unnecessary to repeat the post in L1. Some people who wrote in L1 told us that they only wanted to sell the car to Chinese/Korean friends, either because they did not think their English was good enough to sell a car or because the majority of their intimate friends were Chinese/Korean. The friendship distribution also influenced the birthday scenario and is discussed further in section 5.2.

Although the result for the other scenarios aligned with our hypothesis, there were unanticipated factors that were involved in language choice. We explicitly asked the participants about their language choice for scenario-posts that did not match our hypothesis and found that, although in unexpected ways, they still conformed with our general theme of culture and audience in language choice. For example, we had predicted that the participants would use L1 for the culture-related scenarios, but the desire to share one's culture encouraged some participants to use English. This is explained further in section 5.1.

4.2   Target Audience and Language Obligations

The results from Parts 2 and 4 revealed that the audience is the most influential factor for language choice of multilingual international students on Facebook. The majority (15 out of 18) of the participants said that they had an imagined audience in general, while two participants said that they only had a vague target audience. For each post in Part 2, we asked if the participants remembered the intended audience. They answered positively for 82.5% of the posts that were written in English or L1 and for 92.5% of the posts that were written in both languages. The distribution of the selected intended audience categories is shown in Table 3. People's target audience for the majority of the posts was Friends followed by Work (ex. classmate, professor, colleague, teammate), and then Facebook friends in general. The participants especially targeted their actual friends when they used both languages. People also selected more intended audience categories for posts written in both languages compared to those written in only English or L1. This might suggest that people tried to reach a wider target audience by posting in both languages.

In order to study the social obligation of choosing a language for the audience, we asked participants whether they felt obligated to write in that particular language for their audience. On a scale of 1 (not obligated at all) to 5 (very obligated), the mean obligation score was 2.9 for English posts, 2.6 for L1, and 2.7 for both. People felt most obligated to use English when they were

Table 4. Distribution of responses for Interview Q3c. On the scale from 1 (not obligated at all) to 5 (very obligated), how obligated did you feel to write in English/L1/Both languages for this audience?

	level 1	level 2	level 3	level 4	level 5
Eng	10	3	17	3	7
L1	12	5	14	6	3
Both	16	4	6	6	8

commenting on a post made by an American friend or when American friends were tagged in the original post as illustrated in the introduction.

5   Discussion

5.1   Post Content and Cultural Considerations (RQ1)

We found that culture plays a role in determining the language of a post in two different ways. The collectivistic culture in China and Korea encouraged the multilingual students to write in L1 for emotional or culturally sensitive posts. However, the desire to share one's culture encouraged the students to write posts in English as well. We will look at these seemingly contrasting roles of culture in language choice in the following section.

Collectivistic Culture and L1. A recurring reason for choosing L1, mentioned in the interviews, was its function as a language barrier. The participants wanted to block others from understanding the content of the posts because they were either too personal or contained extreme, usually negative, emotions. P10 commented, "Maybe I'm more used to complain in Chinese. When saying about something, express some feeling not good, I maybe use Chinese more. I think it is also related to the culture. So I think in RenRen, a lot of people express their negative feeling. In English few people complains. It's more common to complain in Chinese." [2] A Korean participant shared a similar feeling that she did not want to share her problems in English, despite having close American friends, because she thought it was more acceptable to do so in Korean [P8]. This aligns with the theory that people are more open to sharing their emotions publicly in a collectivistic culture since the sharing of emotions strengthens the emotional attachment to the ingroup and affirms the cultural value of interdependence [24].

The participants' want to express emotions in L1 was also shown in scenario 5, which was about a professor singing Let It Go in class. Although unexpected, this scenario was affected by culture in two contrasting ways. Since the post tended to express strong emotions, whether of shock, confusion, or excitement, most of the participants wrote in L1. However, one participant explained that she wrote the post in English because American culture was represented through

---

[2] Participant interview quotes are not modified from the original transcriptions.

the song and because it was an event that occurred in a classroom making it public enough to write in English.

Another reason for limiting the audience of a post through language was for potentially offensive and rude posts. Some terms deemed acceptable in some cultures may be taboo in others. For example, one of P11's post was all in English except for a Chinese term referring to African-Americans. He explained, "I think it would be a little uncomfortable [if somebody translated the post] because that's not appropriate for American people. It's only for like, for Chinese people, they don't feel offended. They don't even feel offended to African-American people, but English this is like a sensitive topic. So I don't want them to see it."

For Koreans, the absence of honorifics in English can prompt them to choose L1 if they are addressing people who are older than them. Even if the poster and the intended audience both understand English, it might seem impolite or even rude to write in English because the tone is informal and there is no English term to respectfully address elders (e.g., 'eonni/nuna' referring to an older female and 'oppa/hyeong' referring to an older male). Thus, some posters wrote in Korean to maintain the socioculturally appropriate relationship with older people.

Sharing Cultures in English. Although the cultural differences prompted multilingual students to write some posts in L1 to follow emotion sharing norms in collectivistic cultures, the cultural differences also prompted multilingual students to write certain posts in English to promote their own culture. This point was stressed in our interview with P16, a student in the Intensive English Institution (IEI) program. Most of his Facebook friends were other IEI students from all over the world. When we asked him why he had written all the scenario posts in English, he answered "I just want to introduce some Chinese culture to my American friends and international friends. You know, in English class, we have some classes about different cultures. Everyone introduced his country's special culture to others. It's very interesting. I use English to introduce Chinese culture to my friends." The only time he wrote in Chinese on Facebook was when he shared a link of a TV show called a Bite of China to an American friend who is learning Chinese.

Although his enthusiasm for sharing his culture in English might be extreme, P16 was not alone in his desire to share his culture with others. More examples of sharing one's culture in English were brought up in the interviews as participants explained their language choice for the culture-related posts. We hypothesized that people would use L1 for the Chinese New Year since it was highly cultural and primarily related to the L1-speaking audience. However, we found that people also posted the Chinese New Year scenario in English and in both languages to promote their culture (n=3). P19 explained that she wanted "to share the Korean culture because I remember one of my Chinese friends, she invited us over to celebrate Chinese New Year. So I think it was nice having us also celebrate that culture with her." Another reason for writing the New Year's post in English was the commonness of the phrase "Happy New Year" (n=3). Along with "Happy Birthday" and "Merry Christmas," "Happy New Year" is a commonly used greeting phrase in Korea. As a result, the participants did not feel the need to write, or repeat, the phrase in Korean.

5.2 Division of the Imagined Audience (RQ2)

During the interviews, we noticed that people mentioned their L1 friends as their imagined target audience rather than their friends in general. This suggests that for multilingual users, the imagined audience is segmented into groups based on language. The fragmentation of the imagined audience into L1 and non-L1 imagined audiences prompts bilingual users' to select a specific target group for individual posts based on their content while being considerate of all potential audience.

Imagined Audience and Friendship Distribution. The selection of L1 speakers as the main intended audience is not surprising as nearly 40% of international students reported having no close American friends in a recent survey [8]. The majority of the participants' Facebook friends were L1 speakers, reflecting their actual friend distribution and prompting them to write in L1 for the majority of the posts. In other words, some participants simply did not have enough non-L1 speaking "active friends" to make writing in English worthwhile. P6 pointed out, "10 or 12 international friends [on Facebook out of 400 Facebook friends] are not that active people so I don't really have to, I don't feel the obligation of writing in English." He added that he would definitely write his posts in English if he had more American friends on Facebook and they showed interest in his posts through likes and comments. Similarly, P15 mentioned the lack of American friends even in the "wider friends" circle. She did not think the information in her posts would reach the few American friends she had on Facebook even if she posted in English. She elaborated that she would talk to her American friends in person if she wanted to reach them.

The two scenarios that were highly dependent on the imagined audience and friendship distribution of the participants were the New York scenario and the birthday scenario. People chose to write in L1 for the New York scenario for two main reasons. First, they often travel with close friends who are from the same country as them. Secondly, they felt that Americans might not find their trip to New York as significant or interesting compared to their L1 speaking friends. So their imagined audience for such travel-related posts was primarily L1 speakers. The language choice for the birthday scenario was highly based on the participant's active Facebook friends as well since they were the most likely to have posted a Happy Birthday greeting on the participant's timeline.

Interestingly, most students' imagined English-speaking audience are not only comprised of their American friends because the L1-speaking friends they met in the U.S. also speak English. Subsequently, they can only reach L1 speaking friends if they write a post in L1, but based on their "wider friends," they might reach American friends, other international friends, and some of their L1 speaking friends if they write in English. This is especially true for Chinese students since Facebook is blocked in China, and thus most of their Chinese friends on Facebook are also in the United States. The ability to reach all of their Facebook friends via English is useful in situations when they need to ask for help, promote a petition, sell something, or advertise an event. The use of English for these situations was confirmed in our analysis of participants' actual Facebook posts, and was previous mentioned for the car sale scenario as well.

Language Proficiency of the Poster and the Audience. For some people, it was purely their lack of proficiency in English that led them to write in L1. P5 humorously admitted that he doesn't write English well and did not want to bother people with his "broken English." However, this student was fluent enough in English to pass the mandatory English as Second Language (ESL) course for international graduate students. Thus, it might be the students' perception of their own English that prompts them to use L1, rather than their actual proficiency. Others noted that they could get the basic content across in English, but found it hard to convey subtle nuances such as the accurate expressions of emotions or jokes in English as well as they could in L1. P10 said "I don't know to find the best way to express complaints in a funny way. Not just the complain, I also want it to look interesting. I don't know how to best express it in English." These students chose to write in L1 because they felt that the subtle overtones could make a difference in how people perceived their posts.

Language proficiency has an interesting impact on language choice due to the following dichotomy - the poster's lack of English proficiency prompted writing in L1, yet the presence and the respect for the non-L1 speaking audience prompted writing in English. P8 told us that she used to write most of her posts on Facebook in Korean until her friends in the same program asked her to write in both languages because they didn't know Korean and wanted to know what she was posting. From then on, she consciously wrote all her Facebook posts in English or both languages out of consideration for her friends who do not speak Korean.

One of the participant's Facebook posts about a group project meeting was in English, even though he was writing to a L1-speaking group member. This post was written in English out of respect for the other team members who did not understand L1. We had hypothesized that the posts for the meeting scenario would be in English since the students attend a university in the U.S. and typically write work-related posts in English. Most of the people wrote this post in English, but some people (n=6) still used L1. We believe this is a result of soft priming as participants' interviews revealed that they had read the scenarios in L1 and had continued to think in L1 when they were writing the posts.

5.3   On the Other Side of the Barrier

Choosing L1 can divide the audience into an ingroup that can read a post, and an outgroup that cannot. We wanted to see how international students felt about this division formed by language choices on Facebook. Half of the participants answered that they felt left out when people wrote in languages they did not understand. They pointed out that since they do not know the language and the translator does not work well, they remained curious as to the contents of the post. The students felt especially left out when they knew everyone who was participating in the conversation and they knew that those people also spoke English.

P12 said, "I feel that I'm not the intended reader, the intended audience, and he just ignored me. Or I feel, maybe he thought this is not my business, very sad..." Although P12 would write in Chinese for posts containing extreme emotions that are better expressed in Chinese, he tried to write primarily in English out of respect for the potential audience. P2 took this one step further by creating a list of all the Koreans and only displaying a post to that group whenever he wrote in Korean. Although it took time and effort to create the group of approximately 600 Korean friends, he did not want others to feel "left out" by his Korean posts. It is noteworthy that not all the participants felt left out by the foreign language posts. Approximately half (7 out of 15) of the participants who commented on seeing foreign language posts said that they understand that "everyone has their own small circles." Although they were still curious of the contents of foreign language posts, they were fairly comfortable with the presence of foreign languages on Facebook as they often wrote in L1 themselves.

6   Limitations

There were some limitations of the work that resulted from lack of data. The students who participated in the study attended a university with a large international student body. Therefore, it is likely that the inclination towards writing in L1 might be less consistent for international students in other schools, which limits the generalizability of this study. Studies at other universities with fewer international students are needed to observe international students' language choice in an environment where interaction with American students may be more frequent and necessary.

Some questions in our surveys were hypothetical and self reported, i.e., "In what language would you post a thank you post for birthday wishes?" While our collected data supported our interview findings for many of our questions, a larger collection of longitudinal data would add reliability.

7   Conclusion

Social norms are dynamically evolving in our networked online spaces. Knowing when to use a certain language is not as clear as it is in the physical world. Because one is never certain of the actual audience of one's Facebook post, the factors that determine language choice on Facebook are different from those in face-to-face communication. This study examined multilingual international students' views and motivations for language choice on Facebook. Our findings show that the student's imagined audience and his or her Facebook friend distribution determine the student's dominant language on Facebook. However, other factors such as the cultural content of the post and the language proficiency of the poster and the audience played a role for specific posts. Future work could focus on the other side of the story by surveying American students who see their international friends' posts in L1. It would be noteworthy to see if their views on posts in foreign languages match international students' views. Through this study, we explored the reasons for and the perception of language choice on Facebook. Here we hope to portray language choice not as a tool for exclusion but of consideration for the potential audience and adherence to one's culture, and raise multilingual users' awareness of the impact of language choice on their audience.

References

1. NVivo qualitative data analysis software (2014). http://www.qsrinternational. com/products_nvivo-mac.aspx
2. Acquisti, A., Gross, R.: Imagined communities: awareness, information sharing, and privacy on the facebook. In: Danezis, G., Golle, P. (eds.) PET 2006. LNCS, vol. 4258, pp. 36–58. Springer, Heidelberg (2006)
3. Appel, R., Muysken, P.: Language contact and bilingualism. Amsterdam University Press (2005)
4. Bernstein, M.S., Bakshy, E., Burke, M., Karrer, B.: Quantifying the invisible audience in social networks. In: Proceedings of the SIGCHI Conference on Human Factors in Computing Systems, pp. 21–30. ACM (2013)
5. Choi, J., Jung, J., Lee, S.-W.: What causes users to switch from a local to a global social network site? The cultural, social, economic, and motivational factors of Facebooks globalization. Computers in Human Behavior **29**(6), 2665–2673 (2013)
6. Choy, W.F.: Functions and reasons for code-switching on facebook by UTAR English-Mandarin Chinese bilingual undergraduates. Ph.D. thesis, UTAR (2011)
7. Cunliffe, D., Morris, D., Prys, C.: Young Bilinguals' Language Behaviour in Social Networking Sites: The Use of Welsh on Facebook. Journal of Computer-Mediated Communication **18**(3), 339–361 (2013)
8. Fischer, K.: Many foreign students are friendless in the US, study finds. The Chronicle of Higher Education (2012)
9. García-Gavilanes, R., Kaltenbrunner, A., Sáez-Trumper, D., Baeza-Yates, R., Aragón, P., Laniado, D.: Who are my audiences? A study of the evolution of target audiences in microblogs. In: Aiello, L.M., McFarland, D. (eds.) SocInfo 2014. LNCS, vol. 8851, pp. 561–572. Springer, Heidelberg (2014)
10. Giles, H., Taylor, D.M., Bourhis, R.: Towards a theory of interpersonal accommodation through language: Some Canadian data. Language in Society **2**(02), 177–192 (1973)
11. Gudykunst, W.B., Ting-Toomey, S., Chua, E.: Culture and interpersonal communication. Sage Publications, Inc. (1988)
12. Hofstede, G.: Culture's consequences: International differences in work-related values, vol. 5. Sage (1984)
13. Institute of International Education: A quick look at international students in the U.S. Technical report (2014)

14. Kim, Y., Sohn, D., Choi, S.M.: Cultural difference in motivations for using social network sites: A comparative study of American and Korean college students. Computers in Human Behavior **27**(1), 365–372 (2011)
15. Lanza, E., Svendsen, B.A.: Tell me who your friends are and i might be able to tell you what language (s) you speak: Social network analysis, multilingualism, and identity. International Journal of Bilingualism **11**(3), 275–300 (2007)
16. Lee, C.K.M., Barton, D.: Constructing Glocal Identities Through Multilingual Writing Practices on Flickr.com. International Multilingual Research Journal **5**(1), 39–59 (2011)
17. Litt, E.: Knock, Knock. Who's There? The Imagined Audience. Journal of Broadcasting & Electronic Media **56**(3), 330–345 (2012)
18. Malik, L.: Socio-linguistics: A study of code-switching. Anmol Publications Pvt. Ltd. (1994)
19. Marwick, A.E., Boyd, D.: I tweet honestly, I tweet passionately: Twitter users, context collapse, and the imagined audience. New Media & Society **13**(1), 114–133 (2010)
20. Min, J.-K., Wiese, J., Hong, J.I., Zimmerman, J.: Mining smartphone data to classify life-facets of social relationships. In: Proceedings of the 2013 Conference on Computer Supported Cooperative Work - CSCW 2013, p. 285 (2013)
21. Parveen, S., Aslam, S.: A study on Reasons for code-switching in Facebook by Pakistani Urdu English Bilinguals. Language in India **13**(11), 564–590 (2013)
22. Qiu, L., Lin, H., Leung, A.K.-Y.: Cultural Differences and Switching of In-Group Sharing Behavior Between an American (Facebook) and a Chinese (Renren) Social Networking Site. Journal of Cross-Cultural Psychology **44**(1), 106–121 (2012)
23. Seargeant, P., Ngampramuan, W.: Language choice and addressivity strategies in Thai-English social network interactions. (CMD), pp. 510–531 (2012)
24. Singh-Manoux, A., Finkenauer, C.: Cultural Variations in Social Sharing of Emotions: An Intercultural Perspective. Journal of Cross-Cultural Psychology **32**(6), 647–661 (2001)
25. Smith, C.: By the numbers: 200+ amazing facebook user statistics (June 2015), July 2015. http://expandedramblings.com/index.php/by-the-numbers-17-amazing-facebook-stats/(updated July 21, 2015)
26. Tagg, C., Seargeant, P.: Audience design and language choice in the construction and maintenance of translocal communities on social network sites. In: The Language of Social Media: Identity and Community on the Internet, p. 161 (2014)
27. Tange, H., Lauring, J.: Language management and social interaction within the multilingual workplace. Journal of Communication Management **13**(3), 218–232 (2009)

Analyzing Factors Impacting Revining on the Vine Social Network

Homa Hosseinmardi[1]([✉]), Rahat Ibn Rafiq[1], Sabrina Arredondo Mattson[2],
Richard Han[1], Qin Lv[1], and Shivakant Mishra[1]

[1] Computer Science Department, University of Colorado Boulder, Boulder, USA
{homa.hosseinmardi,rahat.ibnrafiq,richard.Han,
qin.lv,shivakant.mishra}@colorado.edu
[2] Institute of Behavioral Science, University of Colorado Boulder, Boulder, CO, USA
sabrina.mattson@colorado.edu

Abstract. Diffusion of information in the Vine video social network happens via a *revining* mechanism that enables accelerated propagation of news, rumors, and different types of videos. In this paper we aim to understand the revining behavior in Vine and how it may be impacted by different factors. We first look at general properties of information dissemination via the revining feature in Vine. Then, we examine the impact of video content on revining behavior. Finally, we examine how cyberbullying may impact the revining behavior. The insights from this analysis help motivate the design of more effective information dissemination and automatic classification of cyberbullying incidents in online social networks.

Keywords: Information diffusion · Vine · Cyberbullying

1 Introduction

Online social networks such as Facebook, Twitter, YouTube, and Vine are attracting more users every day, and they have become an important source of information sharing and propagation [1]. Vine is a video-based online social network and has become increasingly popular recently in the Internet community. There are almost 40 million registered Vine users as of April 2, 2015 while the total number of Vines played everyday is 15 billion [2]. Using a mobile application, Vine users can record and edit six-second looping videos, which they can share on their profiles for others to see, like, and comment upon. An example of the Vine social network is shown in Figure 1. All user profiles in Vine are public by default unless users change their privacy policies. In the public setting, posts are accessible by all Vine users, not only followers. Using the privacy setting, Vine users can limit the access to their posts to their followers only. Users can also limit who can find them or message them. In the home feed page, featured Vine videos in different categories are provided to a user when logged in.

Due to the wide popularity of online social networks, they have been used for sharing news, art, politics [3–5] and have been researched from different areas,

© Springer International Publishing Switzerland 2015
T.-Y. Liu et al. (Eds.): SocInfo 2015, LNCS 9471, pp. 17–32, 2015.
DOI: 10.1007/978-3-319-27433-1_2

such as marketing and sociology [5]. Vine videos are getting more popular, as they can be easily embedded in Twitter, and the auto-loop feature makes it funny and interesting [6]. One of the most interesting characteristics of Vine is the revining behavior. That is, spreading information in Vine happens via revining a shared 6-second video referred to as a "vine". Users can easily share a video by pressing the revining button provided below the video. Figure 1 also illustrates the revining behavior.

Previous work [7,8] provided detailed characterization of Twitter, looking at the interaction among the users and the temporal behavior of users in Twitter. Other works have looked at the retweeting behavior in Twitter [3,9–12]. Pezzoni et al. measured the influence of a user based on the average number of times that his/her originated tweets have been retweeted [12]. Beside user properties, they also considered the position of the tweet in the feed as a factor impacting retweeting behavior. The authors of [13,14] have considered the number of times a recent seed tweet has been retweeted as the influence of the post, and used this to estimate the total influence and influence score of the user, and then examined a network of influencers. Further, [14] observed that URLs that were rated as having more positive feelings have

Fig. 1. An example of the Vine social network with revining.

been spread more. The authors evaluated the impact of Twitter users in different topics, instead of labeling them as influencer or not influencer. While they built the propagation tree by looking at the followers retweeting, [15] looked at the retweeting methods other than formal retweeting mechanism of Twitter. Looking at different domains, they observed the percentage of retweets coming from non-followers is higher than that from followers. Zhao et al. in [16] considered two types of retweets, coming from direct and indirect followers of the user who originated a tweet. They proposed that looking at the influence of the followers is also important for predicting the number of retweets. There are also considerable amount of works trying to predict the number of retweets [16–20]. In [20], the authors used image features to predict the number of retweets by looking at the image link tweets. Another work [21] analyzed the behavior of the tweets that have been retweeted many times and tried to detect two classes of tweets, first the group of tweets that have been retweeted less than 30 times, and second the group of tweets that have been retweetd more than 100 times. Our work differs from previous works in terms of examining the revining behavior of video-based vines, but leverages prior works in that determining how far a vine has been propagated can be useful as a sense of how influential the post has been.

To date, there has been little prior work examining video-based social networks like Vine. One paper [22] has labeled a small set of about a thousand Vine videos as cyberbullying or not, for the purposes of detecting cyberbullying incidents in Vine. That work does not investigate how vines propagate via revining in the social network, which is the focus of our paper.

This paper makes the following contributions. It is the first paper to provide a detailed characterization of key properties of Vine, a video-based social network. Second, we labeled the content and emotion of a small set of videos and explore their relationship to revining behavior. Third we reveal the difference in revining behavior between vine videos labeled as cyberbullying, and those that are not.

In the following, we first describe our data collection efforts, basic analysis and then provide more characteristics regarding revining. We then analyze the revining behavior of videos both in terms of their labeled video content and cyberbullying content.

2   Dataset

Vine is a 6-second short video sharing platform, launched in 2012. Users can create videos recorded by their mobile phone camera, and apply edits provided by the mobile app. Using snowball sampling, we collected the complete profile information of 55,744 users in Vine. Specifically, starting from 5 random seed users, we collected data for all users within two tops of the seed users, i.e., followers of the seed users, and users who follow those followers. This gives us in total 390,463 unique seed videos generated by these users (approximately 7 vines for each user). For each user we collected their total number of followers, followings, total number of videos created by the user (seed videos) and total

Table 1. Collected features for users and seed vines.

Follower	list of followers of user
Following	list of followings of user
uSeedPost	number of posted videos originated by a user
uPost	number of total posted videos by a user (originated plus revined)
uLoop	number of times all posted videos by a user have been played
uLike	number of times all originated videos by a user have been liked
PostDate	the exact date and time the seed video was originated
Description	the attached caption and tags to the posted video
Revinee	list of all the users who shared the seed video
pLoop	number of times the seed video has been played
pLike	number of times the seed video has been liked
pComment	number of comments the seed video has been liked
pRevine	number of times the seed video has been revined

number of videos revined by the user. We also collected the total number of likes, revines, and loops (# of times video has been played) for each user.

For each of the seed videos, we collected the total number of likes and comments associated with the posed video. Also we collected the creation day, how many times the video has been played (i.e., loop counts) in July 2015. Also we looked at how many times each video has been revined by collecting the user id of all the users who have revined it, along with the time stamp when it has been revined. The complete list of users who have revined a video is accessible directly by collecting information associated to the seed video itself. Table 1 summarizes the features that were collected for users and their seed videos.

3   User Behavior on Vine

3.1   Basic Analysis

We first examine the general behavior of users in Vine, including the number of followers and following, which are user based features, and the number of comments, likes, revines and loops of a video, which are post based features. Figure 2 shows the distributions of the number of followers and following of 55,744 users as complementary cumulative distribution functions (CCDFs). Compared with previous work that reported the following and follower distributions in Twitter [8], there is a much larger gap between the distribution of the number of followers and the number of followings of Vine users. Only 1.3% of the Vine users follow more than 10,000 users, however 19.27% have more than 10,000 followers.

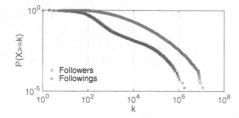
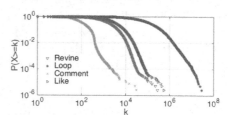

Fig. 2. CCDFs of the number of followers and followings in Vine.

Fig. 3. CCDFs of the number of revines, loops, comments, and likes in Vine.

Figure 3 provides the CCDFs for the number of times a seed video has been played as a loop, has been shared as a revine, has been liked and has been commented upon. The curve for looping is much higher than the other three (i.e., there is a long tail), possibly due to the auto-looping feature. Revining and liking have lower and very similar distributions. Commenting has the distinctly lowest CCDF curve among the features considered for a post.

Table 2 provides the mean, median and maximum values for all user features, including the total number of followers and followings, total number of

Table 2. Statistics on the Vine dataset, for 55,744 users.

	Follower	Following	uLoop	uSeedPost	uLike	uPost
Mean	23,571.14	1,459.0	5.67×10^5	147.2	3,630.0	705.6
Median	1,434.0	99.0	2.97×10^5	52.0	826.0	160.0
Maximum	1.16×10^7	1.83×10^6	2.10×10^9	1.9×10^4	2.10×10^6	7.10×10^4

posts originated by a user (uSeedPost), total number of videos posted by a user (originated+shared), total number of likes a user has received (uLike) and total number of times all his/her posts have been looped (uLoop). We observed that there are outliers that are a large distance from the mean. Namely, the substantial deviation of the mean from the median for nearly every category shows there are a set of Vine users whose feature values are much higher than the behaviors of most of the population. For example, we noticed celebrities provide such a distortive effect, e.g., one celebrity had 12 million followers.

Table 3. Pearson correlation between the collected Vine features. Only for the vines with ∗ is the p-value larger than 0.001, while for the rest the p-value is smaller than 0.001.

	pLike	pComment	pRevine	pLoop	Follower	Following	uLoop	uSeedPost	uLike	uSharedPost
pLike	1.000	**0.672**	**0.840**	**0.596**	**0.473**	-0.009	0.246	-0.018	-0.034	-0.060
pComment	0.672	1.000	**0.692**	0.408	0.209*	0.000	0.131	-0.010	-0.007	-0.03
pRevine	0.840	0.692	1.000	**0.421**	0.295	-0.005	0.144	-0.017	-0.051	-0.046
pLoop	0.596	0.408	0.421	1.000	0.25	-0.006	0.164	-0.033	-0.038	-0.042
Follower	0.473	0.209	0.295	0.256	1.000	0.011	**0.378**	0.141	0.022	0.043
Following	-0.009	0.000*	-0.005	-0.006	0.011	1.000	-0.006	-0.013	0.002*	-0.004*
uLoop	0.246	0.131	0.144	0.164	0.378	-0.006	1.000	0.156	0.030	0.069
uSeedPost	-0.018	-0.010	-0.017	-0.033	0.141	-0.013	0.156	1.000	0.199	**0.512**
uLike	-0.034	-0.007	-0.051	-0.038	0.022	0.002*	0.030	0.199	1.000	0.203
uSharedPost	-0.060	-0.032	-0.046	-0.042	0.043	-0.004*	0.069	0.512	0.203	1.00

Table 3 displays the correlations among different user and post based features. pLike, pComment, pRevine and pLoop are post features and uLoop, uLike, uSeedPost (# posts originated by user) and uSharedPost (# posts revined by user) are user features. The highest correlation is among revines and likes for posts, as was seen in Figure 3 where their CCDFs were also very similar. Revining is also correlated with the number of comments. The correlation of revines for a post with the loop feature is high but is the lowest among post based features. In terms of correlation of the revining with user based features, there is a positive correlation between revining of a post and number of followers that the poster of the seed vine has. There is a very small negative correlation between the number of revines and followings.

There is no correlation between the number of followings a user has and the other user features. The number of followers has considerable correlation with the total number of loops for all seed videos posted by the user (uLoops).

Also the correlation between user's seed post and user's shared post (uShared-Post) is about 0.5, showing users who tend to share more videos also generate more seed videos.

In total, from all posts, there were 366,517 tags in total and 83,147 unique hashtags. As Figure 4 illustrates, only 8.25% of the posts have more than 3 tags and the highest number of tags for a video is 20.

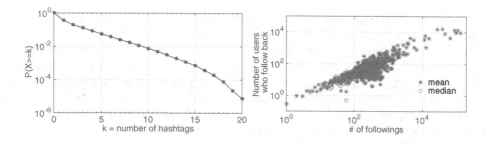

Fig. 4. CCDF of the number of tags.   **Fig. 5.** Mean and median number of users who follow back a user versus number of followings.

As another property of the Vine social network, we investigated the number of users who follow back their following users. Figure 5 shows a positive correlation between the number of followings and the mean number of users who follow back their following users.

3.2   Revining Behavior and Followers and Following

In this subsection, we explore in more detail the relationship between revining behavior and the numbers of followers and following. First, we establish a baseline for originated videos. Figure 6 plots the correlation between the number of originated vine videos and the number of followers. For up to about 1000 followers, there is a linear relation between the log number of followers and log number of mean videos and after that there is no positive trend. We also provide the mean and median for the log scale bins in solid and dash lines, [8]. The mean is clearly above the median, indicating again there are outliers who send original vines much more than other users. In Figure 7, we plot the correlation between the number of originated vines and the number of followings. Up to about a value of 665, the line is linear and after that there is a negative trend between mean value of vine and followings.

In terms of revines, Figure 8 shows the relation between the number of followers and the mean number of revines. Figure 8 demonstrates that as the number of followers increases, the mean number of revinings increases. For up to 245 followers there is positive correlation, and as followers increase the variance of the mean number of revines increase. On Figure 9 it can be seen that as the number

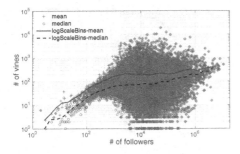

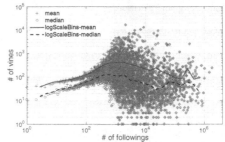

Fig. 6. Mean and median number of original vines versus number of followers.

Fig. 7. Mean and median number of original vines versus number of followings.

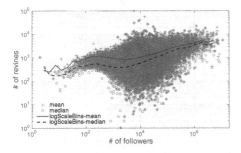

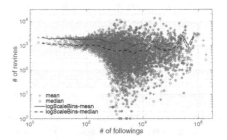

Fig. 8. Mean and median number of revines of the original vines versus number of followers.

Fig. 9. Mean and median number of revines of the original vines versus number of followings.

of following increase the mean number of revines is approximately constant for values close to 245 and after that the variance starts to increase. But there is no positive correlation between number of followings and the number of revines for a post.

3.3 Revining and Temporal Behavior

Next, we explore temporal properties of revining. Figure 10 provides the CCDF of the maximum number of times a seed vine has been revined per day. 19% of the vine videos have reached the maximum revining of more than 1000 times a day. A few seed videos have been revined more than a maximum of 10,000 times in a day.

Figure 11 shows the time difference between the first revine occurring after an original vine has been posted. 48% of the first revines have occurred within one minute and 93% have happened within one hour.

Figure 12 shows the activity of users in terms of posting seed vine videos during the seven days of the week, and across 24 hours. The minimum fraction

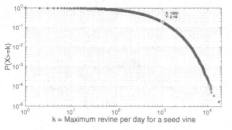
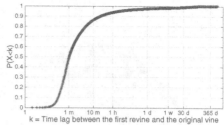

Fig. 10. CCDF of maximum number of times a seed vine has been revined per day.

Fig. 11. The time lag between the first revine and the original vine.

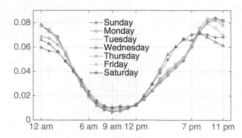
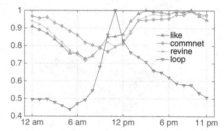

Fig. 12. (left) Percentage of vines posted at time x during a day. (right) Normalized number of likes, revines, comments and loops for posts received at time x in a day. Each graph has been normalized to one by dividing to its highest value.

of the seed videos were posted during the morning, 6am-12pm. The activity increases after that until reaching its peak around 10pm-11pm. The trend of the posting behavior differs somewhat for the weekend, shifting to a higher fraction of posts earlier in the evening, with high proportion of user activity from about 6pm to midnight. On the right side figure, we provide the mean number of likes, comments, revines and loops for the videos posted at time x during a day. It is interesting to see a pattern, showing the videos posted in the evening received on average more likes, comments and are revined more. However, the number of loops shows a completely different behavior and the highest average number of loops belongs to the videos posted at 11am.

3.4   Revining and Hash Tags

Another factor we are interested to explore is the correlation between the post's hash tags and the number of times it has been revined. Table 4 provides the tags with the highest frequency in four different bins of revining. The hash tags are ordered so that the tags on the left have the highest frequency. Each bin contains a quarter of the seed Vine videos. For example 1-bin contains one fourth of the media sessions with the lowest amount of revining. Comedy is the top tag in all four bins. Football is among the top 10 most frequent tags only in the highest

Table 4. The tags with highest frequency in 4 different bins of revining.

1-bin	comedy	funny	remake	6secondcover	lol	loop	revine	music	singing	videoshop
2-bin	comedy	remake	funny	lol	loop	revine	6secondcover	onedirection	teamsour	howto
3-bin	comedy	remake	funny	onedirection	loop	lol	revine	howto	teamsour	6secondcover
4-bin	remake	comedy	funny	onedirection	loop	lol	revine	howto	blackranked	football

Fig. 13. Visualization of the tags associated with the vines

quadrant of the number of revines. Also music and singing only occur in the lowest quadrant of revining. The top ten tags for 2-bin and 3-bin are the same, except reordered.

A visualization of the tags depicted in Figure 13 shows that comedy, funny and remake have the highest frequency and have been seen among the top three tags for all 4 bins.

3.5   Revining and Followers

Figure 14 provides the CCDF for the percentage of revines that have been made by the followers of the user who posted the original Vine video. Since all users in our dataset have public profiles and their profiles and seed vines are accessible by all Vine users, revines are not just limited to their followers. In fact, for 1.67% of the seed vines, none of the revinings have been by the followers of the poster of the original vine. For 58% of the users, more than 90% of the propagation takes place by users who are not direct followers of the users. These include celebrities, sports pages, local singer pages, fun-related pages, etc. Previous work [15] has reported non-follower retweets range from 78.7% to 98.5% for four different domains "Fundraiser", "News", "Petitions" and "YouTube".

Figure 15 displays the relation between the number of additional recipients of the Vine video (users other than the original poster's followers) and the number of followers. There is a positive correlation which shows the higher the number of followers a user has, the greater are the number of users other than followers who access the video post.

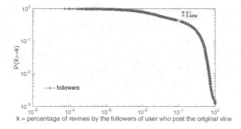

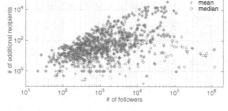

Fig. 14. CCDF of the percentage of the reviners who are the followers of the source of the original vine

Fig. 15. Mean and median number of additional recipients of the vine via revining

3.6   Revine Tree

Figure 16 shows the number of hops a video gets propagated in Vine. For this purpose we collected the user name of all the users who revined a seed video. We then find the ones who are followers of the poster of the seed video and named them as first hop users. At the next step we collected the followers of the first hop users and compared them with the remaining users in the revining list. The common users, if they exist, are named as second hop users. We continued the process until there are no more users

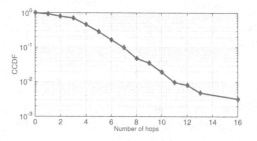

Fig. 16. CCDF of number of hops a video get propagated through the network.

in the x-hop user list. 6.7% of the videos have been propagated 0 hops, meaning none of the revining users have been among their followers. The most common number of hops is 3, i.e., 24.96% of the videos have been propagated 3 hops by their followers. In comparison, for Twitter, the most common number of hops reported is 1 for 85% of the tweets [8]. Also they show the highest number of hops is 10 in tweet propagation, while for Vine the largest number of hops we have discovered is 16. Hence, Vine videos are propagated more on average, and also more in the extreme.

Figure 17 shows an example of a revine tree. The seed video has been revined 8401 times. Looking at the links among these 8401 users, 4501 users have connection with at least one of the 8401 users, with 4595 links. This means that around half of the users that were not among the followers of the original poster have revined the video. As the poster of this Vine video has a public profile, other users who do not follow this user also can see his/her profile content and share the seed video. The dark blue nodes in the right side shows a propagation of video in the network in a community that is not at all connected to the main

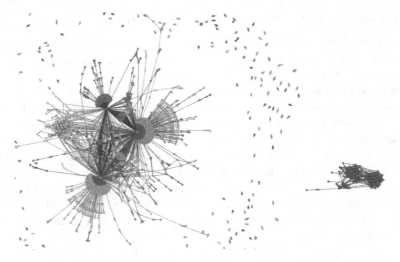

Fig. 17. Revine tree of followers of a seed Vine video who revined the post.

network component. Most of the nodes who are not followers of the original poster are single node components that have not lead to more revining.

The node with highest degree in the pink group is the original poster. Looking at the revining time of the other two nodes with highest in-degree (light blue and green groups), both revined the video in the first 24 hours, one being the first user who revined and the other being the 5th user who revined the video. These later reviners have been more influential than the original poster in the propagation of this vine. Previous work [16] has also observed that real propagation in tweeter happen with a pattern similar to Figure 17.

4   Labeled Data Analysis

In this work, we are specifically interested in how revining behavior is affected by the content of the videos as well as the presence of cyberbullying in the vines. This requires that we label various Vine videos. We selected a set of 983 vines, and collected the appropriate user features of the seed video posters, and the feature set of the posts. Using CrowdFlower, a crowd-sourced website, we labeled the video content and emotions of these vines according to the methodology from [23]. For the video content, People, Person, Indoor, Outdoor, Cartoon, Text, Activity, Animal and Other were provided as categories in multiple choice format to CrowdFlower contributors. Next, the contributors were asked to identify the emotions expressed in the video, given the following options to choose from: Neutral, Joy, Sad, Love, Surprise, Fear and Anger. The same set of videos were labeled for cyberbullying according to the methodology from [22]. Each media session (including a video and its associated comments) was labeled by five different contributors.

4.1   Revining and the Content/Emotion of Vines

We wanted to determine the relationship between the content/emotion of the video and the number of times it gets revined. Figure 18 illustrates the mean and median number of times a seed vine is revined for each of the categories we labeled for the vines. We observed "car", "activity" and "outdoor" have the highest mean. However there is big gap between their mean and median, showing there are outliers with a large amount of revining for these categories. For categories like "food", "fashion" or "cartoon" the mean is close to the median, revealing the lack of significant outliers. For these categories, our analysis found that in terms of the mean number of hops that the "car" category had been propagated on average 5 hops, which was the highest mean among other categories. Next highest was the"text" category, which was propagated with an average 4.4 hops. The category "activity" and "other" had the minimum average number of hops around 2.7.

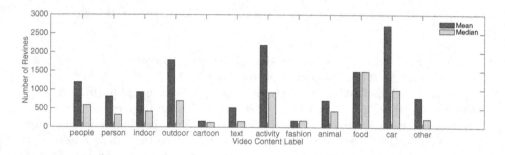

Fig. 18. Mean and median of number of times a video has been revined versus the content of the video.

Figure 19 illustrates the mean and median number of times a seed video has been revined versus the emotional content of the video. We find that "anger" and "sad" have the highest difference between the median and the mean. Positive emotions such as "love", "joy" and "surprised" have higher mean number of revines compared to negative emotion categories such as "sad", "fear" and "anger". "Neutral" behavior seems to behave closer to negative feelings than positive emotions. *In terms of the mean number of hops of revining propagation for different emotions, we found that "love" had the highest mean value 4.4 hops on average and "sad" had the lowest mean value of 2.8 hops.* Previous work [14] also observed that URLs that were rated as having more positive feelings have been spread more. For the rest of the emotion categories the mean value of the category is close to the mean number of hops 3.5.

4.2   Revining and Cyberbullying Vines

Looking at the first revining of the videos for each group, the lag between an original seed vine and its first revine is less than 1 minute for 40% of cyberbullying

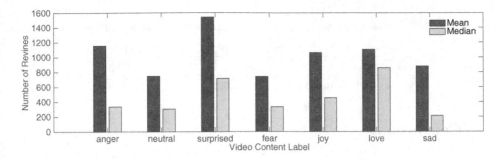

Fig. 19. Mean and median number of times a seed video has been revined versus the emotional content of the video.

vine videos and 50% for non-cyberbullying vine videos. The average time that a cyberbullying vine was revined is 4.01 days versus a mean of 1.27 days for non-cyberbullying vines (p-value 0.05 for t-test). This means the first revines happen faster for the non-cyberbullying group. We also looked at the longest time between two consecutive revining of a seed post for cyberbullying and non-cyberbullying Vine sessions. The average value over the longest inactivity period for cyberbullying is 94.91 days versus a mean of 111.15 days for non-cyberbullying (p-value 0.01 for t-test).

We found that the mean number of revining hops for cyberbullying videos was 3.21 while the mean number of hops for non-cyberbullying videos is 3.78 (p-value 0.003 for t-test). *Not only are non-cyberbullying videos shared more, but they also propagate through the network more deeply compared to cyberbullying videos.*

Tables 5 and 6 provide the mean values for the two different classes of cyberbullying and non-cyberbullying versus user and post features. The owner of cyberbullying video sessions have fewer followers and they also follow fewer users. They have lower number of total likes for their posted videos and the total number of loops is also smaller. However they have shown more activity in terms of posting seed Vine videos compare to non-cyberbullying class users.

Table 5. Mean of user features for cyberbullying and non-cyberbullying class.

	Follower	Following	uLike	uLoop	uSeedPost	uPost
cyberbullying	9.10×10^4	2.27×10^3	5.97×10^3	1.5×10^3	5.12×10^2	1.53×10^3
non-cyberbullying	1.07×10^5	3.03×10^3	7.35×10^3	4.15×10^7	4.44×10^2	1.38×10^3

Table 6 shows that cyberbullying labeled vines have been revined less, though they have been looped more. Also they receive more comments though fewer likes compared to the mean value of the non-cyberbullying labeled vines.

Table 6. Mean of post features for cyberbullying and non-cyberbullying class.

	pRevine	pLoop	pComment	pLike
cyberbullying	719.97	139,534.5	88.5	5,973.3
non-cyberbullying	1,095.6	118,326.2	76.3	7,355.8

5 Conclusions

As far as we are aware, this paper is the first to present a detailed analysis of user behavior in the Vine social network. We analyzed 55,744 profiles of Vine users, along with 390,463 seed Vine videos generated by these users. There are six key findings. First, the number of followers has a positive correlation with the number of seed vines for up to 1000 followers, then the variance of the number of seed vines grows too large. There is also a positive correlation between the number of followers and those revining. Second, for 42% of the users, only 10% of the revines come from the followers, meaning they are responsible for only a small percentage of the propagation of the posts. Third, there is a positive correlation between the number of additional recipients and the followers, supporting the first two statements. Fourth, Vine users are most active in the late evening hours; they are most likely to post original videos after 10pm and receive the greatest number of likes, comments and revines. However the peak of the loops belong to the videos posted at 11am. Fifth, using a smaller set of labeled videos based on their content and emotions, we observed that the videos containing the emotion of "love" have been propagated more deeply, and videos containing the emotion of "sadness" have traveled the lowest number of hops in the network compared with the other emotions. Finally, when examining the differences between videos identified as cyberbullying and those that were not, we found that cyberbullying video sessions were revined less, traveled lower hops in the network, but looped more and receive more comments. Additionally, users who posted cyberbullying videos have less followers and following, but they are more active in posting seed videos. We plan to incorporate these findings into the design of more effective information dissemination and automatic classification of cyberbullying incidences in online social networks.

Acknowledgments. This work was supported in part by the National Science Foundation under awards CNS-1162614 and CNS-1528138.

References

1. Akrouf, S., Meriem, L., Yahia, B., Eddine, M.N.: Social network analysis and information propagation: A case study using flickr and youtube networks. International Journal of Future Computer and Communication **2**, 246 (2013)
2. Craig Smith: By the numbers: 25 amazing Vine statistics, July 12, 2012. http://expandedramblings.com/index.php/vine-statistics/ (accessed August 6, 2015)

3. Boyd, D., Golder, S., Lotan, G.: Tweet, tweet, retweet: conversational aspects of retweeting on Twitter. In: 2010 43rd Hawaii International Conference on System Sciences (HICSS), pp. 1–10. IEEE (2010)
4. Java, A., Song, X., Finin, T., Tseng, B.: Why we twitter: understanding microblogging usage and communities. In: Proceedings of the 9th WebKDD and 1st SNA-KDD 2007 Workshop on Web Mining and Social Network Analysis, pp. 56–65. ACM (2007)
5. Kempe, D., Kleinberg, J., Tardos, É.: Maximizing the spread of influence through a social network. In: Proceedings of the Ninth ACM SIGKDD International Conference on Knowledge Discovery and Data Mining, pp. 137–146. ACM (2003)
6. Nikon Candelle: Why Vine videos are becomming more popular than Instagram videos? January 31, 2015. http://www.rantic.com/articles/vine-videos-becoming-popular-instagram-videos/ (accessed August 16, 2015)
7. Krishnamurthy, B., Gill, P., Arlitt, M.: A few chirps about Twitter. In: Proceedings of the First Workshop on Online Social Networks, pp. 19–24. ACM (2008)
8. Kwak, H., Lee, C., Park, H., Moon, S.: What is twitter, a social network or a news media? In: Proceedings of the 19th International Conference on World Wide Web, pp. 591–600. ACM (2010)
9. Suh, B., Hong, L., Pirolli, P., Chi, E.H.: Want to be retweeted? Large scale analytics on factors impacting retweet in twitter network. In: 2010 IEEE Second International Conference on Social Computing (socialcom), pp. 177–184. IEEE (2010)
10. Taxidou, I., Fischer, P.M.: Online analysis of information diffusion in Twitter. In: Proceedings of the Companion Publication of the 23rd International Conference on World Wide Web Companion, International World Wide Web Conferences Steering Committee, pp. 1313–1318 (2014)
11. Leskovec, J., McGlohon, M., Faloutsos, C., Glance, N.S., Hurst, M.: Patterns of cascading behavior in large blog graphs. In: SDM, vol. 7, pp. 551–556. SIAM (2007)
12. Pezzoni, F., An, J., Passarella, A., Crowcroft, J., Conti, M.: Why do i retweet it? An information propagation model for microblogs. In: Jatowt, A., Lim, E.-P., Ding, Y., Miura, A., Tezuka, T., Dias, G., Tanaka, K., Flanagin, A., Dai, B.T. (eds.) SocInfo 2013. LNCS, vol. 8238, pp. 360–369. Springer, Heidelberg (2013)
13. Mahmud, J.: Why do you write this? Prediction of influencers from word use. In: Eighth International AAAI Conference on Weblogs and Social Media (2014)
14. Bakshy, E., Hofman, J.M., Mason, W.A., Watts, D.J.: Everyone's an influencer: quantifying influence on Twitter. In: Proceedings of the Fourth ACM International Conference on Web Search and Data Mining, pp. 65–74. ACM (2011)
15. Azman, N., Millard, D., Weal, M.: Patterns of implicit and non-follower retweet propagation: Investigating the role of applications and hashtags (2011)
16. Zhao, H., Liu, G., Shi, C., Wu, B.: A retweet number prediction model based on followers' retweet intention and influence. In: 2014 IEEE International Conference on Data Mining Workshop (ICDMW), pp. 952–959. IEEE (2014)
17. Petrovic, S., Osborne, M., Lavrenko, V.: Rt to win! Predicting message propagation in twitter. In: ICWSM (2011)
18. Yu, H., Bai, X.F., Huang, C., Qi, H.: Prediction of users retweet times in social network. International Journal of Multimedia and Ubiquitous Engineering 10, 315–322 (2015)
19. Huang, D., Zhou, J., Mu, D., Yang, F.: Retweet behavior prediction in twitter. In: 2014 Seventh International Symposium on Computational Intelligence and Design (ISCID), vol. 2, pp. 30–33. IEEE (2014)

20. Can, E.F., Oktay, H., Manmatha, R.: Predicting retweet count using visual cues. In: Proceedings of the 22nd ACM International Conference on Conference on Information & Knowledge Management, pp. 1481–1484. ACM (2013)
21. Morchid, M., Dufour, R., Bousquet, P.M., Linarès, G., Torres-Moreno, J.M.: Feature selection using principal component analysis for massive retweet detection. Pattern Recognition Letters **49**, 33–39 (2014)
22. Rafiq, R.I., Hosseinmardi, H., Mattson, S., Han, R., Lv, Q., Mishra, S.: Careful what you share in six seconds: detecting cyberbullying instances in Vine. In: Proceedings of The 2015 IEEE/ACM International Conference on Advances in Social Networks Analysis and Mining (ASONAM 2015) (2015)
23. Hosseinmardi, H., Mattson, S.A., Rafiq, R.I., Han, R., Lv, Q., Mishra, S.: Detection of cyberbullying incidents on the instagram social network. CoRR abs/1503.03909 (2015)

Who Stays Longer in Community QA Media?
- User Behavior Analysis in cQA -

Yoshiyuki Shoji[1]([✉]), Sumio Fujita[2], Akira Tajima[2], and Katsumi Tanaka[1]

[1] Department of Social Informatics, Graduate School of Informatics,
Kyoto University, Kyoto, Japan
{shoji,tanaka}@dl.kuis.kyoto-u.ac.jp
[2] Yahoo Japan Corporation, Tokyo, Japan
{sufujita,atajima}@yahoo-corp.jp

Abstract. Macro and micro analyses of why and when users stop asking and/or answering questions on a community question answering (cQA) site were done for a ten years' worth of questions and answers posted on Yahoo! Chiebukuro (Japanese Yahoo! Answers), the biggest cQA site in Japan. The macro analysis focused on how long participants were active in the QA community from the viewpoints of several user characteristics. In turn, the micro analysis focused on how the participants behaviors changes. The behaviors of both askers and answerers were found to change over the time of their active participation: the askers tended to expand the range of categories for which they asked questions while the answerers tended to contract the range of categories for which they answered questions.

1 Introduction

The popularity of community question answering (cQA) media seems to have declined. More than 15 years have passed since online cQA sites appeared. Such sites quickly became popular and one of the most successful types of consumer generated media, serving as a social knowledge community. Their decline in popularly is attributed to the rapid growth of other social network services (e.g., Twitter and Facebook) and the appearance of specific knowledge services (e.g., Stack Overflow). For instance, Yahoo! Chiebukuro (Japanese Yahoo! Answers)[1], the biggest cQA site in Japan, continues to lose users. As shown in Figure 1, the number of posts and new users (askers + answerers) started decreasing around the beginning of 2012. The decreasing number of new answerers is particularly important as it could indicate the beginning of a negative spiral; no one wants to ask in a community without answerers.

Nevertheless, cQA sites continue to be considered useful information resources: cQA contents frequently appear at the top of search results, and many people read them. For instance, Yahoo! Japan reports that there were 714,000,000 Yahoo! Chiebukuro page views in October 2014 [2]. This number is

---

[1] http://chiebukuro.yahoo.co.jp/

[2] http://i.yimg.jp/images/marketing/portal/paper/media_sheet_open.pdf

© Springer International Publishing Switzerland 2015
T.-Y. Liu et al. (Eds.): SocInfo 2015, LNCS 9471, pp. 33–48, 2015.
DOI: 10.1007/978-3-319-27433-1_3

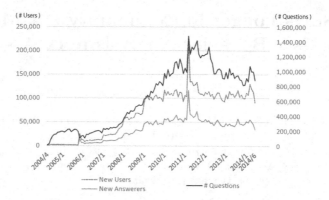

Fig. 1. Number of Newcomers and Posts

much larger than the number of posts in the same month, 524,000. This indicates that cQA contents are still of interest to read-only viewers. Additionally, cQA is a useful information source for researchers. The contents of cQA sites are a good corpus for many research areas, such as question answering systems, information retrieval, and community analysis. For cQA to continue as a useful resource, the decline in the number of users must be reversed.

We analyzed data on cQA users to clarify their reasons for leaving the cQA community. Such understanding should be helpful in stopping them from leaving. We used a dataset obtain from Yahoo! Chiebukuro that contains ten years' worth of data on users, questions, and answers. We conducted basic statistical analysis at the macro level and sequential user behavior analysis at the micro level. The first step was to statistically compare the data on users who soon left with those for users who stayed longer from several aspects. The next step was to model the sequence of actions for users from when they started using the site to when they stopped using the site. To clarify our analysis, we distinguished between long-stay users and short-stay users. Moreover, we compared the user's monthly behavior from joining to leaving the community.

The rest of this paper is organized as follows. Section 2 describes related work. Section 3 explains our method and presents the results of basic statistical analysis. Section 4 explains the comparison of early-phase behavior and late-phase behavior. Section 5 explains our findings and how they might be utilized. We conclude with a summary of the key points in Section 6.

2   Related Work

This section introduces research on cQA service analysis in Section 2.1, research on cQA user motivation on Section 2.2, and research on sequential behavior analysis and phase modeling in Section 2.3.

2.1   cQA Services Analysis

Community question answering is a hot research topic in many areas [22][7]. How to find experts on cQA sites is a typical topic in knowledge community research [23][24][17]. Another typical topic is how to support cQA services. For instance, a question recommendation function helps keep answerers connected to a cQA site [15]. Yang et al. [25] used user lifespan as a tool for comparing cQA communities. User lifespan is tightly related to when and why a user leaves the community. Data on cQA is also used for research in other areas, such as corpus. Several Web information retrieval conferences have included question-answering tracks. For instance, the NTCIR (NII Test Collection for IR Systems) project circulated a cQA dataset as a dataset for a factoid-based retrieval system [6].

There has also been research on the relationship between cQA sites and other social media sites. Morris et al. [21] analyzed data on users who asked questions on an SNS site rather than on a cQA one. One of their findings was that some questions are better suited for a cQA site (e.g., excessively private questions and political questions) while some are better suited for an SNS site (e.g., social invitations and context-dependent questions).

2.2   cQA User Motivation

One of the biggest research topics related to the analysis of cQA services is classification of question types and estimation of the askers' intents. Many researchers have proposed rules for classifying questions and asker intents. Kim et al. [13] classified questions as information, suggestion, opinion, or other for use in estimating the best answer automatically. Harper et al. [10] classified questions into factual, opinion, and advice for use in finding expert answerers on cQA sites. Both classification methods cover opinion. This is because askers on cQA sites often want not only factual information, but also the opinions of others. Liu et al. [19] classified cQA asker intent into navigational, informational, transactional, and social. This classification method supports the conventional Web search query classification [14]. It also highlights a characteristic of cQA sites; that is, the difference between web search users and cQA users is whether they have social intent. Chen et al. [5] also focused on the social intent of the asker, such as subjective, objective, and social. These research efforts generally focused on investigating the reasons that cQA users ask questions. Our work aims to broaden this target; focusing not only on the askers but on all users, including the answerers.

Harper et al. [9] observed that users of cQA sites have two general types of intent. One is an **informational** intent, and the other is a **conversational** intent. Users who have an informational intent tend to ask specific questions that have a clear answer. Those who have a conversational intent want to communicate through cQA sites, and they tend to ask questions that have no clear answer. Harper et al. asserted that users can be classified by using category, text, and user information. They also asserted that conversational questions have low archival quality. Aikawa et al. [2] proposed a method for sorting subjective questions in the same way. Liu et al. [16] discovered the same direction for askers who switch from web search to cQA asking; that is, they ask informational questions more than sticky askers, i.e., those who ask question on a cQA site over a long period of time.

2.3   Sequential Behavior Analysis and Phase Modeling

There has been much research on the analysis of user behavior by using sequential action modeling techniques, not only for cQA but also for other services. One typical example is user modeling for electronic commerce sites. Some researchers have split the total user time into sessions and then created a sequential model for use in estimating whether the user will buy something.

They considered many users actions to be features, such as page transitions, click-through, and search queries [12]. Several modeling methods have been used for this kind of situation. For instance, the hidden Markov model has been widely used for modeling the behaviors of web searchers and web service users [20]. Hassan et al. used relational Markov models [3] to model Web user behavior as a means to enhance Web navigation [11].

Another research analyses the user leaving behavior of online communities, including QA community. Churn Prediction is one of the typical example. Dror et al. predicted how long the new user uses the cQA service [8]. Backstrom et al. analysed users who stay longer in Yahoo! Groups[4] by using machine learning techniques. They pointed the user's staying time in service dipends on the user property but not on the group property.

3   Macro Analysis of cQA User Behaviors

In this section, we explain our macro analysis of users who stop using a cQA service. We compared two types of users: those who use the service for a relatively long time and those who use it for a relatively short time. To do this, we set the following research question and hypotheses:

RQ1: How do users who leave soon differ from those who stay longer?
  H1.1: They make different types of contributions.
  H1.2: The range of categories in which they participate differs.
  H1.3: Their satisfactions and rewards differ.

To answer this research question and to test these hypotheses, we used a massive dataset to analyze user behavior, such as the characteristics of users who leave the QA community. Section 3.1 describes the dataset, and section 3.2 explains our analysis.

3.1   Dataset

The Yahoo! Chiebukuro (Japanese Yahoo! Answers) is the biggest online question answering community in Japan and is freely available to all Yahoo! Japan users. It began operation in 2004. The dataset contained slightly more than ten years' worth of data (March 2004 to December 2014).

- Questions: 84,123,965 questions that received one or more answers
- Answers: 224,969,887 answers to the questions
- Users: 10,391,194 anonymized users who had asked and/or answered one or more questions

Questions that do not receive an answer are deleted from the system, so the dataset contained only questions that received one or more answers. An Asker has the option to specify the best answer (BA). If the asker does not specify a BA, other users can specify one by voting. Yahoo! Chiebukuro has three levels of categories. The top 16 categories are shown in Table 1.

Table 1. Category Detail

Category name	staying	% info	# users	# boards	<0.5	0.5–1	1–3	>3	length
Entertainment and Hobbies	11.6	66%	2,328,386	17,571,681	43%	35%	10%	13%	68.6
Life, Love, and Human Relationship	10.7	0%	1,750,675	8,255,538	46%	34%	9%	11%	181.9
Sports, Outdoor, and Cars	10.3	48%	1,094,271	6,805,630	48%	38%	6%	8%	75.3
Living	9.1	77%	1,092,231	7,303,245	44%	43%	5%	8%	108.9
Kids and Schools	8.9	25%	1,015,797	6,193,570	47%	36%	7%	11%	85.3
Health, Beauty, and Fashion	8.4	68%	843,838	7,185,539	54%	29%	7%	10%	111.7
Internet, PC, and Electronics	7.9	82%	1,204,661	8,348,942	37%	53%	4%	6%	107.9
Local, Travel, and Trip	7.9	56%	707,311	3,887,467	60%	27%	6%	6%	80.7
Knowledge, Education, and Science	7.7	80%	794,644	3,848,875	48%	39%	6%	7%	155.2
News, Politics, and Global	6.8	27%	621,021	2,984,093	62%	25%	6%	8%	96.9
Jobs and Career	6.6	41%	721,055	2,360,837	49%	38%	5%	7%	146.8
Others	6.3	9%	541,983	3,498,145	64%	23%	6%	7%	44.3
Business, Economics, and Money	6.1	51%	495,974	2,065,734	69%	19%	6%	6%	121.7
Manner and Ceremony	6.0	28%	463,642	1,283,267	72%	15%	6%	7%	142.3
Yahoo! Japan	5.7	31%	430,842	2,110,650	75%	11%	7%	8%	85.6
Computer Technology	5.3	92%	358,264	420,537	81%	6%	6%	7%	255.7

Since our research focused on users leaving the cQA community, we neglected users still using the site. It has been reported that a certain percentage of users who leave for a while never come back [25]. We assumed that the threshold is six months; the dataset contained user data from March 2004 to December 2014, so we eliminated users who asked or answered one or more times after June 2014. In this dataset, 87.85% of the users had already stopped using the site. The average staying time was 9.98 months, as shown in Table 2.

3.2   Basic Statistics

To tackle research question 1, to clarify the difference between shot-stay users and long-stay users, we verify three hypotheses.

Table 2. User Staying Time

Staying Time (month)	
Average	9.98
Mode	1
Median	20

Table 3. Only Ask/Answer

%user	
Asking only	52.8%
(10 times +)	2.2%
Answering only	14.9%
(10 times +)	1.9%

Table 4. User Statistics

	Asking	Answering	Both
MAX	29,757	353,126	362,692
Median	8	733	741
Average	7.96	20.36	28.32
Mode	1	0	1
% only 1 times	35.7%	14.9%	37.0%
% 10 times or less	87.8%	86.9%	78.0%
% 50 times or less	97.6%	95.3%	92.9%
% 100 times or more	1.1%	2.8%	4.0%

H1.1: Difference in Contributions. First of all, we set a simple hypothesis: a user's contribution, such as frequency and ratio of asking and answering, reflects his or her length of stay. A breakdown of the user contributions is shown in Table 4. One simple finding is that many Yahoo! Chiebukuro users were light users; they used it once and then stopped using the service. A third or more of the users left the community after asking or answering once. Some were askers who asked only once and left the service. Apparently, they had a problem and they used the service to find a solution. Once they had a solution, they stopped using the service. There were some heavy users as well. They used the service for a long time and frequently post both questions and answers.

Users of cQA sites can contribute as an asker or answerer or as both. The relationship between staying time and the ratio of asking and answering is shown in Figure 3. The horizontal axis represents the average ratio of questions to answers that users posted. The ratio of asking correlates significantly with staying time; the rank correlation coefficient was -0.88. Users who stayed longer tended to answer more than to ask. Moreover, they posted both questions and answers, as opposed to inactive short-stay users who posted either a few questions or a few answers. Few users only asked or answered more than ten times (see Table 3). Another phenomenon that can also be seen is that users in cQA differentiate into two types: asker and answerer. It is unusual to find users who post both questions and answers in similar numbers. The number of posts per month shows the characteristics of askers and answerers. Figure 3 additionally suggests that it is generally difficult to continue posting very frequently for long time. While users who stay longer post at their own pace, the answerers tend to be more frequent posters later in the staying time. The number of heavy answerers who posted more than 100 times per month and stayed 3 years or more was 2011, and the number of users who stayed five years or more was 861. The corresponding numbers for askers were much smaller: 82 and 37, respectively. Thus, users who use cQA sites for a longer period apparently like answering more than asking.

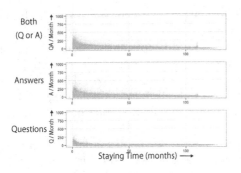

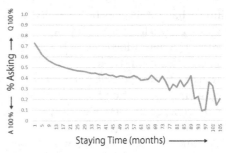

Fig. 2. Number of Posts Per Month vs Staying Time

Fig. 3. Asking Answering Ratio vs Staying Time

H1.2: Difference in Category Participation. The categories on a cQA site directly reflect the users' interests. Moreover, the interests may affect user behavior. To test this, we defined a category for each user on the basis of how many times the user posted answers and/or questions in each category. The defined category was the one with the most posts. If a user posted a similar number of questions and/or answers to two or more categories, we randomly set the user's category to one of them.

Table 1 shows the details for each category. The staying times ranged from 5.3 to 11.6 months. The users who were defined with a daily life category tended to stay longer: the top three categories by staying time were "Entertainment and Hobbies," "Life, Love, and Human Relationship," and "Sports, Outdoor, and Cars." Conversely, the users who were defined with more serious categories, such as "Business, Economics, and Money," "Manner and Ceremony," and "Computer Technology," tended to stay for a shorter time.

Another factor related to staying time was the activeness of the community: the correlation coefficient between staying time and the number of posts was 0.91, and the correlation between staying time and the number of users was 0.88. Each category had a different distribution ratio of long-stay and short-stay users. The columns "< 0.5 (0.5 year or less)," "0.5 – 1 (0.5 – 1 years)," "1 - 3 (1 – 3 years)," and "> 3 (more than 3 years)" shows the component ratios. The categories with many short-stay users seem better suited for askers with sudden problems. The top three categories in this regards were "Computer Technology," "Yahoo! Japan," and "Manner and Ceremony." Each of them contained many questions that needed a quick answer, like "My computer has broken down; what should I do?" and "A relative has died; how much money should I give as a condolence gift."

Moreover, the average length of posts in each category was inversely correlated with staying time. Most of the long-stay users tended to participate in short and frequent communication rather than lengthy discussions. Several researchers have observed that cQA users generally have two types of intent: informational and conversational. This is highly related to the categories. Harper et al. [9]

pointed that intent can be classified into informational and conversational only by category with an accuracy of 70 %. For analysis, we classified 100 randomly selected QA pairs for each category by hand. The ratio of informational questions ("% Info" in Table 1) did not correlate with staying time ($R = -0.00$). Although the component percentages of users in the community correlated with staying time, the ratio of long-stay users was low in the informational categories. The percentage of users who stayed three or more years had a weak inverse correlation with % info ($R = -0.28$).

As another perspective, we focused on category diversity. Adamic et al. pointed that the diversity of categories in which a user posts questions and/or answers indicates whether a user is a specialist or not [1]. That is, the diversity of questions and answers represents the range of the user's interests and knowledge. We calculated the entropy of each user on the basis of the categories in which he or she posted questions and answers:

$$etp(u) = - \sum_{c \in C} \frac{|p \in P(u)|cat(p) = c|}{|P(u)|} log_{|C|} \frac{|p \in P(u)|cat(p) = c|}{|P(u)|}, \qquad (1)$$

where $P(u) = p_1, p_2, p_3\cdot$ is the set of posts by user u, $cat(p)$ is the category of post p, and $C = c_1, c_2, c_3 \ldots$ is the top level category. It drops between zero to one by normalizing by taking logarithm $|C|$. The overall tendency is that a user who stays longer posts in a broader range of categories. We calculated category diversity for both questions and answers. Spearman's rank correlation coefficient for staying time and category diversity for both was 0.99. While both were correlated with staying time, they were markedly different in terms of distribution and detail. As shown in Figure 3.2, there was a sharp contrast between askers and answerers. The most notable feature in this figure is the sparseness at the bottom-right corner of the "Questions" scattergram. Askers who stay longer tend to post questions in a broader range of categories, so it is quite possible that, if an asker continues asking questions in a specific category, he or she will eventually run out of questions. In contrast, answerers who stay longer tend not to post answers to such a broad range of categories. The distribution does not appear to converge or follow a normal distribution. The range of categories in which a user posts answers is assumed to reflect the answerers' field of expertise. Since no one is truly omniscient, no one can continue to answer questions in a wide range of categories. Therefore, answerers who consistently post answers in several specialized categories tend to stay longer.

H1.3: Difference in Satisfaction and Reward. The satisfaction of askers and the reward to answerers have also been used to estimate the motivation of users to continue using a cQA service. The best answer (BA) rate is related to satisfaction and reward [18].

For askers, indicating the BA is a way to indicate satisfaction. Askers typically select the BA when they are satisfied and may not select a BA if all the answers are unsatisfactory. (As mentioned above, other users can then vote for the BA.) Similarly, having his or her answer selected as the BA is rewarding

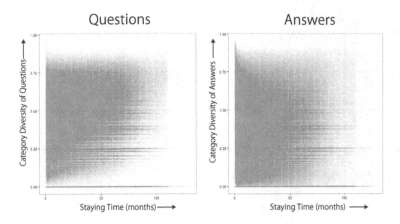

Fig. 4. Category Diversity vs Staying Time

to the answerer. Answerers wanting to get more BA selections tend to answer questions more frequently, and politely.

The BA ratios of askers and answerers are plotted against staying time in Figure 5. The vertical axis represents the ratio of selecting or receiving a BA, and the horizontal axis shows the staying time. For askers, there is no correlation between the rate of selecting the BA and staying time ($R = 0.12$). This is because there are many short stay users, and they tend to select the BA randomly. The graph does show, however, that users who select fewer BAs tend to stay a shorter length of time. The average probability of a user selecting the BA is 0.46. For answerers, the correlation is not much better ($R = 0.17$), indicating that they do not seem particularly concerned about the BA rate.

In our dataset, the average number of answers for a question was 2.7. It is thus natural for the BA rate to converge to 0.37. In other words, people who answer questions over a long period of time do not seem eager for their answers to be selected as the BA.

4   Micro Analysis of cQA User Behaviors

To analyze user behavior from the micro view, we compared the users' actions during the user action phase, from the month in which the user joined the service to the month in which the user left the service. Again we set a research question and three hypotheses.

RQ2: Do users change their actions during the period from when they start using the site to when they stop using the site?
H2.1: The contribution type changes.
H2.2: The category in which they participate changes.
H2.3: The satisfaction and reward change.

To test these hypotheses, we compared users' monthly behaviors from when they started using the site to when they stopped using it.

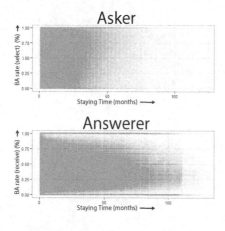

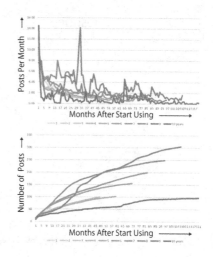

Fig. 5. Ratio of Giving or Receiving
Best Answer vs Staying Time

Fig. 6. Posts frequency vs n-th month

4.1   User Phase Analysis

Before analyzing the user action phase, we removed the data on users who posted less than five questions and/or answers. The number of users who posted 5 times or more was 3,256,402, i.e., 31.24% of all users. We compared the monthly activity of users from two viewpoints: staying time and period of time from start using or stopped using. For instance, we focused on what users who stayed m years did in the n-th month after starting to use the site or on the last action of long-stay users before they left.

H2.1: Difference in Contributions. How users commit to the site may vary depending on the their period of use. We focused on changes of their contribution, such as the frequency of posts and ratio of asking and answering.

Figure 6 shows the relation between their period of use and frequency of posts. The upper graph shows the number of post per month, and the lower graph shows the cumulative number of posts. The horizontal axis means how many months have passed after users start using the site. It shows the behavior of ten user groups divided by their staying time, from a year to 10 years. The graph shows the overall trend, that is, users start using the cQA site in the active state. They frequently post questions and answers. After that, they decrease the frequency of posts, and finally they leave. For most users, the period when they post with high frequency is right after joining. Both short-stay users and long-stay users post about ten times in a month when they start using the service. Their participation decreases as time goes on, although short-stay users are easily excited but quick to cool down again (see Table 5). The long-stay users tend to continuously use cQA even after their participation decreases, in contrast to the short-stay users who post more frequently and leave abruptly.

Table 5. Frequency of Post and Time Prsiod

Staying years	Period of use		
	starting 1 year	3 years	5 years
1	12.53	3.25	-	-
2	10.98	2.36	-	-
3	10.21	2.31	1.15	-
4	9.23	2.75	1.31	-
5	9.63	2.19	1.27	0.68
6	12.39	3.92	1.74	1.10
7	12.56	3.82	1.80	1.70
8	10.47	4.98	2.19	1.60
9	7.55	1.09	2.40	0.71
10	14.50	0.89	2.50	0.39

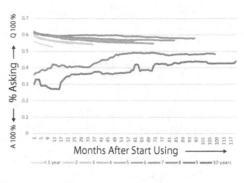

Fig. 7. QA ratio vs n-th month

Asking and answering also depends on the period of use. Figure 7 shows the ratio of asking and answering in n-th month from joining the service. The vertical axis shows the users' asking ratio up to that time. As explained above, most users start using cQA as askers, and increase their percentage of answering during their using. Except the remarkable long-stay users (those who stay nine or more years), users start using the cQA site by posting 60 % questions and 40% answers, and leave the site if the ratio of asking becomes 55%. It is possible that they exhaust the questions and lose interest in the site. For the users who leave after 1 year to 9 years, the speed of decreasing of asking ratio is slower. On the other hand, users who use the cQA site for nine or more years are more heterogeneous than the others. What we can say is that the users who use Yahoo! Chiebukuro for 9 years or more are Early Adopters; because the service started in 2004. They answered more frequently than most of the people joining later, especially in the early period of the service. The amount of users were smaller at that time, therefore they may retained communication by posting many answers for rare questions.

We found that 67.76% of the users who posted five times or more, started by asking questions. As mentioned in section 3.2, many users stop using a cQA site after they post a few questions. Others apparently started to like the site and become regular visitors, and some of the regular visitors become long-staying users. There was only a small difference between the first action of short-stay users (stay less than a year) and that of long-stay users (stay 5 or more years). The percentage of short-stay users whose first post was a question was 62.48 % and that for long-stay users was 70.10%. That is, more long-stay users start by asking a question. The last posts of most users before they leave the site are also asking, as 67.76% of users stop using the service after they post a question. These facts suggest two factors: a great multitude of users post answer in their middle period, and some particular users who are enthusiastic answerers post a large amount of answers. It is possible that users who leave by asking a question

Table 6. first Post and Last Post

Category / Staying Years	First Post						Last Post					
	Asking			Answering			Asking			Answering		
	<1	1–5	5–10	<1	1–5	5–10	<1	1–5	5–10	<1	1–5	5–10
Entertainment and Hobbies	12.6%	13.1%	8.3%	8.9%	7.7%	4.6%	10.4%	10.3%	7.7%	10.7%	7.8%	5.1%
Internet, PC, and Electronics	6.7%	9.3%	11.0%	2.0%	1.6%	2.0%	7.5%	11.4%	13.5%	2.4%	2.0%	2.2%
Life, Love, and Human Relationship	7.1%	4.9%	2.8%	5.7%	4.4%	3.1%	5.3%	4.4%	3.5%	7.0%	5.7%	5.7%
Living	5.0%	6.8%	8.4%	2.6%	2.3%	3.0%	4.8%	7.1%	9.3%	3.3%	2.6%	2.9%
Health, Beauty, and Fashion	6.2%	6.6%	6.5%	3.0%	2.4%	2.9%	5.7%	6.3%	5.5%	3.5%	2.3%	2.1%
Sports, Outdoor, and Cars	4.0%	5.5%	6.2%	3.0%	2.8%	3.4%	3.5%	5.1%	6.0%	3.5%	3.0%	3.2%
Knowledge, Education, and Science	3.9%	4.1%	4.0%	1.8%	1.5%	1.8%	3.5%	4.2%	4.0%	2.1%	1.5%	1.4%
Local, Travel, and Trip	2.4%	3.8%	4.1%	1.0%	1.0%	1.3%	2.5%	4.8%	6.4%	1.2%	1.2%	1.5%
Kids and Schools	3.8%	3.8%	3.6%	2.4%	2.3%	2.6%	3.0%	3.4%	2.6%	2.9%	2.5%	2.4%
Others	2.2%	1.2%	0.9%	1.8%	0.7%	0.6%	1.8%	1.0%	0.7%	2.2%	0.9%	0.6%
News, Politics, and Global	1.0%	0.7%	0.7%	1.3%	0.9%	0.9%	1.0%	0.8%	0.7%	1.6%	1.2%	1.2%
Jobs and Career	2.2%	2.6%	2.5%	0.9%	0.8%	0.9%	1.8%	2.6%	2.5%	1.0%	0.9%	0.9%
Yahoo! JAPAN	2.4%	2.6%	4.6%	1.6%	0.9%	1.6%	1.9%	1.0%	0.8%	1.8%	0.5%	0.3%
Business, Economics, and Money	1.7%	2.5%	3.4%	0.7%	0.6%	0.8%	1.6%	2.6%	3.6%	0.8%	0.6%	0.7%
Manner and Ceremony	0.9%	1.3%	1.9%	0.7%	0.6%	0.8%	0.8%	1.1%	1.5%	0.8%	0.6%	0.7%
Computer Technology	0.4%	0.6%	0.8%	0.1%	0.1%	0.1%	0.4%	0.6%	0.8%	0.1%	0.1%	0.1%

are losing their interest in the site and they just do not want to search a question relevant to their interest. They may be tired to be committed to the cQA site.

H2.2: Difference in Categories. The users must change the category which they commit in response to current time period. First, we focused on the category which the user committed at the first posting. Table 6 shows the category and the type of first posts. As explained above, user tend to start using the cQA site as an asker. Regardless of users' staying time, a lot of user start using cQA in entertainment and advice categories. These categories are good appeal to get new users. They are originally popular categories. Let us compare the last category they commit before leave. The ratio of answering increases especially in counseling categories and Entertainment categories. This may reflects that users who stays longer in cQA tend to love communication but not serious arguments.

Category diversity which users ask or answer also change with their phase. Figure 8 shows the relation between users' phase and the diversity of committing categories. Diversity generally increases when a user posts in more categories in a balanced manner. Therefore, it has correlation to the length of using. Users post questions and answers in new categories for a while after joining the service. On some level, increasing of the diversity stops because they stop penetrating new categories or they posting their own specific categories disproportionately.

H2.3: Difference in Satisfaction and Reward. Users' behavior on getting or giving BA also depends on their phase and staying time. The simplest hypothesis that the asker who got good answer must stay, and answerer who got BA must stay. To test it, we focus on their first actions. Figure 9 shows the relation between first action and staying time by the BA. Surprisingly, in the case of the user who post a question first, the ratio of giving BA has inverse correlation to

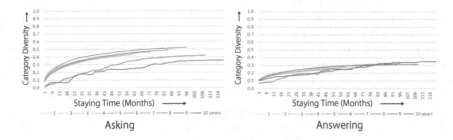

Fig. 8. Category-diversity vs Time Period

the staying time (R = -0.82). Through the whole data, probability of giving BA at first asking time is 33.54%. One of long-stay users is lower, especially in the case of staying months is longer than 90 months. One reason is that they are early adopters and stay the site since it was immaturity. In that time, the quality of answer might be lower, and the rule of giving BA was not enough spread.

In the case of users who join the cQA as answerers, user who got BA at first post tend to stay linger(R=0.45). To get BA on first post is not easy; the probability if 10.88%. Such answerers might imbibe a taste for getting BA as an incentive.

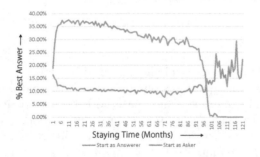

Fig. 9. BA Rates on First Posts

5 Discussion

We identified several tendencies of long-stay and short-stay users of cQA sites from our analysis that should be useful in keeping users active on cQA sites for a longer period of time. First of all, many users use cQA only once as an asker and leave. Keeping such users is an important issue. Related to this, we have that answerers tend to stay longer than askers. It may be possible to recommend a catchy question for askers that would cause them to unconsciously switch from being an asker to being an answerer. Most users start using cQA as askers; then, as they stay longer, they differentiate into askers and answerers over time.

For user who regularly uses cQA for a reasonable amount of time, navigation or recommendation should be personalized to their stance.

Our findings related to categories may also be useful for keeping users. Askers who post questions in a wide range of categories tend to stay longer, while answerers who post in a narrow range of categories tend to stay longer. For askers, personalized navigation may help them to expand the range of categories in which they ask questions. For answerers, it is better to let them focus on their field of expertise. cQA sites are also used to get information about various types of amusement. Users who mainly post in entertainment-related categories tend to stay longer. The categories in which the posts are short tend to be more popular and attract more long-stay users. Some desultory exchanges of words in short sentence are not useful for improving the information quality of cQA sites, but it is useful to sustain the community in active state.

Askers who tend to select the BA tend to stay longer. In other words, askers who feel that they did not receive a good answer tend to leave sooner. It is safe to assume that good answers make askers more satisfied, and this encourages them to stay longer. In contrast, answerers who stay longer are apparently less concerned about their BA rate.

Our phase analysis revealed that most users of a cQA site are not always active. While users who stay longer are also not always active, they tend to come back sooner to answer or to ask. One simple way to keep users active is to send them reminders or advertisements.

6   Conclusion

We have conducted both macro and micro analysis to clarify why and when users stop asking or answering questions on cQA sites. We used a dataset obtained from the Yahoo! Chiebukuro site and analyzed it both statistically and sequentially. For the macro analysis, we set three simple hypothesis related to contribution, category, and reward and satisfaction and analyzed how long users participate in the QA community. For the micro analysis, we analyzed how cQA users behave by comparing users' actions in early-phase and late-phase. We found that askers who stay longer in cQA sites tend to ask questions in a wider range of categories while answerers who stay longer tend to answer questions in a narrower range of categories.

Acknowledgments. This work was supported in part by the following projects: Grants-in-Aid for Scientific Research (Nos. 15H01718) from MEXT of Japan.

References

1. Adamic, L.A., Zhang, J., Bakshy, E., Ackerman, M.S.: Knowledge sharing and yahoo answers: everyone knows something. In: Proceedings of the 17th International Conference on World Wide Web, pp. 665–674. ACM (2008)
2. Aikawa, N., Sakai, T., Yamana, H.: Community qa question classification: Is the asker looking for subjective answers or not? IPSJ Online Transactions **4**, 160–168 (2011)
3. Anderson, C.R., Domingos, P., Weld, D.S.: Relational markov models and their application to adaptive web navigation. In: Proceedings of the Eighth ACM SIGKDD International Conference on Knowledge Discovery and Data Mining, pp. 143–152. ACM (2002)
4. Backstrom, L., Kumar, R., Marlow, C., Novak, J., Tomkins, A.: Preferential behavior in online groups. In: Proceedings of the 2008 International Conference on Web Search and Data Mining, pp. 117–128 (2008)
5. Chen, L., Zhang, D., Mark, L.: Understanding user intent in community question answering. In: Proceedings of the 21st International Conference Companion on World Wide Web, pp. 823–828. ACM (2012)
6. Sakai, T., Ishikawa, D., Kando, N.: Overview of the ntcir-8 community qa pilot task (part i): the test collection and the task. In: Proceedings of NTCIR-8 Workshop Meeting, pp. 421–432 (2010)
7. Dearman, D., Truong, K.N.: Why users of yahoo!: answers do not answer questions. In: Proceedings of the SIGCHI Conference on Human Factors in Computing Systems, pp. 329–332. ACM (2010)
8. Dror, G., Pelleg, D., Rokhlenko, O., Szpektor, I.: Churn prediction in new users of yahoo! answers. In: Proceedings of the 21st International Conference on World Wide Web, pp. 829–834. ACM (2012)
9. Harper, F.M., Moy, D., Konstan, J.A.: Facts or friends?: distinguishing informational and conversational questions in social q&a sites. In: Proceedings of the SIGCHI Conference on Human Factors in Computing Systems, pp. 759–768. ACM (2009)
10. Harper, F.M., Raban, D., Rafaeli, S., Konstan, J.A.: Predictors of answer quality in online q&a sites. In: Proceedings of the SIGCHI Conference on Human Factors in Computing Systems, pp. 865–874. ACM (2008)
11. Hassan, A., Jones, R., Klinkner, K.L.: Beyond dcg: user behavior as a predictor of a successful search. In: Proceedings of the Third ACM International Conference on Web Search and Data Mining, pp. 221–230. ACM (2010)
12. Iwata, T., Watanabe, S., Yamada, T., Ueda, N.: Topic tracking model for analyzing consumer purchase behavior. In: Proceedings of the 21st International Joint Conference on Artifical Intelligence, pp. 1427–1432. Morgan Kaufmann Publishers Inc. (2009)
13. Kim, S., Oh, J.S., Oh, S.: Best-answer selection criteria in a social q&a site from the user-oriented relevance perspective. Proceedings of the American Society for Information Science and Technology **44**(1), 1–15 (2007)
14. Lee, U., Liu, Z., Cho, J.: Automatic identification of user goals in web search. In: Proceedings of the 14th International Conference on World Wide Web, pp. 391–400. ACM (2005)
15. Li, B., King, I.: Routing questions to appropriate answerers in community question answering services. In: Proceedings of the 19th ACM International Conference on Information and Knowledge Management, pp. 1585–1588. ACM (2010)

16. Liu, Q., Agichtein, E., Dror, G., Maarek, Y., Szpektor, I.: When web search fails, searchers become askers: understanding the transition. In: Proceedings of the 35th International ACM SIGIR Conference on Research and Development in Information Retrieval, pp. 801–810. ACM (2012)

17. Liu, X., Croft, W.B., Koll, M.: Finding experts in community-based question-answering services. In: Proceedings of the 14th ACM International Conference on Information and Knowledge Management, pp. 315–316. ACM (2005)

18. Liu, Y., Bian, J., Agichtein, E.: Predicting information seeker satisfaction in community question answering. In: Proceedings of the 31st Annual International ACM SIGIR Conference on Research and Development in Information Retrieval, pp. 483–490. ACM (2008)

19. Liu, Y., Li, S., Cao, Y., Lin, C.-Y., Han, D., Yu, Y.: Understanding and summarizing answers in community-based question answering services. In: Proceedings of the 22nd International Conference on Computational Linguistics, vol. 1, pp. 497–504. Association for Computational Linguistics (2008)

20. Miller, D.R.H., Leek, T., Schwartz, R.M.: A hidden markov model information retrieval system. In: Proceedings of the 22nd Annual International ACM SIGIR Conference on Research and Development in Information Retrieval, SIGIR 1999, New York, NY, USA, pp. 214–221. ACM (1999)

21. Morris, M.R., Teevan, J., Panovich, K.: What do people ask their social networks, and why?: a survey study of status message q&a behavior. In: Proceedings of the SIGCHI Conference on Human Factors in Computing Systems, pp. 1739–1748. ACM (2010)

22. Nam, K.K., Ackerman, M.S., Adamic, L.A.: Questions in, knowledge in?: a study of naver's question answering community. In: Proceedings of the SIGCHI Conference on Human Factors in Computing Systems, pp. 779–788. ACM (2009)

23. Pal, A., Konstan, J.A.: Expert identification in community question answering: exploring question selection bias. In: Proceedings of the 19th ACM International Conference on Information and Knowledge Management, pp. 1505–1508. ACM (2010)

24. Riahi, F., Zolaktaf, Z., Shafiei, M., Milios, E.: Finding expert users in community question answering. In: Proceedings of the 21st International Conference Companion on World Wide Web, pp. 791–798. ACM (2012)

25. Yang, J., Wei, X., Ackerman, M.S., Adamic, L.A.: Activity lifespan: an analysis of user survival patterns in online knowledge sharing communities. In: ICWSM (2010)

Analyzing Labeled Cyberbullying Incidents on the Instagram Social Network

Homa Hosseinmardi[1]($\boxtimes$), Sabrina Arredondo Mattson[2], Rahat Ibn Rafiq[1],
Richard Han[1], Qin Lv[1], and Shivakant Mishra[1]

[1] Computer Science Department, University of Colorado Boulder, Boulder, CO, USA
{homa.hosseinmardi,rahat.ibnrafiq,
richard.han,qin.lv,shivakant.mishra}@colorado.edu
[2] Institute of Behavioral Science, University of Colorado Boulder, Boulder, CO, USA
sabrina.mattson@colorado.edu

Abstract. Cyberbullying is a growing problem affecting more than half of all American teens. The main goal of this paper is to study labeled cyberbullying incidents in the Instagram social network. In this work, we have collected a sample data set consisting of Instagram images and their associated comments. We then designed a labeling study and employed human contributors at the crowd-sourced CrowdFlower website to label these media sessions for cyberbullying. A detailed analysis of the labeled data is then presented, including a study of relationships between cyberbullying and a host of features such as cyberaggression, profanity, social graph features, temporal commenting behavior, linguistic content, and image content.

1 Introduction

As online social networks (OSNs) have grown in popularity, instances of cyberbullying in OSNs have become an increasing concern. In fact more than half of American teens have reported being the victims of cyberbullying [1]. Moreover, research has found links between experiences of cyberbullying and negative outcomes such as decreased performance in school, absenteeism, truancy, dropping out, and violent behavior [2], and potentially devastating psychological effects such as depression, low self-esteem, suicide ideation, and even suicide [3–6], that can have long term effects in the future life of victims [7]. Incidents of cyberbullying with extreme consequences such as suicide are now routinely reported in the popular press. For example cyberbullying of Jessica Logan via her image shared in Facebook and MySpace and of Hope Sitwell with her image shared in MySpace is attributed to their suicides [8], [9].

Given the gravity of the consequences cyberbullying has on its victims and its rapid spread among middle and high school students, there is an immediate and pressing need for research to understand how cyberbullying occurs in OSNs today, so that effective techniques can be developed to accurately detect cyberbullying. In [6], it is reported that experts in the field of cyberbullying could favor automatic monitoring of cyberbullying on social networking sites and propose effective follow-up strategies.

© Springer International Publishing Switzerland 2015
T.-Y. Liu et al. (Eds.): SocInfo 2015, LNCS 9471, pp. 49–66, 2015.
DOI: 10.1007/978-3-319-27433-1_4

Our work makes the important distinction between cyberaggression and cyberbullying. Cyberaggression is defined as aggressive online behavior that uses digital media in a way that is intended to cause harm to another person[10]. Examples include negative content and words such as profanity, slang and abbreviations that would be used in negative posts such as hate, fight, wtf. Cyberbullying is one form of cyberaggression that is more restrictively defined as (1) an act of aggression online with (2) an imbalance of power between the individuals involved and (3) repetition of the aggression [2,10–15]. Similar to traditional bullying, it is the combination of the aggressive behavior, repeated acts, and the victim's inability to defend himself or herself that severely impacts many teens [2]. Particularly important in the context of cyberbullying, is the permanent nature of the online posts (until they're removed), the ease and wide distribution in which aggressive posts can be made,

Fig. 1. An example of comments posted on Instagram. To give more room for the text, we have moved the associated image to overlay some of the text.

the difficulty of identifying the behavior, the ability to be connected and exposed to online interaction 24/7, and the growing number of potential victims and perpetrators [4]. The power imbalance can take on a variety of forms including physical, social, relational or psychological [14,16–18], such as a user being more technologically savvy than another [2], a group of users targeting one user, or a popular user targeting a less popular one [19]. Repetition of cyberbullying can occur over time or by forwarding/sharing a negative comment or photo with multiple individuals [19].

Facebook, Twitter, YouTube, Ask.fm, and Instagram have been listed as the top five networks with the highest percentage of users reporting experience of cyberbullying [20]. Instagram is of particular interest as it is a media-based social network, which allows users to post and comment on images. An example of an Instagram media session is shown in Figure 1. Cyberbullying in Instagram can happen in different ways, including posting a humiliating image of someone else by perhaps editing the image, posting mean or hateful comments, aggressive captions or hashtags, or creating fake profiles pretending to be someone else [21].

The main goal of this paper is to study cyberbullying in Instagram. To do so, we first collected a large sample of Instagram data comprised of 3,165K media sessions (images and their associated comments) taken from 25K user profiles. Next, we provided labeling instructions and Instagram media sessions to human labelers at the crowd-sourced CrowdFlower website to identify occurrences of cyberbullying and cyberaggression in Instagram. We then analyzed the labeled data set, reporting the relationship of different features of these media sessions to cyberbullying. This paper make the following important contributions:

- We provide a clear distinction between cyberbullying and general cyberaggression. Cyberbullying is a form of cyberaggression that can have devastating effects on the victims and perpetrators. Most of the prior research in this area is more appropriately described as investigating cyberaggression.
- We obtain ground truth cyberbullying behavior in Instagram by instructing human crowd-sourcers to label Instagram images and their associated comments according to both the more restrictive definition of cyberbullying and the more general definition of cyberaggression. Labelers are provided with the image and its associated comments at the same time to be able to understand the context and label accordingly.
- We present a novel detailed analysis of the labeled media sessions, including the relationships between cyberbullying and a host of factors, such as cyberaggression, profanity, social graph properties (liking, followers/following), the interarrival time of comments, the linguistic content of comments, and labeled image content.

2   Related Works

Prior works that investigated cyberbullying [22–32] are more accurately described as research on cyberaggression, since these works do not take into account both the frequency of aggression and the imbalance of power. These works have largely applied a text analysis approach to online comments, since this approach results in higher precision and lower false positives than simpler list-based matching of profane words [33]. Previous research [27, 29, 34, 35] applied text based cyberbullying on Formspring.me and Myspace dataset. Dinakar *et al.* investigated both explicit and implicit cyberbullying by analyzing negative text comments on YouTube and Formspring profiles [25]. Sanchez and Kumar proposed using a Naive Bayes classifier to find inappropriate words in Twitter text data for bullying detection [26]. They tracked potential bullies, their followers, and the victims. Also some researchers tried to detect bullies and victims by looking at the number of received and sent, beside detecting aggressive comments [36] and [32] . Dadvar *et al.* investigated how combining text analysis with MySpace user profile information such as gender can improve the accuracy of cyberbullying detection in OSNs [23] . Huang *et al.* [37] has consider some graph properties besides text features, however they also worked only over comment-based labeled data. Another work has looked at the time series of posted comments of Formepring dataset, in which each question answer pair was labeled separately as cyberaggression and then their severity predicted [38]. These works largely focus on text-based analysis and unlike our work do not examine the features associated with the media objects such as images or videos belonging to those comments, as in Instagram. Kansara *et al.* [39] suggest only a framework for using images beside text for detecting cyberbullying.

Other work analyzed profanity usages in Instagram [40] and Ask.fm [41] comments, but did not label the data in terms of cyberbullying. Additional research investigated aspects of the Instagram social network, but not in the context of cyberbullying. For example, [42] explored users' photo sharing experience in a museum. Silva *et al.* [43] considered the temporal photo sharing behavior of Instagram users and Hu *et al.* [44] categorized Instagram images into eight popular image categories and the Instagram users into five types in terms of their posted images. [45] concluded that Instagram users tend to be more active during weekends and at the end of the day, and that Instagram users are more likely to like and comment on the medias that are already popular, thereby inducing the rich get richer phenomenon.

3   Data Collection

Starting from a random seed node, we identified 41K Instagram user ids using a snowball sampling method from the Instagram API. Among these Instagram ids, 25K (61%) users had public profiles while the rest had private profiles. Due to the limitation on the private profiles' lack of shared information, the 25K public user profiles comprise our sample data set. For each public Instagram user, the collected profile data includes the media objects (videos/images) that the user has posted and their associated comments, user id of each user followed by this user, user id of each user who follows this user, and user id of each user who commented on or liked the media objects shared by the user. We consider each media object plus its associated comments as a *media session*.

Labeling data is a costly process and therefore in order to make the labeling of cyberbullying more manageable, we sought to label a smaller subset of these media sessions. To have a higher rate of cyberbullying instances, we considered media sessions with at least one profanity word in their associated comments. We tag a comment as "negative" using an approach similar to [41]. For this set of 25K users, 3,165K unique media sessions were collected, where 697K of these sessions have at least one profane word in their comments by users other than the profile owner, where a profane word is obtained from a dictionary [46], [47].

In addition, we needed media sessions with enough comments so that labelers could adequately assess the frequency or repetition of aggression, which is an important part of the cyberbullying definition. We selected a threshold of 15 as a lower bound on the number of comments in a media session, considering that the average ratio of comments posted by users other than friends to comments posted by the profile owner in an Instagram profile is around 16 [40]. At the end 2,218 media sessions (images and their associated comments) were selected randomly for the task of labeling.

4   Cyberbullying Labeling

In this section, we explain the design and methodology for labeling the selected set of media sessions. In Instagram, each media session consists of a media object posted by the profile owner and the corresponding comments for the media object. For example, Figure 1 illustrated a media session in which hateful comments were posted for an Instagram image on the profile of the owner. Such a media session was used in the labeling process, in which labelers were shown both the image and the associated text comments in order to make determinations for cyberaggression and cyberbullying.

With input from a social science expert, co-author Mattson, we designed simple instructions to help human contributors identify whether the media session constituted an act of cyberaggression or cyberbullying. During the instructional phase prior to labeling, contributors were given the aforementioned definitions of cyberaggression and cyberbullying along with related examples. The example questions provided more details to help contributors accurately label the online behavior and distinguish between cyberaggression and cyberbullying based on the social science definitions they were provided.

In order to provide quality control, we only permitted the highest-rated contributors on CrowdFlower to have access to our job. Next, a mentoring phase was provided for the potential contributors that included instructions and a set of example media sessions with the correct label. Further, to monitor the quality of the contributors and filter out the spammers, potential contributors were asked to answer a set of test questions in two phases: quiz mode and work mode. Potential contributors needed to answer correctly a minimum number of test questions to pass the quiz mode and qualify as a contributor for the job. We also incorporated quality control checks during the labeling process (work mode) by inserting random test questions. A contributor was filtered out if he/she failed this work mode.

Finally, a minimum time threshold was set to filter out contributors who rushed too quickly through the labeling process. The minimum number of test questions to get back high-quality data was recommended by Crowd-Flower. More details about the labeling process statistics have been provided in the appendix.

An example of a media session that each crowd-sourcer was asked to label is shown in Figure 2. Each media session was then labeled by five contributors that asked them to use the instructions and definitions we provided to determine whether the post included cyberaggressive behavior or

Fig. 2. An example of the labeling study, showing an image and its corresponding comments, and the study questions.

cyberbullying behavior. Specifically we asked 1) Is there any cyberaggressive behavior in the online posts? Mark yes if there is at least one negative word/comment and or content with intent to harm someone or others and 2) Is there any cyberbullying behavior in the online posts? Mark yes if there are negative words and repeated negativity against a victim that cannot easily defend him or herself.

We conducted a separate second phase of labeling focused only on the image contents in order to identify the content and category of the image. We provided separate instructions to label the image contents of media sessions, so that we could investigate the relationship between cyberbullying and cyberaggression and image content. We reasoned that the content or category of an image may help identify incidents of cyberbullying and cyberaggression. More detail regarding image labeling has been explained in the appendix.

5 Analysis and Characterization of Ground Truth Data

We submitted our first study with 2,218 media sessions (images and their associated comments) to CrowdFlower. CrowdFlower assesses a degree of trust for each contributor based on the percentage of correctly answered test questions, as explained in Section 4. This trust value is incorporated by CrowdFlower into a weighted version of the majority voting method called a "confidence level" for each labeled media session. We decided to keep media sessions whose weighted trust-based metric was equal to or greater than 60%. We deemed them to be strong enough support for majority voting from contributors with higher trust. Overall, 1,954 (88%) of the original pure majority-vote based media sessions wound up in this higher-confidence cyberbullying-labeled group. For this higher-confidence data set, 29% of the media sessions belonged to the "bullying" group while the other 71% were deemed to be not bullying.

5.1 Labeling and Negativity Analysis

The distribution of the media sessions based on the number of votes (out of five votes) received for cyberaggression and cyberbullying respectively has been provided in Figure 3. The left chart shows the fraction of samples that have been labeled as cyberaggression k times, and the right chart shows the fraction of samples that have been labeled as cyberbullying k times. The higher the number of votes for a given media session, the more confidence we have that the media session contains an incident of cyberaggression or cyberbullying, with five votes means unanimous agreement. Similarly, the lower the number of votes for a given media session, the more confidence we have that the media session *does not* contain an incident of cyberaggression or cyberbullying, with zero votes means unanimous agreement.The inter-rater agreement Fleiss-Kappa value for cyberbullying is 0.5 and for cyberaggression is 0.52.

We notice that for both cyberaggression and cyberbullying, most of the probability mass is around media sessions labeled by all four or five contributors the same, i.e. either 0 or 1 votes (about 50% for cyberaggression and about 62% for cyberbullying), or 4 or 5 votes (about 31% for cyberaggression and about 23% for cyberbullying). *Thus, a key finding is that the contributors are mostly in agreement about what behavior constitutes cyberaggression, and what behavior constitutes cyberbullying in Instagram media sessions.* Only about 13–17% of the media sessions have two or three votes, which indicates that there is some disagreement in a small fraction of media sessions about whether the session contains an incident of cyberaggression or cyberbullying. This disagreement can be attributed to the fact that different people have different levels of sensitivity and a conversation may seem normal to one person and hurtful to another.

Next, we observe that about 30% of the media sessions have not been labeled as cyberaggression by any of the five contributors. Since all media sessions contained at least one comment with one or more profane word, this suggests that only employing a profanity usage threshold to detect cyberaggression can produce many false positives. We make a similar observation for cyberbullying. We notice about 40% of the media sessions have not been labeled as cyberbullying by any of the five contributors. Applying a majority voting criterion to a binary label as cyberbullying or not, 30% of the samples have been labeled as cyberbullying. This is despite the fact that all the media sessions contain at least one profane word. This leads us to our second important finding. *A classifier design for cyberbullying detection cannot solely rely on the usage of profanity among the words in image-based discussions, and instead must consider other features to improve accuracy.*

In order to understand the relationship between cyberaggression and cyberbullying, we plotted in Figure 4 a two-dimensional heat map that shows the distribution of media sessions as a function of the number of votes each media session received for cyberaggression and cyberbullying. We observe that a

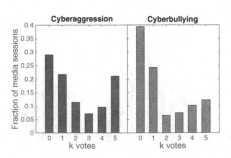

Fig. 3. Fraction of media sessions that have been voted k times as cyberaggression (left) or cyberbullying (right).

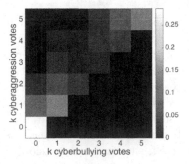

Fig. 4. Two-dimensional distribution of media sessions as a function of the number of votes given for cyberaggression versus the number of votes given for cyberbullying, assuming five labelers.

significant fraction of the sessions exhibit strong agreement in terms of either receiving high numbers of votes for both cyberbullying and cyberaggressions, or receiving low numbers of votes for both cyberbullying and cyberaggression. This can be inferred from the high energy in the upper right and lower left part of the diagonal. In addition, it is promising that the area below the diagonal is essentially zero, meaning no session has received more votes for cyberbullying than for cyberaggression. This conforms with the definition that cyberbullying is a subset of cyberaggression. The Pearson's correlation between number of votes for cyberbullying and number of votes for cyberaggression is 0.9.

We see that the remaining significant energy in the distribution appears in the area above the diagonal. Media sessions in this area exhibit the property that if they receive N_1 cyberbullying votes and N_2 cyberaggression votes, then $N_2 \geq N_1$. The area where $N_1 \leq 2$ and $N_2 \geq 3$ corresponds to cases where there is cyberaggression but not cyberbullying. These observations lead us to our third important finding. *A media session that exhibits cyberaggression does not necessarily exhibit cyberbullying, and a classifier design for cyberbullying detection must go much beyond merely detecting cyberaggression.* This is a very important finding, because as we noted in Section 2, prior work on detecting cyberbullying has mainly focused on detecting cyberaggression as they do not take into account the frequency of aggression or imbalance of power, which are crucial features of cyberbullying.

Finally, we are interested in understanding the relation between cyberbullying/cyberaggression and the percentage of negativity in the comments. We divided all the media sessions into nine different bins based on the percentage of negativity in their comments. Bin $(n_1 - n_2]$ contains all media sessions with bigger than $n_1\%$ and smaller than or equal to $n_2\%$ negativity. None of the media sessions contained more than 90% negative comments. Next, we calculated percentage media sessions for each bin that can be identified as cyberaggression or cyberbullying based on majority of votes, i.e. where the number of votes is 3 or higher.

Figure 5 shows these fractions, left figure for cyberaggression and right figure for cyberbullying. We observe that as the percentage of negativity increases, so does the fraction of media sessions up until 50% negativity for cyberaggression and 60% for cyberbullying. This increase is as expected, since cyberaggression or cyberbullying is typically accompanied with negativity in the postings. However, we notice that the percentage of cyberaggression or cyberbullying starts decreasing after these peaks as the percentage of negativity

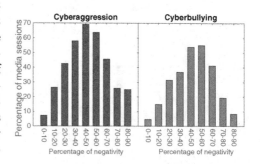

Fig. 5. Percentage of media sessions that have been labeled as cyberaggression (left) and cyberbullying (right) versus their negativity percentages.

increases. This is quite an unexpected result and seems counter-intuitive. To understand this, we examined closely the media sessions that have very high negativity. We noticed that these media sessions typically involved discussions about sports, politics, tattoos, or were just friendly talks. People tend to use lots of profanity words in such discussions, even though they are not insulting any one person in particular. This leads us to our final important finding about negativity analysis. *A media session with a significantly high percentage of negativity (more than 60-70%) typically implies a low probability that the session contains a cyberbullying incident.*

5.2   Temporal and Graph Properties Analysis

Since different comments in a media session are posted at different times, it is important to understand the relationship between the temporal nature of comment postings and cyberbullying/cyberaggression. We define the strength of cyberbullying/cyberaggression as the number of votes received for labeling a media session as cyberbullying/cyber-aggression, and explore the Pearson's correlation of cyberbullying/cyberaggression strength and temporal behavior comment arrivals. We would like to understand how human contributors incorporated the definition of cyberbullying, which includes the temporal notion of repetition of negativity over time, into their labeling. Given the time stamps on the collected comment, we compute the interarrival time between two consequent comments. We then count the number of interarrival times of comments in a media session that have a value less than $x = 1$min, 5 min, ..., 6 months.

Figure 6 illustrates the correlation between the number votes and the number of comments arrive with $\leq x$ seconds after their previously received comment. We see that there is a correlation of about 0.3 between the strength of support for cyberbullying and media sessions in which there are frequent postings within one hour of previous post. Further, we find that as we expand the allowable interarrival times between comments, the correlation weakens considerably. A similar pattern was observed for cyberaggression. In fact, on average around 40%

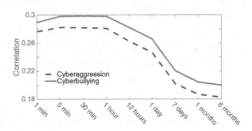

Fig. 6. Pearson's correlation between the number of votes and the number of comments that arrive in $\leq x$ seconds after their previously received comment.

of the comments arrive in less than 1 hour after previously received comments in cyberbullying media sessions, however only 30% of the comments have been received with the same interarrival time in non-cyberbullying samples ($p < 0.001$, based on t-test). *A key finding here is that media sessions that contain cyberbullying have relatively low comment interarrival times, that is the comments in these media sessions are posted quite frequently.*

Table 1. Mean values of social graph properties for cyberbullying versus non-cyberbullying samples and aggression versus non-cyberaggression. ($*p < 0.05$).

Label	*Likes	*Media objects	Following	Followers
Non-cyberbullying	9,684.4	1,145.7	668.1	415,676.2
Cyberbullying	7,029.0	1,198.3	626.7	463,073.1
Non-cyberaggression	9,768.6	1,133.7	665.9	421,075.3
Cyberaggression	7,551.3	1,204.3	640.3	440,403.6

Next, we examine the relationship between cyberbullying/cyberaggression and the social network graph features such as the number of likes for a given media object, number of comments posted for a media object, number of users a user is following, and the number of followers of a user. Table 1 shows these numbers, for categories of non-cyberbullying sessions, cyberbullying sessions, non-cyberaggression sessions and cyberaggression sessions. We observe that media sessions that contain cyberbullying/cyberaggression share more media objects than media sessions that do not contain cyberbullying/cyberaggression, but on average receive lower number of likes. Souza *et al.*'s [45] analysis of Instagram users shows there is a positive correlation between number of followers and number of likes for typical Instagram users. Users who receive cyberbullying do not follow the same pattern. In fact, the average number of likes per post for non-cyberbullying sessions is 4 times the average number of likes for cyberbullying sessions, and the average number of likes per post for non-cyberaggression sessions is 4.5 times the average number of likes for cyberaggression sessions. In terms of number of following and followers, the distinction is not as pronounced, although we see that the media sessions with cyberbullying/cyberaggression incidents have more followers and less following compared to the media sessions without cyberaggression/cyberbullying. *The key finding here is that the users of media sessions with cyberbullying/cyberaggression have lower number of likes per post while have more followers.*

5.3   Linguistic and Psychological Analysis

We now focus on the pattern of linguistic and psychological measurements of cyberbullying/cyberaggression media sessions versus non-cyberbullying/non-cyberaggression. For this purpose, we have applied Linguistic Inquiry and Word Count (LIWC), a text analysis program to find which categories of words have been used for cyberbullying/cyberaggression labeled media sessions. LIWC evaluates different aspects of word usages in psychologically meaningful categories, by counting the number of the words across the text for each category [48]. LIWC has often been used for studies on variations in language use across different people. Published papers show that LIWC have been validated to perform well in studies on variations in language use across different peoples [49]. We first analyze the number of words, and usage of pronouns, negations and swear words

(Figure 7). Next, we look at some of the personal concerns such as work, achievements, leisure, etc. (Figure 7). Finally, we investigate some of the psychological measurements such as social, family, friends, etc. (Figure 8). For each of these cases, we first obtain the LIWC values for each media session comment set. We then calculate the average LIWC value for each of the four classes: media sessions with cyberbullying, media sessions with no cyberbullying, media sessions with cyberaggression, and media sessions with no cyberaggression. The bars shown in Figures 7-8 represent the ratio of average LIWC value for cyberbullying class to that of non-cyberbullying, and the ratio of average LIWC value for cyberaggression class to that of non-cyberaggression.

In Figure 7, we first notice that the word count for media sessions with cyberbullying/cyberaggression is significantly higher than for media sessions with no cyberbullying/cyberaggression ($p < 10^{-5}$). Next, as expected, for swear words (e.g., damn, piss) and negations (e.g., never, not), the ratio is higher for cyberbullying/cyberaggression category ($p < 10^{-5}$). It is interesting to note that the ratios for the third person pronouns (she, he, they) are more than 1.3 ($p < 10^{-5}$), the ratio for the first person singular pronoun (i) is 0.85, and the ratios for first person plural and second person pronouns (we, you) is closer to 1. This leads us to our first key finding with respect to the linguistic features. *A user is less likely to directly refer to himself/herself and more likely to refer to other people in third person in postings involving cyberbullying or cyberaggression.*

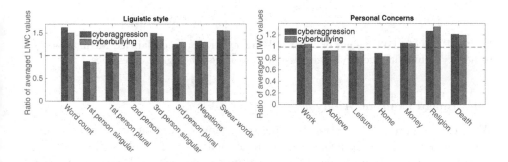

Fig. 7. (left) Ratio of LIWC values of cyberbullying/cyberaggression labeled media sessions to non-cyberbullying/non-cyberaggression class in Linguistic categories. (right)Ratio of LIWC values of cyberbullying/cyberaggression labeled media sessions to non-cyberbullying/non-cyberaggression class in Personal Concerns categories.

For personal concerns (Figure 7), "religion" (e.g., church, mosque) and "death" (e.g., bury, kill) categories have higher ratios (more than 1.2, p < 0.1). This is in line with our findings in our previous work on profanity usage analysis in ask.fm social media, where we observed that there is high profanity usage around words like "muslim" [41]. This suggests that religion-based cyberbullying may be quite prevalent in social media. On the other hand, ratios for personal categories like "work", "money" and 'achieve" are much closer to 1.

For psychological measurements (Figure 8), we notice that the ratios for "negative emotion", "anger", "body", and "sexual" categories are significantly higher than 1 (more than 1.4, $p < 10^{-5}$), and the ratio for "positive emotion" category is significantly lower than 1 (0.76, $p < 10^{-5}$). Higher ratios for "body" (e.g. face, wear) and "sexual" (e.g. slut, rapist) categories provide evidence for appearance-based and sexual-based cyberbullying in social media. For other psychological measurement categories, such as "social", "friend", etc., the ratios are closer to 1. Based on our observations from Figures 7 and 8, our final important finding with respect to linguistic features is as follows: *There is a higher probability of cyberbullying in postings involving religion, death, appearance and sexual hints, and cyberbullying posts typically have higher occurrences of negative emotions and lower occurrences of positive emotions.*

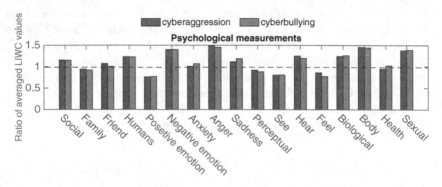

Fig. 8. Ratio of LIWC values of cyberbullying/cyberaggression labeled media sessions to non-cyberbullying/non-cyberaggression class in Psychological categories.

5.4 Image Content Analysis

We now explore the relationship between image content and cyberbullying/cyberaggression in a media session. If the majority of labelers chose a given content category for an image, then that image was counted as belonging to that category. Note that it was possible for contributors to place an image in more than one category. More than 70% of the images were labeled with only one category, and around 20% were labeled with two categories. However, there were a few images that were labeled with up to eight unique categories.

Figure 9 shows the fraction of the contents for all labeled data in the green bar. The "dont know" choice was given as we realized that for some images it is hard to figure out what is in the image. As some images belong to more than one category, the bars will sum up to more than one. First, we observe that the most common labels for image content are Person/People, Text and Sports.

Next, the heights of the blue and red bars embedded inside each green bar relative to the height of the green bar indicate the fraction of images belonging to the media sessions that contained cyberaggression and cyberbullying respectively. For example, for the "Text" category, about 1/3 of the images with "Text"

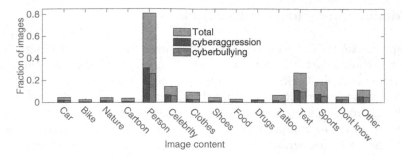

Fig. 9. Fraction of image categories for all media sessions, cyberbullying and cyberaggression classes.

were associated with media sessions containing cyberaggression and cyberbullying. We observe that for some content categories such as "Drugs", the overall fraction is quite small (green bar height is low), but most of the images in those categories do belong to media sessions with cyberaggression/cyberbullying in them. To see this more clearly, Figure 10 plots the fraction of images labeled as cyberaggression/cyberbullying for each content category. We notice that for content category "Drugs", 75% of the images belong to media sessions containing cyberbullying, while for content categories like "Car", "Nature", "Person", "Celebrity", "Text" and "Sport", 30%-40% of the images belong to media sessions containing cyberbullying/cyberaggression. Also, whenever images contain bike, food, tattoo, etc., there is little cyberbullying occurring. *The key finding here is that certain image contents such as Drug are strongly related with cyberbulllying, while some other image contents such as bike, food, etc. have a very low relationship with cyberbullying.*

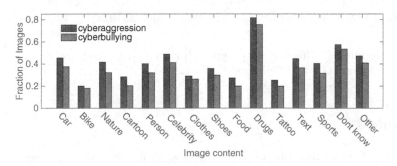

Fig. 10. Fraction of images which have been labeled as cyberbullying and cyberaggression for each content category.

6   Conclusions and Future Work

This paper makes the following major contributions. First, an appropriate definition of cyberbullying that incorporates both frequency of negativity and imbalance of power is applied in large-scale labeling, and is differentiated from cyberaggression. Second, cyberbullying is studied in the context of a media-based social
network, incorporating both images and comments in the labeling. We found
that labelers are mostly in agreement about what constitutes cyberbullying and
cyberaggression in Instagram media sessions. Third, a detailed analysis of the
distribution results of labeling of cyberbullying incidents is presented, including
a correlation analysis of cyberbullying with other factors derived from images,
text comments, and social network meta data. We found a significant number
of media sessions containing profanity and cyberaggression were not labeled as
cyberbullying, suggesting that detection of cyberbullying must be more sophisticated than merely looking for profanity. We observed that media sessions with
very high percentage of negativity above 60-70% actually correspond to a lower
likelihood of cyberbullying. Also, media sessions with cyberbullying exhibit more
frequent commenting. We found that users of media sessions containing cyberbullying demonstrate a lower number of likes per post. Finally, cyberbullying has
a higher probability of occurring when media sessions contain certain linguistic
categories such as death, appearance, religion and sexuality content. Similarly,
certain image contents such as "drug" are highly related to cyberbullying while
other image categories such as "tattoo" or "food" are not.

In the future, we hope to build upon this analysis. We hope to examine more
features for their correlation with cyberbullying, such as new image features,
mobile sensor data, etc. Such features should be auto-generated by software
rather than requiring human labeling. We also wish to obtain greater detail from
the labeling process. Streamlining down to two labeling questions improved the
response rate, quality and speed, but limited our ability to ask more detailed
questions about other aspects of cyberbullying, such as different types and roles.

Acknowledgments. This work was supported in part by the National Science Foundation under awards CNS-1162614 and CNS-1528138.

References

1. National Crime Prevention Council: National Crime Prevention Council (2011).
   http://en.wikipedia.org/wiki/Cyberbullying (accessed July 6, 2011)
2. Kowalski, R.M., Giumetti, G.W., Schroeder, A.N., Lattanner, M.R.: Bullying in
   the digital age: A critical review and meta-analysis of cyberbullying research among
   youth (2014)
3. McNamee, D.: Cyberbullying 'causes suicidal thoughts in kids more than traditional bullying (2014). http://www.medicalnewstoday.com/articles/273788.php
   (accessed May 31, 2015)

4. Hinduja, S., Patchin, J.W.: Cyberbullying research summary, cyberbullying and suicide (2010)
5. Menesini, E., Nocentini, A.: Cyberbullying definition and measurement. some critical considerations. Journal of Psychology **217**, 320–323 (2009)
6. Van Royen, K., Poels, K., Daelemans, W., Vandebosch, H.: Automatic monitoring of cyberbullying on social networking sites: From technological feasibility to desirability. Telematics and Informatics **32**, 89–97 (2015)
7. Strickland, A.: Bullying by peers has effects later in life (2015). http://www.cnn.com/2015/05/08/health/bullying-mental-health-effects/index.html (accessed May 2015)
8. cbcNews: Jessica Logan - Victims of bullying (2008). http://www.cbsnews.com/pictures/victims-of-bullying/11/ (accessed May 31, 2015)
9. NoBullying.com: The top six cyberbullying case ever (2015). http://nobullying.com/six-unforgettable-cyber-bullying-cases/ (accessed May 31, 2015)
10. Kowalski, R.M., Limber, S., Limber, S.P., Agatston, P.W.: Cyberbullying: Bullying in the digital age. John Wiley & Sons (2012)
11. Patchin, J.W., Hinduja, S.: An update and synthesis of the research. Cyberbullying prevention and response: Expert perspectives, p. 13 (2012)
12. Hunter, S.C., Boyle, J.M., Warden, D.: Perceptions and correlates of peer-victimization and bullying. British Journal of Educational Psychology **77**, 797–810 (2007)
13. Olweus, D.: Bullying at school: What we know and what we can do. Blackwell (1993)
14. Olweus, D.: School bullying: Development and some important challenges. Annual Review of Clinical Psychology **9**, 751–780 (2013)
15. Smith, P.K., del Barrio, C., Tokunaga, R.: Definitions of bullying and cyberbullying: how useful are the terms? routledge. In: Principles of Cyberbullying Research. Definitions, Measures and Methodology (2012)
16. Dooley, J.J., Pyżalski, J., Cross, D.: Cyberbullying versus face-to-face bullying. Zeitschrift für Psychologie/Journal of Psychology **217**, 182–188 (2009)
17. Monks, C.P., Smith, P.K.: Definitions of bullying: Age differences in understanding of the term, and the role of experience. British Journal of Developmental Psychology **24**, 801–821 (2006)
18. Pyżalski, J.: Electronic aggression among adolescents: An old house with. Youth culture and net culture: Online social practices, p. 278 (2010)
19. Limber, S.P., Kowalski, R.M., Agatston, P.A.: Cyber bullying: A curriculum for grades 6–12. Hazelden, Center City (2008)
20. Ditch the Label Anti Bullying Charity: The annual cyberbullying survey 2013 (2013). http://www.ditchthelabel.org/annual-cyber-bullying-survey-cyber-bullying-statistics/
21. Hinduja, S.: Cyberbullying on Instagram (2013)
22. Ptaszynski, M., Dybala, P., Matsuba, T., Masui, F., Rzepka, R., Araki, K., Momouchi, Y.: In the service of online order tackling cyberbullying with machine learning and affect analysis (2010)
23. Dadvar, M., de Jong, F.M.G., Ordelman, R.J.F., Trieschnigg, R.B.: Improved cyberbullying detection using gender information. In: Proceedings of the Twelfth Dutch-Belgian Information Retrieval Workshop (DIR 2012), Ghent, Belgium, pp. 23–25. University of Ghent, Ghent (2012)
24. Reynolds, K., Kontostathis, A., Edwards, L.: Using machine learning to detect cyberbullying. In: Fourth International Conference on Machine Learning and Applications, vol. 2, pp. 241–244 (2011)

25. Dinakar, K., Jones, B., Havasi, C., Lieberman, H., Picard, R.: Common sense reasoning for detection, prevention, and mitigation of cyberbullying. ACM Trans. Interact. Intell. Syst. **2**, 18:1–18:30 (2012)
26. Sanchez, H., Kumar, S.: Twitter bullying detection. In: NSDI 2012, Berkeley, CA, USA, p. 15. USENIX Association (2012)
27. Kontostathis, A., Reynolds, K., Garron, A., Edwards, L.: Detecting cyberbullying: query terms and techniques. In: Proceedings of the 5th Annual ACM Web Science Conference, pp. 195–204. ACM (2013)
28. Xu, J.M., Jun, K.S., Zhu, X., Bellmore, A.: Learning from bullying traces in social media. In: Proceedings of the 2012 Conference of the North American Chapter of the Association for Computational Linguistics: Human Language Technologies, pp. 656–666. Association for Computational Linguistics (2012)
29. Nahar, V., Al-Maskari, S., Li, X., Pang, C.: Semi-supervised learning for cyberbullying detection in social networks. In: Wang, H., Sharaf, M.A. (eds.) ADC 2014. LNCS, vol. 8506, pp. 160–171. Springer, Heidelberg (2014)
30. Nahar, V., Li, X., Pang, C.: An effective approach for cyberbullying detection. In: Communications in Information Science and Management Engineering, CISME 2013 (2013)
31. Dinakar, K., Reichart, R., Lieberman, H.: Modeling the detection of textual cyberbullying. In: The Social Mobile Web (2011)
32. Nahar, V., Unankard, S., Li, X., Pang, C.: Sentiment analysis for effective detection of cyber bullying. In: Sheng, Q.Z., Wang, G., Jensen, C.S., Xu, G. (eds.) APWeb 2012. LNCS, vol. 7235, pp. 767–774. Springer, Heidelberg (2012)
33. Sood, S., Antin, J., Churchill, E.: Profanity use in online communities. In: Proceedings of the SIGCHI Conference on Human Factors in Computing Systems, pp. 1481–1490. ACM (2012)
34. Nandhini, B., Sheeba, J.: Cyberbullying detection and classification using information retrieval algorithm. In: Proceedings of the 2015 International Conference on Advanced Research in Computer Science Engineering & Technology (ICARCSET 2015), p. 20. ACM (2015)
35. Nandhini, B.S., Sheeba, J.: Online social network bullying detection using intelligence techniques. Procedia Computer Science **45**, 485–492 (2015)
36. Nalini, K., Sheela, L.J.: Classification of tweets using text classifier to detect cyber bullying. In: Satapathy, S.C., Govardhan, A., Raju, K.S., Mandal, J.K. (eds.) Emerging ICT for Bridging the Future - Volume 2. AISC, vol. 338, pp. 637–645. Springer, Heidelberg (2014)
37. Huang, Q., Singh, V.K., Atrey, P.K.: Cyber bullying detection using social and textual analysis. In: Proceedings of the 3rd International Workshop on Socially-Aware Multimedia, pp. 3–6. ACM (2014)
38. Potha, N., Maragoudakis, M.: Cyberbullying detection using time series modeling. In: 2014 IEEE International Conference on Data Mining Workshop (ICDMW), pp. 373–382. IEEE (2014)
39. Kansara, K.B., Shekokar, N.M.: A framework for cyberbullying detection in social network (2015)
40. Hosseinmardi, H., Rafiq, R.I., Li, S., Yang, Z., Han, R., Mishra, S., Lv, Q.: Comparison of common users across Instagram and Ask.fm to better understand cyberbullying. In: The 7th IEEE International Conference on Social Computing and Networking (SocialCom) (2014)

41. Hosseinmardi, H., Ghasemianlangroodi, A., Han, R., Lv, Q., Mishra, S.: Towards understanding cyberbullying behavior in a semi-anonymous social network. In: Advances in Social Networks Analysis and Mining (ASONAM 2014), pp. 244–252 (2014)
42. Weilenmann, A., Hillman, T., Jungselius, B.: Instagram at the museum: communicating the museum experience through social photo sharing. In: Proc. of the SIGCHI Conf. on Human Factors in Computing Systems, CHI 2013, pp. 1843–1852 (2013)
43. Silva, T.H., de Melo, P.O.S.V., Almeida, J.M., Salles, J., Loureiro, A.A.F.: A picture of Instagram is worth more than a thousand words: workload characterization and application. In: DCOSS, pp. 123–132. IEEE (2013)
44. Hu, Y., Manikonda, L., Kambhampati, S.: What we instagram: a first analysis of instagram photo content and user types. In: Proc. of the 8th International AAAI Conference on Weblogs and Social Media (ICWSM 2014) (2014)
45. Araujo, C.S., Correa, L.P.D., da Silva, A.P.C., Prates, R.O., Meira Jr., W.: It is not just a picture: revealing some user practices in instagram. In: 2014 9th Latin American Web Congress (LA-WEB), pp. 19–23. IEEE (2014)
46. NoSwearing.com: (Bad word list and swear filter) (accessed November 10, 2014)
47. von Ahn's Research Group, L.: Negative words list form. Luis von Ahn's Research Group (2014)
48. Pennebaker, J.M., Francis, M.E., Booth, R.J.: Linguistic Inquiry and Word Count. Lawerence Erlbaum Associates, Mahwah (2001)
49. De Choudhury, M., Gamon, M., Counts, S., Horvitz, E.: Predicting depression via social media. In: International AAAI Conference on Weblogs and Social Media, Boston, MA (2013)

Appendix

Labeling Statistics

Overall, 176 potential contributors worked on the quiz questions, 144 passed the quiz mode, while 31 contributors failed and 1 gave up. The labeled data that we finally obtained were from 139 *trusted* contributors, while the the rest were filtered out during the work mode. Table 2 provides the number of trusted judgments and the contributors' accuracy for 11,090 total judgments.

Table 2. Labeling process statistics. Trusted judgments are the ones made by trusted contributors.

Trusted Judgments	10987
Untrusted Judgments	103
Average Test Question Accuracy of Trusted Contributors	89%
Labeled Media Sessions per Hour	6

Image Labeling

Human contributors were given detailed instructions for identifying image's content. We first sampled 1,200 images from the selected subset of media sessions to determine a suitable set of representative categories to be used in the labeling. A graduate student examined all the images and classified them to different possible categories. Then, a social science expert checked the categories again and revised them. Some of the dominant categories identified were the presence of a human in the image, as well as text, clothes, tattoos, sports and celebrities. We then asked contributors to identify which of the aforementioned categories were present in the image. Multiple categories could be selected for a given image. Each media session was labeled by three different contributors. At the end, our social science expert checked a set of random media sessions and images to confirm the quality of the labeled data for both studies.

Uncovering Social Media Reaction Pattern to Protest Events: A Spatiotemporal Dynamics Perspective of Ferguson Unrest

Jiaying He[1(✉)], Lingzi Hong[2], Vanessa Frias-Martinez[2], and Paul Torrens[1]

[1] Department of Geographical Sciences, University of Maryland,
College Park, MD, USA
{hjy0608,ptorrens}@umd.edu
[2] School of Information Studies, University of Maryland, College Park, MD, USA
{lzhong,vfrias}@umd.edu

Abstract. Social platforms like Twitter play an important role in people's participation in social events. Utilizing big social media data to uncover people's reaction to social protests can shed lights on understanding the event progress and the attitudes of normal people. In this study, we aim to explore the use of Twitter during protests using Ferguson unrest as an example from multiple perspectives of space, time and content. We conduct an in-depth analysis to unpack the social media response and event dynamics from a spatiotemporal perspective and to evaluate the social media reaction through the integration of spatiotemporal tweeting behavior and tweet text. We propose to answer the following research questions. (1) What is the general spatiotemporal tweeting patterns across the US? (2) What is the spatiotemporal tweeting patterns in local St. Louis? (3) What are the reaction patterns in different US urban areas in space, time and content?

1 Introduction

Social movements such as protests have significant impacts on public policies and political decisions. Understanding the social movements is important because engaging in this kind of collective efforts is one of the limited methods for ordinary people to pursue their political goals [10]. In recent decades, the development of new information and communication technologies have led to increasing political participation through information spread, opinion expression and activism [9,11,16]. Specifically, social media platforms can provide the public with easier access to protest plan details and affect people's motivation to participate, thus facilitating protests both locally and globally [15]. Compared to traditional media, they can deliver protest information at a high speed, allow active participation in event organization, search information as needed, and bring pre-vetted information through personal social networks [13,16]. Hence social media can provide users enriched experiences involving themselves actively in the events rather than simply receiving news. For example, during the Arab Spring, digital media, especially Twitter,

T.-Y. Liu et al. (Eds.): SocInfo 2015, LNCS 9471, pp. 67–81, 2015.
DOI: 10.1007/978-3-319-27433-1_5

has prompted the protest mobilization through reporting real-time event magnitudes and providing basis for collaboration and emotional mobilization [2,11].

In recent years, Twitter has rapidly become one of the most important media for information dissemination and communication. Users can use Twitter to identify interesting topics, express opinions and share news with the public: traditional media can use Twitter to quickly transmit the latest news, while normal users can exchange viewpoints with friends or strangers. Due to the large amount of users conveying their thoughts on Twitter, the great quantity of newly released tweets can lead to information burst at an astonishing rate through the social media network in response to social issues. People can thus generate concrete understanding of movement process in multiple dimensions by analyzing large amount of related tweets and exploring spatiotemporal patterns through Twitter. Especially for researchers, tweets are important crowd sourced data to learn the reaction patterns and real thoughts in social movements. In this way, we can fully understand the ongoing trends and objective opinions of protests in time through the detailed tweet information including time, location and content.

On August 9th, 2014, an unarmed black teenager, Michael Brown, was shot to death by a white police officer, Darren Wilson, in Ferguson, a suburban area of St. Louis, Missouri. This incident quickly spread on Twitter, Facebook and other platforms. Meanwhile tensions between the public and police in Ferguson boiled over which led to riots. Several nights of protests were held to support transparent investigation while state of emergency and curfews were implemented there. More and more related reports were posted on Twitter. Hashtags like "#Ferguson and #TheyGunnedMeDown" became popular. The tweet count spiked right after the shooting, indicating the growing attention among people from local to global. Protesters also utilized Twitter to organize events by spreading plans, sending announcements and collecting donations [7].

Though many studies use social media data to study social movements, most of them examine the large-scale pattern in social networks or focus on the content in small-scale data. In this study, we are motivated by the lack of efforts in comprehensively understanding the social media usage and reaction pattern towards protests integrating space, time and content. Our main contributions are: (1) a detailed analysis to unpack the social media response and event dynamics from a spatiotemporal perspective; (2) an in-depth study of the social media reactions through the integration of space, time and content. We aim to answer the following questions using the Ferguson unrest as an instance. (1) What is the general spatiotemporal tweeting patterns across the US? (2) What is the spatiotemporal tweeting patterns in local St. Louis? (3) What are the reaction patterns in different US urban areas in space, time and content?

2   Related Work

Existing research has examined how social media influenced the information spread and user reactions in social movements. Gaby and Caren's study showed that social media platforms like Facebook helped spread social movement information quickly and reach large amount of audiences [8]. During the 2011 Tunisian and Egyptian Revolutions, Twitter has supported conversations among different types of users including normal people, activists, bloggers and journalists by examining news dissemination and user activities [12]. In addition to information dissemination, social media have played an important role in increasing normal people's participation in social protest events. Poell and Borra [14] assessed user participation in social media protest reporting based on a set of tweets, videos and photos related to the Toronto G20 protest event from Twitter, YouTube and Flickr. They found that YouTube and Flickr did not facilitate users' participation in reporting while Twitter to some extent has more alternative reporting. Valenzuela's study of protests in Chile also suggested that frequent use of social media for opinion expression has significantly positive relationship with protest participation [17]. Tufekci and Wilson [16] further pointed out that social media use could greatly increase the odds for respondents to attend protests on the first day, while traditional media did the opposite. In addition, Earl et al.[6] identified that Twitter created a new dynamic in the protester and police interaction as it has been frequently used by protesters to share information about protest details and the actions of police.

Furthermore, previous studies have integrated spatial analysis methods to explore the social movements by analyzing the social and information network structures. Conover et al.[4] tried to understand if the spatial patterns of communication networks could reflect the goals and needs of protest movement Occupy Wall Street. They reached the conclusion that the network had high levels of locality while non-local attention mainly focused on high-profile locations. Croitoru et al.[5] examined the spatial footprint, social network structure and content of both Occupy Wall Street and Boston Bombing Twitter datasets in both physical and cyber spaces to understand the information exchange during social movement events. Bastos et al.[1] investigated the relationship between the locations of protesters attending demonstrations and those of the users who tweeted the protests during the 2013 Vinegar protests in Brazil. Their study indicated that users tweeting the protests were geographically distant from the street protests and that users from isolated areas relied on Twitter hashtags to remotely engage in the demonstrations.

As ordinary users commonly have different activity patterns compared to media outlet accounts [18], our work differs from earlier studies by emphasizing on the spatiotemporal reaction patterns of ordinary people to the protest events. We start with analyzing the general tweeting patterns towards the Ferguson protest event. Next we explore the reaction patterns in space, time and content. Further we compare the different reactions among different urban areas.

3   Dataset

From August 10th, 2014 to August 27th, 2014, we collected 13,238,863 tweets mentioning "Ferguson" using the Twitter Streaming API. In total, 2,052,364 unique users are included in this dataset. Among all the 13 million tweets, 72.87% of them are retweeting tweets while 3.47% are in reply to others, indicating that retweeting is the major behavior in information diffusion. Around 0.833% of all tweets are tagged with geographical coordinates. In addition, we obtained multiscale geographical data of US from 2014 TIGER/Lines Shapefiles to assist our analysis.

4   Data Preprocessing

As we are interested in the spatiotemporal behaviors of ordinary Twitter users, we focus on the geo-tagged tweets to study the users' behaviors from social type and initiative perspectives. First we compare the time series trend and behavior patterns of geo-tagged and non-tagged tweets. Then we distinguish the media outlet accounts and eliminate out their tweets to avoid potential bias in analyzing local people's reactions.

4.1   Measurement of Reaction Types

Existing studies have utilized different measurements to evaluate users' influence and activity considering following and retweeting relationships [3,19]. Here we evaluate the reaction patterns of Twitter users from social type and initiative perspectives.

Social type can be identified by one's social network built on Twitter. It can be a result of users' response to certain events on Twitter, or an indication of activity, authority or influence in real life so that they tend to react more in Twitter. We categorize users into different social types based on the numbers of their followers and friends, which represent their past activities. Based on the two features, we use K-means clustering method to classify the users to three levels: influencer, intermediary and acceptor. Within each level, users have similar overall prestige. The users with followers far more than friends act as the influencers in the social type, the users with friends far more than followers act as the acceptors, while users who have similar numbers of followers and friends play the part of medium.

Initiative specifically reflects users' preferences of posting original tweets. To observe whether users tend to post new tweets or retweet to spread information, we evaluate the initiative by calculating each user's retweeting ratio with the number of retweets divided by the number of all tweets one posted. Based on the ratio distribution, we group the users to three classes with different initiative levels. If the ratio is less than 0.2, the user mainly publishes new tweets and contributes new content to Twitter. Then we label the user as high initiative. Users with retweeting ratio higher than 0.8 are categorized as low initiative with passive action because they mainly retweet from others. Then the rest of the users are considered to have medium initiative. In general, 21.4% of all users mainly publish new tweets with retweets ratio of almost 0, while 58.4% mainly retweet.

4.2   Geo-tagged and Non-tagged Tweets

After defining the reaction types, we clean our dataset prior to any data analysis. We first remove the tweets related to other topics such as Alex Ferguson or Connie Ferguson and the retweeted tweets posted before August 10th. According to Morstatter et al.'s study [13], geo-tagged tweets collected from Streaming API can evenly represent all tweets' locations in North America. Thus using geo-tagged tweets to represent the spatial distribution of all tweets is plausible. In addition, we check the distribution of daily tweet count for both geo-tagged tweets and all tweets and computed their correlation to ensure that the temporal patterns are consistent in general (Fig. 1). The Pearson's r correlation of the daily tweet count reaches 0.9985, indicating that the frequency distribution of the two datasets are highly linearly correlated. Hence using geo-tagged tweets for analysis will not introduce bias to the temporal data distribution.

We further compare the reaction of users who add geo-tags and who do not. The result shows no significant difference among users of all social types (Fig. 2(a)). The percentage of influencers among the users who post geo-tagged tweets is relatively higher than that among those who do not, though the difference is not obvious. This suggests that tweeting with or without geo-location is not influenced by one's social network structure. For different initiative levels (Fig. 2(b)), however, the result shows that users who tweet with geo-tags have significantly higher initiative of publishing more new tweets compared to those

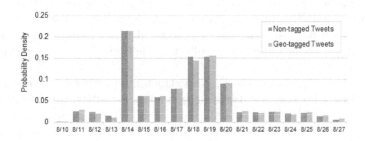

Fig. 1. Histogram of non-tagged tweets and geo-tagged tweets in US by dates

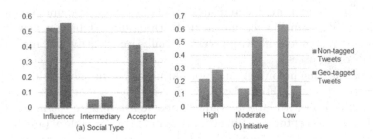

Fig. 2. Percentage of different social type and initiative levels for users who post geo-tagged and non-tagged tweets

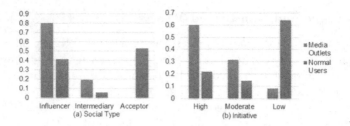

Fig. 3. Percentage of different social type and initiative levels for media outlets and normal users

who do not. The user group with geo-tagged tweets has a larger proportion of people with high and moderate initiative compared to the group with non-tagged tweets, indicating users who tweet with geo-tags tend to have higher initiative to post original tweets.

4.3   Media Outlets and Normal Users

Playing a more influential and active role in investigating and disseminating news, media outlet accounts usually act differently from normal social media users [18]. With large number of followers, tweets posted by these accounts always spread quickly and have broader audiences. Thus including them in our spatiotemporal analysis may lead to bias in understanding normal users' reactions. Here we apply the Support Vector Machine (SVM) classification method incorporated with manual processing to distinguish media outlet accounts and further examine if media outlets have different reaction patterns when compared to ordinary users.

To characterize media accounts, we extract nine features regarding users' social networks, influences and tweeting behaviors including: number of followers, number of friends, status count, number of tweets, number of retweets, number of replies, average retweeted times and average replied times during this event. For classification training, we randomly sample 200 accounts and label 30 of them as media outlets with instances of TV news (e.g. CBSEveningNews), newspapers (e.g. Washingtonpost) and self-media accounts (e.g. JMitchellNews). We use SVM to train our model, and then manually review accounts labeled with media to identify real media account next. With modified labels, next we retrain the model and repeat the above process. Finally we identify 108 media outlet accounts out of all users.

With classified media outlets and normal users, we further evaluate the two groups' reaction patterns. From Fig. 3, we find out that about 80% of media outlets act as influencers with large amount of people listening to them. None of the media outlets is an acceptor, meaning that they mainly play a role of publishing and disseminating information. About 60% media outlets have high or middle initiative, indicating that they actively publish new tweets in general. However, there are still a minor proportion of media outlets that prefer to retweet rather than publish new tweets. On the contrary, normal users in this event are more passive to publish original tweets as around 63.89% of them have low initiative. Regarding the social types, more than half of normal users act as acceptors. Still, there are almost 40% of normal users act as influencers in this event. The comparison of media outlets and normal users shows that media accounts in Twitter still work as traditional media with high initiative and strong influence. Even though many normal users participate in the Ferguson event and post new tweets on Twitter, their influence is not comparable to the media outlets. Most normal users still act as channels to spread information created by others.

In summary, geo-tagged tweets can represent the spatial and temporal distribution of the whole tweet dataset, while the users who post geo-tagged tweets tend to have higher initiative than those who do not. In addition, media accounts turn out to have quite different reaction behaviors, which could lead to bias when analyzing spatiotemporal behaviors of normal users. As a result, we eliminate tweets by media users out for our further study.

5   Research Questions

In this paper, we focus on analyzing the spatiotemporal dynamics of the Ferguson event in detail from different spatial scales by addressing answers to three research questions. Research question 5.1 examines the general spatiotemporal patterns across the whole US. Research question 5.2 digs into the spatiotemporal tweeting patterns and specific events happened in local St. Louis. While research question 5.3 further compares the reaction patterns among four major US urban areas in space, time and content.

5.1   What is the General Spatiotemporal Tweeting Patterns Across the US?

Spatial Distribution. To examine the general tweeting patterns in US, we apply the Average Nearest Neighbor (ANN) and Kernel Density Estimation (KDE) methods to evaluate the general spatial distribution of tweets in US. ANN compares the average distance between each feature and all nearest neighbors to that of a hypothetical random distribution. Then it determines whether the point features is clustered, dispersed, or randomly distributed in space. Different from ANN, KDE is a density-based method to calculate the feature density in a neighborhood around each of these features. In detail, it estimates a density at each feature location by counting the feature number in a small region defined by kernel value centered at the feature location. The ANN result shows that the significance level p-value is smaller than 0.05, suggesting that geo-tagged tweets are highly clustered in US. The Kernel Density map (Fig. 4) shows that there are obvious high density of tweets in St. Louis near the incident in the Missouri and Illinois boarder. Other high density areas can be easily identified in main urban areas like New York and Washington DC, while the tweet density in other areas is generally low.

After examining the general spatial tweeting pattern, we utilize a Density-Based Spatial Clustering of Applications with Noise (DBSCAN) method to accurately identify the most highly clustered tweeting areas. DBSCAN groups closely packed points with plenty of nearest neighbors together and makes them as outlier points compared to low-density regions. In DBSCAN, we set the minimum sample size in a neighborhood to 5000 and identify four largest clusters in St. Louis, New York, Washington DC and Los Angeles urban areas (Table 1). Among all the four urban areas, St. Louis has the highest tweet count, though the other three areas are large urban areas with higher population than St. Louis. Thus we can infer that local St. Louis people have more intense reaction to the event. By integrating all above methods together, we can generate a thorough understanding of the spatial tweeting pattern in this case.

Table 1. Number of tweets in the four urban areas

Location	St. Louis	Washington DC	Los Angeles	New York
Number of tweets	12,434	4,657	4,329	7,628

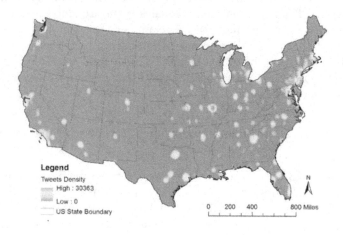

Fig. 4. Kernel density map of tweet counts in the US

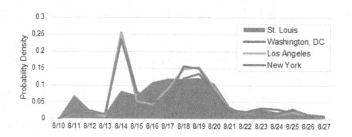

Fig. 5. Histogram of daily tweet counts in the four urban areas

Temporal Trend. To understand the usage trend in time, we examine the daily tweet counts from August 10th to August 27th, 2014. Similarity of temporal pattern among New York, Washington D.C. and Los Angeles can be found easily, while St. Louis shows a completely different trend (Fig. 5). St. Louis tend to have relatively even response during the whole period, while other three areas have a peak on August 14th and then some more response during August 18th to 19th. Specifically in St. Louis, the first peak emerges on August 11th right after the event happened. Then the count keeps increasing from 12th till 18th, during which time Governor Nixon declared emergency state in Ferguson and imposed a curfew. However for the other areas, the highest peak exists on August 14th, when President Obama addressed the nation on the situation in Ferguson, saying there was no excuse for protesters to turn to violence or for excessive force by police. During 17th and 18th, another peak occurs in other areas related to the Governor's declaration and curfew. Then the percentage for all places became low after the 21st.

To quantitatively compare the temporal trend, we calculate Root-mean-square Deviation (RMSE) and Pearson's r correlation of the daily tweets distribution densities from the four areas to those of the whole geo-tagged tweet dataset. In addition, we compute the skewness to measurement the asymmetry of daily tweets probability distribution. The results (Table 2) suggest that New York, Washington DC and Los Angeles are more positively skewed compared to St. Louis. The RMSE value of St. Louis is around 0.04 while the values of the other cities are only around 0.01. The correlation between St. Louis and the whole geo-tagged tweets is only 0.77 while the correlations between the other cities and the whole geo-tagged tweets are over 0.99. These indicators suggest that St. Louis tweets have obviously different temporal pattern compared to other places in US in general.

In summary, tweets are highly clustered in US, especially in large urban areas such as St. Louis and New York. In particular, local St. Louis has the most concentrated tweets. Also, the temporal trend of St. Louis is different compared to other areas in US.

Table 2. Skewness, RMSE and correlation results for four urban areas

Urban Area	Skewness	RMSE	Correlation
St. Louis	0.22787	0.03706	0.77001
New York	0.55577	0.00947	0.99019
Washington D.C.	0.57499	0.01451	0.97174
Los Angeles	0.50537	0.01145	0.99185

5.2 What is the Spatiotemporal Tweeting Patterns in Local St. Louis?

For this question, our purpose is to focus on the local St. Louis area and identify where, when and what are local people tweeting about during the protest.

Spatiotemporal Hotspots. To explore the small-scale pattern within St. Louis, we apply a spatiotemporal clustering method Space-Time Permutation Scan statistic to identify significant tweeting hotspots in St. Louis. Originally designed for detecting disease outbreaks, this method uses only time and location data to identify potential spatiotemporal clusters. It makes minimal assumptions about the time and location, and adjusts for natural purely spatial and purely temporal variation.

Fig. 6 shows the four identified clusters in St. Louis. The western part of St. Louis has a cluster covering a large area with 1453 tweets from August 11th to 14th, corresponding to citizens' first reaction peak to this event. Then in the Ferguson area near the incident, there are two smaller tweeting clusters during August 15th to 17th and August 20th to 21st. These two clusters match well in space and time with the local protests happened on West Florissant Avenue and the burned Quiktrip store in Ferguson. Then at the end of our data collection period, there is another small tweeting cluster from August 25th to 26th located close to the Friendly Temple Missionary Baptist Church, at which the funeral of Michael Brown held on August 25th.

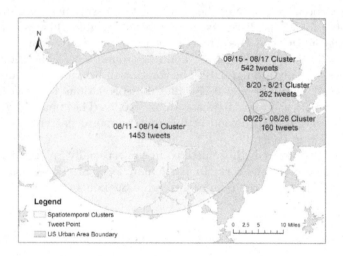

Fig. 6. Spatiotemporal clusters identified in St. Louis urban area

Text Analysis. In addition, we dig into the tweet content to examine how people responded to the hot issues in this event on Twitter. As the general sentence length in our tweet dataset is not suitable for commonly used topical modeling algorithms, in this study we calculate and compare the frequencies of popular terms tweeted in our dataset.

We start with the tweets extracted from the four clusters in St. Louis to identify the topics that people are mostly interested in. First we calculate the frequency of each term considering its different forms. For example, "Peaceful" is regarded the same as "Peace". "Governor" and "Nixon" all refer to the Governor Nixon of Missouri. We then remove the most tweeted terms that appear commonly in all clusters, such as "Ferguson", "Michael Brown", "Police", etc. For the cluster from 25th to 26th, during which time Michael Brown's funeral was held, the word "funeral" has a high occurrence frequency. While almost nobody talks about it in tweets from other clusters. In the cluster from 15th to 17th, "Quiktrip" is among the highest frequency words. This is corresponding to the fact that the Quiktrip store at which the shot happened, was the epicenter of the protest and was looted and burned by the angry protesters during the unrest in those days. Similarly, during 20th to 21st, the "Florissant" has high frequency, in accordance with the protests held on the West Florissant Avenue at that time.

Thus by extracting the local tweeting clusters and identifying the tweet content, it is easy to find out that the local spatiotemporal tweeting patterns strongly reflected the real-time events happened in local St. Louis area.

5.3   What are the Reaction Patterns in Different US Urban Areas in Space, Time and Content?

After exploring the spatiotemporal tweeting patterns in two different spatial scales, we further study the reaction patterns of ordinary Twitter users in space, time and content by comparing the four representative urban areas including St. Louis, Washington DC, New York and Los Angeles.

Twitter User Reaction in Space. To analyze whether geography plays a role in user reaction we mainly looked at the data from two perspectives: 1) Users in and out of St. Louis. 2) Users in four cities: St. Louis, Washington DC, New York and Los Angeles.

We first separate users into two groups: in and out of St. Louis. By comparison of the users' reaction types (Fig. 7), we find out that for both groups, more than 50% the users who respond to the Ferguson event are influencers. The percentage of influencers is relatively higher outside St. Louis. Around 30% of them also act as acceptors to receive information from others. While more users in St. Louis play the role of intermediary to spread news and opinions. We further conduct an independent-samples t-test to examine the difference of social types in and out of St. Louis. With p-value equals to 2.2e-16. The result suggests that the users' social types are significantly different in and outside St. Louis. We also look into the percentage of people who have high, moderate or low initiative in Ferguson event. Regarding initiative, users in St. Louis tend to publish original tweets more while users out of St. Louis retweeted tweets more. Comparison of social types shows that tweets published by users in St. Louis are retweeted more than other tweets. The possible reason is that users in St. Louis report the events with twitter, then the news is spread to other places.

For the second analysis, we focus on comparing the reaction patterns of users in the four different urban areas. Result (Fig. 8) shows that in St. Louis more ordinary users are active, because active users with fewer followers are more than other areas. Meanwhile, the Twitter users in St. Louis tend to have a larger percentage of high and medium initiative than those in the other three cities. There is

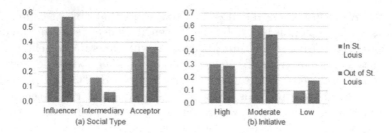

Fig. 7. Percentage of different social types and initiative levels for users in and out of St. Louis

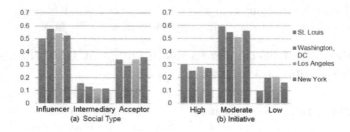

Fig. 8. Percentage of different social types and initiative levels for users in the four urban areas

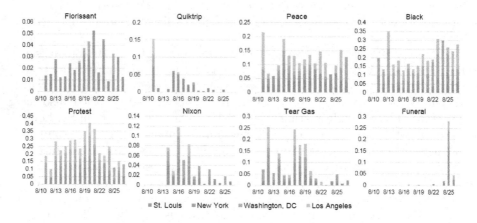

Fig. 9. Popularity trends of the selected eight terms in the four urban areas

not much difference in user types among Washington DC, New York and Los Angeles. We further compare the being retweeted times for tweets in the four cities and find out that tweets posted in St. Louis are retweeted 8.55 times on average, which is several times higher than that of the three other cities, which are separately 4.17, 3.84 and 1.97. This result also suggests that tweets generated by users in St. Louis have higher influence and are more valued than those generated in other areas.

Text Analysis in Space and Time. Based on the text analysis session in 5.2, we identify eight terms with different context to further compare their daily popularity trends in different areas: "Florissant", "Quiktrip", "Peace", "Black", "Protest", "Nixon", "Tear" "Gas" and "Funeral". "Florissant" and "Quiktrip" are corresponding to the local protest events happened in St. Louis. "Peace", "Black", and "Protest" stand for the common topics that people will talk about regarding social movements related to racial problems. While "Nixon", "Tear Gas" and "Funeral" are more about the news commonly reported about the protest status. For each term, we compute the rate of term frequency to the total tweet number during each day for all the four cities. We then use the Analysis of Variance (ANOVA) test to quantitatively measure the difference of the trends. Specifically, we conduct two groups of ANOVA test: (1) all four cities; (2) the three cities other than St. Louis.

Table 3. ANOVA test result for the selected eight terms

Words	p-value (4 cities)	p-value (3 cities)
Florissant	2.26e-15*	0.088669
Quiktrip	0.000774*	0.676055
Peace	0.024167*	0.105103
Black	3.75e-05*	0.008304*
Protest	0.24671	0.402924
Nixon	0.273281	0.203327
Tear Gas	0.828834	0.748623
Funeral	0.742757	0.884234

From Fig. 9, it is easy to find that words representing local event, such as Florissant and Quiktrip, were mainly discussed in local St. Louis area. ANOVA result (Table 3) shows that for both words, there are significant differences among all four cities while no difference among the cities outside St. Louis. This indicates that local people care more about the local event and may use Twitter to exchange details for the protest plans, such as the tweets "Marching up and down W Florissant" and "Group of men in purple robes and shirts marching north on Florissant Ave in #Ferguson anyone know what organization they're with?" Even though we exclude big media outlets, smaller media accounts may also get into St. Louis to report the latest progress on Twitter, such as "Protests are peaceful right now on west Florissant. We'll bring you an update from Ferguson tonight on 41ActionNews". For the words about popular social movement topics, our results suggest that local people talk more about "Peace" or "Justice", while words related to race such as "Black" is more popular outside St. Louis and have various popularity trends among different cities. Moreover, for the words about the general public events and announcements made by politicians such as "Tear Gas" and "Nixon", they tend to have similar popularity patterns among users in different areas.

Combining the previous analysis in space, time and content, we can identify that Twitter users in different areas have different reactions towards the protest events. People in St. Louis show higher initiative for attending the online discussion and are more active in spreading and accepting the information about the protests. Local people tend to respond to and publish tweets about the issues related to event details, while people from other areas have response to the general topics related to the events. Besides, announcements from politicians such as the President and Governor can attract public attention in large spatial coverage, which can be reflected on social media platforms like Twitter.

6   Conclusion

In our study, we have conducted an in-depth analysis to understand the social media response and protest event dynamics from a spatiotemporal perspective using Ferguson unrest as an example. Also, we have explored normal people's reactions through the integration of space, time and tweet content. Specifically, we have focused on measuring how people react on Twitter from the social type and initiative aspects.

Our results have shown that media outlets have much higher influence and initiative compared to normal users. Normal users who post geo-tagged tweets tend to have higher influence and initiative than those who do not, though the difference of influence is not obvious. In general, tweets in US are highly clustered in urban areas, especially St. Louis and large cities like New York, Washington D.C. and Los Angeles. The daily temporal patterns vary between St. Louis and other urban areas. The spatiotemporal clustering analysis indicates that tweets inside St. Louis have significant spatiotemporal clusters, suggesting the existence of hot issues in local area in correspondence to the real protest events happened in St. Louis. Though users in St. Louis have less followers, they tend to publish more original tweets and have higher retweeted times than users in other areas, suggesting that their words have higher popularity on Twitter. In addition, text analysis indicates that people in various areas have different responses to different terms. More local people tend to respond to local event details, while people from other areas have responses to the general topics related to the events. Besides, politicians' announcements to the public have broad audiences during the protests. Future work will focus on the network structure, sentiment analysis and socioeconomic factors in detail to fully understand the mechanisms of information spread and user interaction in social protests.

Acknowledgments. This research is funded by the NSF #14-524 . We thank Ed Summers from Maryland Institute for Technology in the Humanities (MITH) for data collection.

References

1. Bastos, M.T., da Cunha Recuero, R., da Silva Zago, G.: Taking Tweets to the Streets: A Spatial Analysis of the Vinegar Protests in Brazil. First Monday **19**(3) (2014)

2. Breuer, A., Landman, T., Farquhar, D.: Social Media and Protest Mobilization: Evidence from the Tunisian Revolution. Democratization, 1–29 (2014)
3. Cha, M., Haddadi, H., Benevenuto, F., Gummadi, P.K.: Measuring User Influence in Twitter: The Million Follower Fallacy. ICWSM **10**(10–17), 30 (2010)
4. Conover, M. D., Davis, C., Ferrara, E., McKelvey, K., Menczer, F., Flammini, A.: The Geospatial Characteristics of a Social Movement Communication Network. PloS One **8**(3) (2013)
5. Croitoru, A., Wayant, N., Crooks, A., Radzikowski, J., Stefanidis, A.: Linking Cyber and Physical Spaces through Community Detection and Clustering in Social Media Feeds. Computers, Environment and Urban Systems (2014)
6. Earl, J., McKee Hurwitz, H., Mejia Mesinas, A., Tolan, M., Arlotti, A.: This Protest will be Tweeted: Twitter and Protest Policing during the Pittsburgh G20. Information, Communication and Society **16**(4), 459–478 (2013)
7. Fillion, R. M.: How Ferguson Protesters Use Social Media to Organize. The Wall Street Journal, November 24, 2014
8. Gaby, S., Caren, N.: Occupy Online: How Cute Old Men and Malcolm X Recruited 400,000 US Users to OWS on Facebook. Social Movement Studies **11**(3–4), 367–374 (2012)
9. Garrett, R.K.: Protest in an Information Society: A Review of Literature on Social Movements and New ICTs. Information, Communication and Society **9**(02), 202–224 (2006)
10. Giugni, M.: Social Protest and Policy Change: Ecology, Antinuclear, and Peace Movements in Comparative Perspective. Rowman and Littlefield (2004)
11. Gonzlez-Bailn, S., Borge-Holthoefer, J., Rivero, A., Moreno, Y.: The Dynamics of Protest Recruitment through an Online Network. Scientific Reports **1** (2011)
12. Lotan, G., Graeff, E., Ananny, M., Gaffney, D., Pearce, I.: The Revolutions were Tweeted: Information Flows during the 2011 Tunisian and Egyptian Revolutions. International Journal of Communication **5**, 31 (2011)
13. Morstatter, F., Pfeffer, J., Liu, H., Carley, K.M.: Is the sample good enough? Comparing data from twitter's streaming API with twitter's firehose. In: Seventh International AAAI Conference on Weblogs and Social Media (2013)
14. Poell, T., Borra, E.: Twitter, YouTube, and Flickr as Platforms of Alternative Journalism: The Social Media Account of the 2010 Toronto G20 Protests. Journalism **13**(6), 695–713 (2012)
15. Shirky, C.: The Political Power of Social Media. Foreign Affairs **90**(1), 28–41 (2011)
16. Tufekci, Z., Wilson, C.: Social Media and the Decision to Participate in Political Protest: Observations from Tahir Square. Journal of Communication **62**(2), 363–379 (2012)
17. Valenzuela, S.: Unpacking the Use of Social Media for Protest Behavior: the Roles of Information, Opinion Expression, and Activism. American Behavioral Scientist **57**(7), 920–942 (2013)
18. Wu, S., Hofman, J. M., Mason, W. A., Watts, D. J.: Who says what to whom on twitter? In Proceedings of the 20th International Conference on World Wide Web, pp. 705–714 (2011)
19. Romero, D.M., Galuba, W., Asur, S., Huberman, B.A.: Influence and passivity in social media. In: Gunopulos, D., Hofmann, T., Malerba, D., Vazirgiannis, M. (eds.) ECML PKDD 2011, Part III. LNCS, vol. 6913, pp. 18–33. Springer, Heidelberg (2011)

Detecting Opinions in a Temporally Evolving Conversation on Twitter

Kasturi Bhattacharjee[✉] and Linda Petzold

Department of Computer Science, University of California, Santa Barbara, USA
{kbhattacharjee,petzold}@cs.ucsb.edu

Abstract. The immense growth of online social networks from simply being a medium of connecting people to assuming a variety of roles has led to a massive increase in their use and popularity. Today, networks like Facebook and Twitter act as news sources, mediums of advertising and facilitators of socio-political revolutions. In such a scenario, it is of vital importance to be able to detect the opinions of social network users in order to study the opinion flow processes that unfold in these networks. For many topics, the focus of the conversation evolves over time based on the occurrence of real-world events, which makes opinion detection challenging. Since it is not practical to label samples from every point in time, a general supervised learning approach is infeasible. In this work we propose a temporal machine-learning model that has its underpinnings in social network research conducted by sociologists over the years, to detect user opinions in evolving conversations. It uses a combination of hashtags and n-grams as features to identify the opinions of Twitter users on a topic, from their publicly available tweets. We use it to detect temporal opinions on Obamacare and U.S. Immigration Reform, for which it is able to identify user opinions with a very high degree of accuracy for a randomly chosen set of users over time.

1 Introduction

Online social networks were initially developed as a means of connecting people from different parts of the world, by facilitating communication between them. However, in today's day and age, they have grown to assume various other roles, including news sources, platforms for users to voice their opinions on current events, mediums for viral marketing, and facilitators of socio-political revolutions [5,7]. This variety of roles has led to the tremendous increase in the use and popularity of these networks in general [3,8], and makes them an interesting subject of study.

To understand the opinion flow processes that unfold in these networks, the detection of opinions and sentiments of users is of great importance. Moreover, owing to the volume of posts being generated daily, it is important to be able to perform this task in an automated fashion. In this paper we present a method for detecting the opinions of Twitter users on a given topic *over time*, using data mining and machine learning techniques.

© Springer International Publishing Switzerland 2015
T.-Y. Liu et al. (Eds.): SocInfo 2015, LNCS 9471, pp. 82–97, 2015.
DOI: 10.1007/978-3-319-27433-1_6

Recently proposed methods in the field of opinion detection in general are based either on machine learning, or lexicons of words. There is no temporal aspect to these approaches. They are trained on labeled data and/or use a predetermined lexicon of words. However, in the case of *temporal opinion detection*, which is the problem we address here, the focus of the conversation shifts from one sub-topic to another, thus new textual features emerge at every time point. The lack of training data at every timestep renders general supervised approaches infeasible.

The method we propose in this paper for *temporal opinion detection* borrows from social network research conducted by sociologists over the years [16,19]. A key observation from social network research is that temporal evolution of user opinions is a *slow* process. People are inherently resistant to changing their opinions. We propose a regularized supervised approach that requires labeled data only at the initial time, and enables us to use opinions detected in a previous timestep when performing predictions for the future. Additionally, the method can capture relevant textual features over time, thus highlighting the conversational sub-topics that emerge at every timestep.

We select Twitter as the source of data for our experiments, and Obamacare as the primary topic of interest. *Obamacare* is a popular term coined to represent the Affordable Care Act (ACA) which was signed into law by President Barack Obama on March 23, 2010 [1]. Since its inception, it has garnered much political and social attention in the US, and has emerged as one of the most popular topics of discussion in social media platforms [6]. The Act also underwent several reforms over time, each addressing a different issue. This led to an *evolving* online conversation on the topic, since the focus of the discussions would *shift* from one sub-topic to another over time. The above characteristic makes this topic interesting and challenging for opinion detection, as we shall illustrate in the later sections.

In order to demonstrate the generality of our method, we selected another topic for our experiments, namely, the U.S. Immigration Reform bill (the Border Security, Economic Opportunity, and Immigration Modernization Act of 2013) that was introduced in the US Senate in April, 2013. The bill would allow for many undocumented immigrants to gain legal status and become U.S. citizens. Additionally, it would make the border more secure by adding up to 40,000 border patrol agents [4]. This topic was also extensively discussed on Twitter. The details about the data collection process for both topics are elaborated in Section 4.1.

Contributions of the Paper

1. This work proposes a machine-learning model to accurately detect opinions of Twitter users over time using their tweets, even when the topic of conversation is evolving in nature. Training is required only at the initial time.
2. The proposed method also showcases the textual features that are most effective at identifying the opinions at different time points. These features aid in identifying the most popular sub-topics that emerge at every time point.

The remainder of the paper is organized as follows. Section 2 discusses the existing literature in this field. Section 3 describes in detail our proposed method for solving the problem at hand. Section 4 contains details on the data we collected and the techniques we used for pre-processing the data. Section 5 discusses the implementation of the method. Section 6 elaborates on the experiments conducted to validate the method and the results obtained.

2   Related Work

The prior research on opinion detection or sentiment analysis can be broadly classified into two groups: lexicon-based methods and machine learning-based methods. The lexicon-based methods work by using a predefined collection (lexicon) of words, where each word is annotated with a sentiment. Various publicly available lexicons are used for this purpose, each differing according to the context in which they were constructed. Examples include the Linguistic Inquiry and Word Count (LIWC) lexicon [30,31] and the Multiple Perspective Question Answering (MPQA) lexicon [25,37,38]. The LIWC lexicon contains words that have been assigned into categories, and matches the input text with the words in each category [24]. The MPQA lexicon is a publicly available corpus of news articles that have been manually annotated for opinions, emotions, etc. These lexicons have been widely used for sentiment analysis across various domains, not just specifically for social networks [9,13,21]. Other popular sentiment lexicons that have been designed for short texts are SentiStrength [35] and SentiWord-Net [10,17]. These lexicons have been extensively used for sentiment analysis of social network data, online posts, movie reviews, etc. [20,23,32,34]. However, as seen in [12], they do not perform well for opinion detection on Twitter users.

Machine learning techniques for sentiment analysis include classification techniques such as Maximum Entropy, Naive Bayes, SVM [22], k-NN based strategies [15], and label propagation [36]. These require labeling of data for training, which is accomplished either by manually labeling posts [36], or through the use of features specific to social networks such as emoticons and hashtags [15,22]. Some of the existing research combines lexicon-based methods and machine-learning methods [33]. None of the above methods address the problem of temporal opinion detection that is the topic of this paper.

In prior work [12], we addressed the problem of opinion detection on Twitter users over a fixed period of time. There was no temporal aspect to the problem. We developed a supervised learning approach using a regularized logistic regression model. We used textual features, namely hashtags and n-grams, to detect user opinions on two topics: U.S. Politics and Obamacare, with a high accuracy. The Obamacare dataset used in that work contained tweets over a short time period and hence did not capture the evolving nature of the conversation. However, when we applied the same method to the current dataset that spans a larger timeline, it failed to detect user opinions accurately (details in Section 6), thus leading us to the development of the proposed model for temporal opinion detection.

3 Temporal Opinion Detection Over an Evolving Conversation

In this section we describe the problem at hand and discuss the social network research that our proposed model is based on. Thereafter, we delve into the details of the proposed model.

3.1 Opinion Change Processes Over Time

The key point of our proposed opinion detection model is that users tend to change their opinions very slowly. This forms a basis of the seminal opinion change models from sociology [16,19]. We present three factors owing to which transition to a different opinion takes place gradually. First, people vary in their readiness to be influenced by their neighbors. Every person has some amount of stubbornness and attachment to their own opinions and beliefs. This is a factor that most models of opinion change consider. For example, a widely-used opinion change model arises from the Social Influence Network Theory of Friedkin and Johnson [19], and is given by

$$\mathbf{y}^{(t)} = \mathbf{A}\mathbf{W}\mathbf{y}^{(t-1)} + (\mathbf{I} - \mathbf{A})\mathbf{y}^{(1)}, \tag{1}$$

where $\mathbf{y}^{(t)}$ is a vector of the users' opinions at time t, $\mathbf{W} = [w_{ij}]$ is the matrix of interpersonal influences, which stores the amount of influence user j has on user i. $\mathbf{A}$ is a diagonal matrix of the users' susceptibilities to interpersonal influence. As is evident from (1), $\mathbf{A}$ determines how anchored the users remain to their initial opinions $\mathbf{y}^{(1)}$, which regulates how much they are influenced by their network neighbors to change their opinions.

Second, we treat the responses of all users as homogeneous from the point of view of opinion change. Thus the opinion of any user, as well as the opinions of all the users she is influenced by, evolve over time. The influenced user slowly changes her opinion in response to the changing opinions of her influencers.

Third, multiple neighbors influence each user. Most opinion models, including Social Influence Network Theory (1) and the DeGroot model [16], assume that a user's opinion is the average of the opinions of her neighbors and her own opinions. This averaging effect tends to dampen dramatic changes [19], making opinion change a slow process. This key observation leads to the main assumption in our proposed model. *For a sufficiently large set of users, most users are not likely to change their opinions drastically over a short period of time.*

3.2 Opinion Detection Models

In this section we discuss our previous model on opinion detection for Twitter users (with no temporal aspect) [12]. Thereafter, we present our proposed model for temporal opinion detection.

Static Opinion Detection Model. In previous work [12], we assumed user opinion to be a distribution over positive and negative types, and used textual features derived from the tweets to learn a weighted combination of the features that would best classify the opinions.

We begin with training data $(\mathbf{x}_i, y_i)$, $i = 1, ...n$, where n is the number of users, $\mathbf{x}_i$ is the ith data vector of size $k \times 1$, with k number of features, and y_i is the ith user's discrete opinion value in $\{-1, 1\}$. For the ith user, the probability that she has a positive opinion is given by:

$$P(y_i = 1 | \mathbf{x}_i, \beta) = \frac{1}{1 + \exp\left(-\beta^T \mathbf{x}_i\right)}, \tag{2}$$

where β is a $k \times 1$ feature weight vector. Note that there is no concept of time in this model.

We minimized an l_2-regularized logistic loss function to learn β:

$$L(\beta) = -\log\left(\prod_{i=1}^{n} P(y_i | \mathbf{x}_i, \beta)\right) + \lambda \|\beta\|_2^2$$

$$= \sum_{i=1}^{n} \log\left(1 + \exp\left(-y_i(\beta^T \mathbf{x}_i)\right)\right) + \lambda \|\beta\|_2^2,$$

where λ is the regularization parameter. Thus, given a set of features $\mathbf{x}$ and a set of known outputs y in the training data, the logistic regression model learns the parameter β that determines the relationship between $\mathbf{x}$ and y. Once the model has been learned, it can then be used to predict the outcomes of the test data, given their features $\mathbf{x}$.

Temporal Opinion Detection Model. In this work, we extend the above regularized logistic regression model, with an added element of time. As in the previous work, user opinions are classified as positive and negative types. Here, we have data samples $\mathbf{x}_i^{(t)}$, $i = 1, ...n$ and $t = 1, 2, ...$ Further, we have labels only for the first timestep, i.e., $y_i^{(1)}$, $i = 1, ...n$. Labeled samples are required for the first timestep, but not for the subsequent timesteps.

Now, extending (2) for any t^{th} timestep for user i, we obtain

$$P(y_i^{(t)} = 1 | \mathbf{x}_i^{(t)}, \beta^{(t)}) = \frac{1}{1 + \exp(-\beta^{(t)T} \mathbf{x}_i^{(t)})} \tag{3}$$

where $y_i^{(t)}$ is the discrete opinion value in $\{-1, 1\}$ in timestep t, $\mathbf{x}_i^{(t)}$ is a $k \times 1$ data vector and $\beta^{(t)}$ is a $k \times 1$ feature weight vector for timestep t.

We do not have labels on the samples for timestep $t+1$, as previously stated. Hence, to predict the opinions for timestep $t + 1$, we apply the key observation from Section 3.1 that *most* users do not change their opinions drastically in a single timestep. Thus, we assume that *most* users hold the same opinion as in the previous timestep. Most of the opinions in the previous timestep will therefore

Table 1. Examples of hashtags and n-grams over time on Obamacare

Feature type	Timestep 1	Timestep 5	Timestep 8
Hashtags	#obamacare, #koch, #getcovered, #cvs, #gop	#obamacare, #fullrepeal, #dontfundit, #aca, #trainwreck	#obamacare, #irs, #koch, #debtceiling, #gop
Unigrams	obamacare, gop, health, republicans, healthcare	obamacare, website, insurance, fix, coverage	obamacare, enrollment, work, hhs, job
Bigrams	obamacare will, the gop, benefits to, howard dean, fund obamacare	obamacare enrollment, signed up, fix obamacare, website failed, obamacare promises	3.3 million, signed up, million jobs, the koch, the irs

be the same as those in the next timestep, i.e. $y_i^{(t)}$ is the same as $y_i^{(t+1)}$ for most users. Following this assumption, we use $y_i^{(t)}$ from the previous timestep, and new textual features $\mathbf{x}_i^{(t+1)}$ from the current timestep to learn $\beta^{(t+1)}$.

Thus, we minimize the following l_2-regularized logistic loss function over consecutive timesteps t and $t + 1$:

$$L(\beta^{(t+1)}) = -\log\left(\prod_{i=1}^{n} P\left(y_i^{(t)}|\mathbf{x}_i^{(t+1)}, \beta^{(t+1)}\right)\right) + \lambda\|\beta^{(t+1)}\|_2^2 \qquad (4)$$

$$= \sum_{i=1}^{n} \log\left(1 + \exp\left(-y_i^{(t)}(\beta^{(t+1)^T}\mathbf{x}_i^{(t+1)})\right)\right) + \lambda\|\beta^{(t+1)}\|_2^2 \qquad (5)$$

The regularization helps to avoid overfitting [26] and to take care of the fact that this is an underdetermined system since $n \ll k$. Thus, by minimizing (4), we learn $\beta^{(t+1)}$ even in the absence of labeled samples at time $t + 1$. We use the open-source machine learning tool scikit-learn [28] to implement logistic regression with l_2 regularization.

4 Data Collection and Preprocessing

In this section we describe the method used to collect the dataset for this work, and the data pre-processing steps involved.

4.1 Data Collection

To crawl tweets on a topic of interest, we randomly selected users and collected their tweets over a period of time using the Twitter Streaming API.

Table 2. Examples of hashtags and n-grams over time on Immigration

Feature type	Timestep 1	Timestep 3
Hashtags	#immigration, #takeit-tothehouse, #weall-shallovercome, #movefor-ward, #immigrationen-forcement	#immigration, #immi-grationnews, #protests, #deport
Unigrams	immigrants, taxes, system, reform, drafted	gop, population, reforms, senator
Bigrams	million people, to diver-sity, immigration reform, require immigration	gop is, for immigration, need jobs, domestic issue, immigration reform

For Obamacare, tweets were crawled over a period of 8 months from July 2013 to February 2014. We have 757,960 users and 4,203,900 tweets in our dataset. For the topic of Immigration, tweets were crawled over the months of July, August and September, 2013, yielding a total of 15,001 users and 44,626 tweets. We consider each month to be 1 timestep for the sake of our experiments. On the topic of Obamacare, we selected 936 users that have tweets every month on which to test our model, and for the topic of Immigration, we picked 111 users.

4.2 Data Cleaning and Preprocessing

Twitter data is inherently noisy and filled with abbreviations and informal words. We clean and pre-process the dataset in the following manner to enable a better extraction of features.

- **URL removal:** In our method, URLs would not contribute to the feature extraction and were therefore removed.
- **Stopword removal:** Stopwords such as "a", "the", "who", "that", "of", "has" , etc. were removed from the tweets before extracting n-grams, which is a common practice.
- **Punctuation marks and special character removal:** Punctuation marks such as ":", "" etc. and special characters such as "[]", ",", "?", etc. were removed before extracting n-grams.
- **Additional whitespace removal:** Multiple white spaces were replaced with a single whitespace.
- **Conversion to lowercase:** Tweets are not generally case-sensitive owing to the informal language used. For instance, for our method, the word "Obama" should be considered the same as "obama" when parsing through a tweet. We converted the tweets to lowercase to preserve uniformity in feature extraction.
- **Tokenization:** The tweets were tokenized into words to extract n-grams from them.

5   Implementation Details

In this section we describe the features we chose to use in the model, and also explain the steps taken to implement the model.

5.1   Feature Engineering

As mentioned in the Introduction, we used textual features extracted from the tweets for opinion detection in [12]. The features used were hashtags and n-grams. Apart from highlighting the topic of a tweet, hashtags have been found to carry some additional information regarding the bias of the tweet itself [15,36]. For example, on the topic of Obamacare, *#defundobamacare, #getcovered, #fullrepeal* are examples of hashtags that clearly portray the opinion of the person that uses them.

However, at times, hashtags by themselves are not sufficient to capture the opinion, for instance,

"Let's abolish the IRS before it enforces #obamacare! please sign and rt this petition if you agree"

In the above tweet, the hashtag *#obamacare* is not sufficient to capture the opinion of the tweet. The entire tweet needs to be considered to get the actual opinion. For this purpose, we use the n-gram model which is considered a powerful tool for sentiment extraction [11,14,27]. We extract n-grams out of the tweets to capture the bias from the tweet itself.

At every timestep, we order the features according to the number of users that use them. We use the 1000 most popularly used hashtags, 2000 most popularly used unigrams and 2000 most popularly used bigrams from each timestep for our experiments. The choice of the number of features was governed by the usage of the features. For instance, after the first 1000 hashtags, the usage of the hashtags drops significantly, thus motivating us to use the most popular 1000 tags as our features. Similar reasons led to the use of the top 2000 unigrams and bigrams. Thus we had 5000 features at every timestep.

For every user i at time t in (4), $\mathbf{x}_i$ contains the number of times user i uses each of the 5000 features at that timestep. Owing to the evolving nature of the conversation, this set of features changes over time. However, using our model described in Section 3.2, we can *automatically* learn a new β at every timestep for a new set of features by minimizing (4). Tables 1 and 2 show a few examples of features found on several timesteps.

5.2   Implementation

In our experiments, we consider each month to be a timestep, and study the same set of n users across all timesteps. The following provides a detailed description of the steps taken at every timestep.

- At timestep 1:
  - We begin by labeling a subset of the users such that those with a positive opinion on the given topic are assigned a label +1 and those with a negative opinion are assigned a label -1. Let d be the number of users that are labeled at timestep 1. The data matrix is built using the 5000 textual features (as described in Section 5.1), thereby leading to a $d \times 5000$ matrix, $X^{(1)}$. We use this data to train the model (4) to learn $\beta^{(1)}$. For Obamacare, $d = 201$ (89 positive, 112 negative), and for the Immigration dataset, $d = 30$ (24 positive, 6 negative).
  - We then assign opinion labels to the larger unlabeled set of $n - d$ users using the learned $\beta^{(1)}$. This step is performed to get the opinion labels for all n users at this timestep. We now proceed with the entire set of n users for the subsequent steps.
- For each subsequent timestep, $t + 1$:
  - We minimize the regularized logistic loss function (4) between the opinions of users at t and $t + 1$ to learn $\beta^{(t+1)}$.
  - We then use the learned $\beta^{(t+1)}$ to predict opinions at time $t + 1$. This forms $y_i^{(t+1)}$.

6   Experimental Results

In this section, we outline in detail the experiments we conducted on the dataset, and the metrics we used to evaluate it. Further, we report the insights that the method provided with respect to the sub-topics that were being discussed at every timestep.

6.1   Temporal Opinion Detection Results

To evaluate the model on our primary topic of interest, *Obamacare*, we label the opinions of a random group of users on some of the key timesteps to test whether our model captures their opinions correctly. We were particularly interested in determining whether the model detects the opinions correctly after the occurrence of a significant event with respect to Obamacare. One such event occurred on October 27, 2013, when the main website for the Affordable Care Act, *Healthcare.gov* crashed. This created a great deal of chatter on Twitter (see Figure 3 for a plot of the number of users that mentioned the website crash over time. As is evident, the number of users goes up significantly towards the end of October which was when the website crash occurred, and continues to be a focus of conversation during November as well.) To determine whether our model captures the opinions being echoed right after this occurrence, we focus on Timestep 5 which contains tweets from the beginning of November 2013, and throughout the rest of the month. We select 88 users at random from that timestep, for testing our model.

The other timestep that we pick for these tests was timestep 5, which was the month of February 2014. In that month, the Department of Health and Human

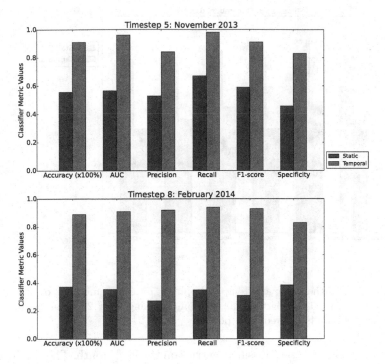

Fig. 1. Comparison of the static and temporal opinion detection methods on Obamacare. The methods are compared on two timesteps of interest across all classifier metrics.

Services (HHS) announced the signing up of 3.3 million people for Obamacare, which was a significant event in the Obamacare timeline. Another event that generated a large volume of tweets at that time was that some firms were firing employees to avoid Obamacare costs, but were certifying to the IRS that the firings were not on the grounds of Obamacare, to avoid penalty of perjury. We label 43 randomly selected users from this timestep.

To validate the usefulness and the need for our method, we first present the results obtained by simply using the *Static Opinion Detection Model* described in Section 3.2 for temporal opinion detection. Thus we used the β learned from the training samples at timestep 1 to predict opinions for later timesteps. As seen in Figure 1, the accuracies achieved using the static method on timesteps 5 and 8 are 55.68% and 37.2% respectively, while our proposed temporal method yields accuracies of 90.9% and 89.0% respectively for the two timesteps. Moreover, the temporal method outperforms the static method across all popularly-used classifier metrics [18] such as AUC, F1-score, etc.

To demonstrate the generality of our method, we also conducted experiments on the topic of *U.S. Immigration Reform bill*. Since we only have 3 months' data

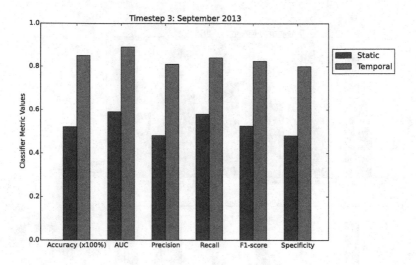

Fig. 2. Comparison of the static and temporal opinion detection methods on Immigration. The static method is no better than random guessing after 2 timesteps, but the temporal method shows high predictive power.

on the topic, we evaluated the classifier metrics on the last month. The results are reported in Figure 2. The temporal method yields better performance than the static method in this case as well. As is evident, the static method yields about 50% accuracy, which can simply be obtained by random guessing. However, using the temporal method yields a significantly higher accuracy of 85%. The temporal method also performs much better in comparison to the static method across all classifier metrics as well.

6.2   Significant Feature Detection and Emergence of Temporal Sub-Topics

Out of the 5000 features used at every timestep, some of the textual features are more informative in detecting opinions than others. To determine this set of informative features over time, we evaluate the statistical significance of each feature of the Obamacare dataset for predicting user opinion. We follow the technique described in Section 5, Algorithm 3 of [29] for significance testing, which we describe here for the sake of completeness. For the timestep of interest, we run our l_2 regularized temporal model on the data, and store the weights that each of the features are assigned by the model. Then we randomize the labels on the samples and run our model on the randomized data. Let $\hat{\theta}$ be the coefficient obtained from this set. For each randomized run m, let $\tilde{\theta}$ be the random coefficient vector obtained from the fixed feature vector $\mathbf{x}$ and the randomized response $\tilde{y}$. For ν randomized runs, we obtain ν coefficient vectors $\tilde{\theta}$.

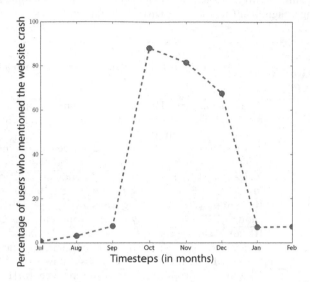

Fig. 3. Mentions of Obamacare website crash over time

For each dimension, the coefficient value in each $\tilde{\theta}$ represents a random statistical relationship between the feature and the response. Then the p-value of the lth dimension is computed as

$$\frac{\text{Count}(|\tilde{\theta}_l| > |\hat{\theta}_l|)}{\nu + 1} \tag{6}$$

where "Count" represents the number of times the absolute value of the random coefficient for the lth dimension exceeded the absolute value of the same coefficient obtained from the training set. This is a commonly used permutation test for statistical hypothesis testing [2]. Features that had a p-value less than 0.05 are selected as the most significant features with a confidence of at least 95%.

Using the significant features obtained at each timestep, we examine the dataset for tweets carrying these features. This led to the discovery of the various sub-topics of conversation (related to the main topic of Obamacare), that users participated in over time. Most of the sub-topics can be tied to real-world events that aligned with the timestep under consideration. This further reflects the evolving nature of the topics of conversation. Table 3 illustrates the sub-topics of interest that were detected over the various timesteps. For example, in July 2013, we find that the IRS emerged as an important sub-topic of discussion. Similarly, Obama's apology and a count of how many people were enrolling in Obamacare were popular sub-topics in November 2013. In February 2014, the 3.3 million enrollment mark and Megyn Kelly (a Fox News anchor who covered a great deal of negative news related to Obamacare) were sub-topics that emerged as being popular. Thus our method is able to detect evolving sub-topics of conversation among users over time.

Table 3. Significant features (95% statistical significance) on Obamacare at three different time steps. Significant features capture the temporally evolving sub-topics.

Time step	Significant Features	Temporal sub-topics inferred from tweets
Jul 2013	*braveheart, gifs*	The Washington Examiner publishes funny series of gifs from movie Braveheart depicting Republicans' failed attempts at defunding Obamacare.
	employees	News sources report that Obamacare call center employees were not being offered healthcare benefits.
	kyle	News report by reporter Kyle Cheney on Politico.com stating that CVS was going to publicize Obamacare.
	irs	IRS employees unwilling to sign up for Obamacare, although IRS was heavily involved in enforcing Obamacare.
	howard	Howard Dean, former Democratic National Committee Chairman, comments that Independent Payment Advisory Board will be unable to keep costs down.
	premiums	Obamacare premiums are lowered even further in eleven states.
	empire	Cited article discussing civil lawsuits, environmental damage caused by the output from industries, etc. of the Koch brothers' empire and related controversies.
Nov 2013	*warning*	Republicans "warning" people of Obamacare, and that the website crash is a "warning" in itself.
	case	Blog by Peter Suderman ("Time To Start Considering Obamacares Worst-Case Scenarios") discussing failure of online enrollment system negatively affecting Obamacre.
	apology	– Obama apologizing to people whose insurance plans were being canceled, even though he said that people could keep their existing coverage if they liked. – Ed Schultz demands that Republicans, rather than the President, should apologize "for not having any plan".
	scorecard	Obamacare scorecard: how many actually enrolled, and how a larger number of people lost their insurance.
Feb 2014	*@megynkelly*	Megyn Kelly, a Fox news anchor who covered (negative) news related to Obamacare.
	wednesday	Dept. of Health and Human Services announces on a Wednesday (Feb 12, 2014) that 3.3 million people signed up for Obamacare, but it includes hundreds of thousands of individuals defaulting their first premium payment.
	firings	Firms required to certify to the IRS that Obamacare was not a factor in their firing their employees (although it was).
	tgdn	New hashtag (Twitter Gulag Defense Network) started in January 2013 to counter Twitter Gulag, a way to trick Twitter systems into thinking that live profiles are actually spambot profiles. Apparently, many conservative profiles were being shut down by leftists employing this policy.

7   Conclusion

In this work we have proposed a novel temporal opinion detection method that can successfully detect the opinions of Twitter users engaging in an evolving conversation. Our primary topic of interest is Obamacare, for which the focus of conversation shifted from one sub-topic to another due to the various events associated with the event that occurred over time. We also selected the topic of U.S. Immigration Reform to demonstrate the generality of our method. Our proposed temporal machine-learning method performs well across all classifier metrics of importance. Additionally, it leads to automatic detection of informative features that point to important, and changing sub-topics.

Acknowledgment. This work has been supported by the Institute for Collaborative Biotechnologies through grant W911NF-09-0001 from the U.S. Army Research Office. The content of the information does not necessarily reflect the position or the policy of the Government, and no official endorsement should be inferred. The authors would like to thank Prof. Noah Friedkin of the Sociology Department at the University of California Santa Barbara for his insightful comments during the research.

References

1. Obamacare facts. http://obamacarefacts.com/obamacare-facts/
2. p-value. http://en.wikipedia.org/wiki/P-value
3. Twitter usage statistics. http://www.internetlivestats.com/twitter-statistics/
4. U.S. immigration reform. https://en.wikipedia.org/wiki/Border_Security,_Economic_Opportunity,_and_Immigration_Modernization_Act_of_2013
5. Revolutionizing revolutions: Virtual collective consciousness and the arab spring (2012). http://www.huffingtonpost.com/yousri-marzouki/revolutionizing-revolutio_b_1679181.html
6. Obamacare sees social media surge ahead of deadline, March 2014. http://www.nextgov.com/health/2014/03/obamacare-sees-social-media-surge-ahead-deadline/81625/
7. Social networks and social media in ukrainian "euromaidan" protests (2014). http://www.washingtonpost.com/blogs/monkey-cage/wp/2014/01/02/social-networks-and-social-media-in-ukrainian-euromaidan-protests-2/
8. The top 20 valuable facebook statistics (2015). https://zephoria.com/social-media/top-15-valuable-facebook-statistics/ (updated February 2015)
9. Akkaya, C., Wiebe, J., Mihalcea, R.: Subjectivity word sense disambiguation. In: Proceedings of the 2009 Conference on Empirical Methods in Natural Language Processing, vol. 1, pp. 190–199. Association for Computational Linguistics (2009)
10. Baccianella, S., Esuli, A., Sebastiani, F.: Sentiwordnet 3.0: an enhanced lexical resource for sentiment analysis and opinion mining. In: LREC, vol. 10, pp. 2200–2204 (2010)
11. Bespalov, D., Bai, B., Qi, Y., Shokoufandeh, A.: Sentiment classification based on supervised latent n-gram analysis. In: Proceedings of the 20th ACM International Conference on Information and Knowledge Management, pp. 375–382. ACM (2011)

12. Bhattacharjee, K., Petzold, L.: Probabilistic user-level opinion detection on online social networks. In: Aiello, L.M., McFarland, D. (eds.) SocInfo 2014. LNCS, vol. 8851, pp. 309–325. Springer, Heidelberg (2014)
13. Bono, J.E., Ilies, R.: Charisma, positive emotions and mood contagion. The Leadership Quarterly **17**(4), 317–334 (2006)
14. Dave, K., Lawrence, S., Pennock, D.M.: Mining the peanut gallery: opinion extraction and semantic classification of product reviews. In: Proceedings of the 12th international conference on World Wide Web, pp. 519–528. ACM (2003)
15. Davidov, D., Tsur, O., Rappoport, A.: Enhanced sentiment learning using twitter hashtags and smileys. In: Proceedings of the 23rd International Conference on Computational Linguistics: Posters (2010)
16. DeGroot, M.H.: Reaching a consensus. Journal of the American Statistical Association **69**(345), 118–121 (1974)
17. Esuli, A., Sebastiani, F.: Sentiwordnet: a publicly available lexical resource for opinion mining. Proceedings of LREC **6**, 417–422 (2006)
18. Fawcett, T.: An introduction to ROC analysis. Pattern Recognition Letters **27**(8), 861–874 (2006)
19. Friedkin, N.E., Johnsen, E.C.: Social influence networks and opinion change. Advances in Group Processes **16**(1), 1–29 (1999)
20. Garas, A., Garcia, D., Skowron, M., Schweitzer, F.: Emotional persistence in online chatting communities. Scientific Reports **2** (2012)
21. Gilbert, E., Karahalios, K.: Predicting tie strength with social media. In: Proceedings of the SIGCHI Conference on Human Factors in Computing Systems (2009)
22. Go, A., Bhayani, R., Huang, L.: Twitter sentiment classification using distant supervision. CS224N Project Report, Stanford, 1–12 (2009a)
23. Kucuktunc, O., Cambazoglu, B.B., Weber, I., Ferhatosmanoglu, H.: A large-scale sentiment analysis for Yahoo! answers. In: Proceedings of the Fifth ACM International Conference on Web Search and Data Mining (2012)
24. LIWC: LIWC software (2001). http://www.liwc.net/index.php
25. MPQA: MPQA (2005). http://mpqa.cs.pitt.edu/lexicons/
26. Ng, A.Y.: Feature election, L 1 vs. L 2 regularization, and rotational invariance. In: Proceedings of the Twenty-First International Conference on Machine Learning (2004)
27. Pak, A., Paroubek, P.: Twitter as a corpus for sentiment analysis and opinion mining. In: LREC (2010)
28. Pedregosa, F., et al.: Scikit-learn: Machine learning in Python. Journal of Machine Learning Research **12**, 2825–2830 (2011)
29. Pendse, S.V., Tetteh, I.K., Semazzi, F.H., Kumar, V., Samatova, N.F.: Toward data-driven, semi-automatic inference of phenomenological physical models: application to eastern sahel rainfall. In: SIAM International Conference on Data Mining (2012)
30. Pennebaker, J.W., Chung, C.K., Ireland, M., Gonzales, A., Booth, R.J.: The development and psychometric properties of LIWC 2007. Austin, TX, LIWC. Net (2007)
31. Pennebaker, J.W., Francis, M.E., Booth, R.J.: Linguistic inquiry and word count: LIWC 2001 (2001)
32. Tan, C., Lee, L., Tang, J., Jiang, L., Zhou, M., Li, P.: User-level sentiment analysis incorporating social networks. In: Proceedings of the 17th ACM SIGKDD International Conference on Knowledge Discovery and Data Mining, pp. 1397–1405. ACM (2011)

33. Tan, S., Li, Y., Sun, H., Guan, Z., Yan, X., Bu, J., Chen, C., He, X.: Interpreting the public sentiment variations on twitter. IEEE Transactions on Knowledge and Data Engineering **26**(5), 1158–1170 (2014)
34. Thelwall, M., Buckley, K., Paltoglou, G.: Sentiment in twitter events. Journal of the American Society for Information Science and Technology **62**(2), 406–418 (2011)
35. Thelwall, M., Buckley, K., Paltoglou, G., Cai, D., Kappas, A.: Sentiment strength detection in short informal text. Journal of the American Society for Information Science and Technology **61**(12), 2544–2558 (2010)
36. Wang, X., Wei, F., Liu, X., Zhou, M., Zhang, M.: Topic sentiment analysis in twitter: a graph-based hashtag sentiment classification approach. In: Proceedings of the 20th ACM International Conference on Information and Knowledge Management (2011)
37. Wiebe, J., Wilson, T., Cardie, C.: Annotating expressions of opinions and emotions in language. Language Resources and Evaluation **39**(2–3), 165–210 (2005)
38. Wilson, T.: Fine-grained subjectivity analysis. Ph.D. thesis, Doctoral Dissertation, University of Pittsburgh (2008)

Identifying Similar Opinions in News Comments Using a Community Detection Algorithm

Jonathan Scott[1]([⊠]), David Millard[1], and Pauline Leonard[2]

[1] Web and Internet Science, School of Electronics and Computer Science,
University of Southampton, Southampton, UK
js3g10@soton.ac.uk
[2] Sociology, School of Social Sciences, University of Southampton, Southampton, UK

Abstract. Despite playing many important roles in society, the news media have been frequently criticised for failing to represent a wide range of viewpoints. Online news systems have the potential to allow readers to add additional information and perspectives. However, due to the simplicity of the filtering mechanisms typically employed, these systems can themselves be prone to over-promoting popular viewpoints at the expense of others. Previous research has attempted to diversify news comments through the use of content similarity, sentiment analysis, named entity recognition, and other factors. In this paper we propose the use of a commonly used community detection algorithm on a network of voting data to identify sentiment groups in news discussion threads, with the eventual goal that these groups may be used to present diverse content. In a controlled experiment with 154 participants, we verify that the Louvain Community Detection algorithm is able to group users with accuracy comparable to an average human. This produces groups containing users who share similar sentiment on a given topic. This is an important step towards ensuring that each group is represented, as by using this method future news systems can ensure that more diverse views are represented in open comment threads.

Keywords: Community detection · News · Comments · Discussion · Sentiment · Viewpoint

1 Introduction

In 2008, Stromback claimed that the media have become "the most important source of information for most people in advanced democracies around the world" [29]. They fill many roles in modern democratic society: As an agenda-setter the media influences the focus of public opinion [18,30], as the "fourth estate" they are expected to hold the powerful to account [27], and as an information provider they are tasked with ensuring the population are informed about the processes and decisions which concern their lives [2,16].

Criticism of the news media has been frequent however, with many complaints of an over-reliance on elite sources (e.g. [4,8,13]). They have also been

© Springer International Publishing Switzerland 2015
T.-Y. Liu et al. (Eds.): SocInfo 2015, LNCS 9471, pp. 98–111, 2015.
DOI: 10.1007/978-3-319-27433-1_7

shown to be failing in their role as information provider. In 2013, an Ipsos MORI survey [14] found that, among other misperceptions, British people estimated the amount of benefit fraud as 34 times higher than official estimates, that 24% of the population were Muslim (compared to the official figure of 5%) and 31% of the population were immigrants (compared to the official figure of 13%). A similar report in 2014 found very similar figures [24]. With welfare and immigration regularly ranking high on election priorities [31] this shows that the media have not adequately informed the public of the information they need to properly participate in democracy.

Part of this issue may be explained by the lack of perspectives represented in the media. In 1972, McCombs found that amongst local daily newspapers, national newspapers, and national news broadcasts, there existed a high degree of similarity of news agenda [19]. More recently, Nick Davies found that 60% of the stories in four chosen "quality newspapers"[1] comprised wholly or mainly of material from newswires and public relations groups, and only 12% could be confidently attributed to a named reporter [8]. This reveals a situation where much of the information the public receive is provided by very few sources.

The rise of online news has allowed for new methods of citizen involvement in the news process. Once restricted to passive consumption of journalist-produced content, readers now regularly contribute to news, through online commenting and discussion systems, submitting media directly to journalists, and on some systems through the creation of their own stories. In 2014 it was found that half of social network users shared news on their social network accounts, 46% of users discussed news on these sites, and roughly 10% of users had published news videos they made themselves [22].

However, with this much increased public contribution comes the difficult problem of sorting and filtering this content into a form that can be easily consumed, a job performed in traditional media by a news editor. One common solution is to allow other readers to vote for high quality content and then show the highest rated prominently. This is the solution employed by a number of social media platforms, as well as the websites of many newspapers [28].

One issue with this method of filtering is that it can encourage groupthink [21]. Jokingly referred to as the "hive mind" within Reddit communities, this refers to the common phenomenon whereby members of a group tend to reach a consensus and avoid dissenting opinions. Muchnik et al. performed an experiment on an un-named social news aggregation website and found that both prior positive votes and prior negative votes increased the number of future positive votes [23], and Mills found that on Reddit, minority opinions are "slightly marginalised but not excluded" [21].

There is evidence that being responded to is a key determinant in the decision to continue contributing to a community [3,6]. Interfaces which make less highly rated comments less visible can lead to discouraging future contributions from those who do not agree with the majority viewpoints. Over time this will lead to less diversity in news discussions, undermining the potential of online news to

---

[1] The Times, The Guardian, The Independent, and The Daily Telegraph.

solve the issues identified by critics of traditional journalism. To avoid this, it is important to ensure that news discussions present diverse comments and fairly represent the viewpoints of the different groups involved.

This paper provides an overview of existing attempts at diversifying comments online, proposes a method of diversification using voting data and community detection algorithms, and presents an experiment to test the ability of community detection algorithms to automatically detect groups of commenters with similar opinions and views. If comment systems could automatically group users in this way then it would be possible to design new commentary systems that move beyond simple chronology or popularity in order to highlight a more diverse set of comments and opinions to users.

2   Background

Previous work by Giannopoulos, et al. attempted to diversify comments by building on textual diversity algorithms used in other domains, and added additional criteria specific to comments. They produced a system for selecting diverse comments using measures of content similarity, sentiment analysis, named entity recognition, and comment quality [11]. We will instead approach the problem by concentrating on groups of users rather than individual comments, relying on homophily and an assumption that people will tend to maintain a consistent viewpoint within the timeframe of a single discussion thread.

Previous research has attempted to identify communities on social networks and in online discussions without the focus on diversification. This has primarily used explicit friend/follower relationships or text-mining. Jaffali, et al. look at community identification using tweets in the context of friend recommender systems. They propose an algorithm which performs text-mining and sentiment analysis, producing communities of users which share sentiment towards a given entity [15]. Abu-Jbara, et al. utilise opinion mining and sentiment analysis on online Arabic discussions to identify groups who share an opinion by their use of subjective language [1]. Parau, et al. investigate the use of multidimensional data in identifying sentiment communities [26]. Each of these approaches has merit, but they do not leverage the very common interaction of users voting for content they agree with.

Voting is one of the more common interactions in online communities, including Reddit, Facebook [12], and Twitter [20], and we propose the use of this data to identify users who share sentiment towards a topic. It should be noted that by "sentiment" we mean "an attitude toward something; opinion"[2]. For this voting data to be useful for our purposes, we require voting be performed for a particular reason: to indicate agreement or disagreement with a piece of content. Kriplean, et al. view the "like" buttons on social websites as overloading two functionalities: providing a way to recognise and appreciate a speaker, but also including an implicit agreement with the content [17]. Using the buttons in this

---

[2] "sentiment" definition from dictionary.com. Accessed 14/10/15.

way is contrary to the rules of many popular social news systems[3] but as mentioned, Reddit users regularly discuss the "hive mind" and it has been shown that there is at least some marginalisation of minority views in social news [21]. This indicates that to some extent, users are voting for content with which they agree rather than only for content which they think is of high quality.

In this paper we will use this voting data to create a network of users with edge weights representing the number of times they've mutually "liked" a piece of content. Once the network is created, it can be partitioned using one of many existing community detection algorithms.

2.1   Community Detection Algorithms

A community is "a subgraph of a network whose nodes are more tightly connected with each other than with nodes outside the subgraph" [10]. Community detection techniques are commonly used for analysing networks such as community organisations and scientific collaborations [9]. They have also been used for tag disambiguation, user profiling, and event detection [25] and to improve friend recommendation systems and collaborative filtering techniques.

There are some specific requirements of a community detection algorithm to be used with online discussions. First, it is required that the algorithm be able to run in near-real-time on average news discussions. This is so that the results can be immediately presented to users to maximize the impact. Second, due to the nature of online discussions, there is no way of predicting in advance how many communities will exist or what the size of those communities will be, techniques that require this information to be provided will not be suitable for the task.

Community detection algorithms can be broadly grouped into three categories: divisive algorithms, which detect and remove inter-community links; agglomerative algorithms, which recursively merge communities; and optimization algorithms, which attempt to maximize some function [5]. They are typically judged on the "modularity" of the communities detected, a value between -1 and 1 that represents the density of links within communities compared to the density of links between communities.

There are a number of commonly used algorithms which optimize for modularity, though Fortunato and Barthelemy identified a problem with these techniques, showing that modularity "contains an intrinsic scale that depends on the total number of links in the network", and that communities which are smaller than this scale may not be detected at all [10]. This is particularly a problem in very large networks where smaller communities will not be detected.

The method used in this paper is the Louvain Community Detection algorithm, as proposed in 2008 by Blondel, et al. [5]. It is a heuristics-based method which optimises for modularity. It is simple to implement, performs well (achieving high modularity in low computing time), and runs well even on large net-

---

[3] e.g. Reddit "Moderate based on quality, not opinion" from http://www.reddit.com/wiki/reddiquette or Slashdot "simply disagreeing with a comment is not a valid reason to mark it down" from http://slashdot.org/faq

works. It also partially deals with the resolution limit problem identified by Fortunato and Barthelemy.

2.2 Louvain Community Detection

The Lovain Community Detection algorithm is very fast, with linear complexity on typical data [5]. The algorithm begins with each node assigned to its own community, and repeatedly executes two phases. The first phase is described by Blondel, et al. as:

> For each node i we consider the neighbours j of i and we evaluate the gain of modularity that would take place by removing i from its community and by placing it in the community of j. The node i is then placed in the community for which this gain is maximum (in case of a tie we use a breaking rule), but only if this gain is positive. If no positive gain is possible, i stays in its original community. This process is applied repeatedly and sequentially for all nodes until no further improvement can be achieved and the first phase is then complete.

The second phase involves creating a new network whose nodes are the communities created during the first phase. The weights of edges between the new nodes are the sum of the edge weights between the two communities. These two phases are repeated until there are no more changes.

Due to the recursive nature of the algorithm, it provides intermediate stages which allow for different levels of granularity in the communities. This feature helps to partially avoid the resolution limit identified in by Fortunato and Barthelemy as selecting different levels will result in communities of different sizes appearing. These intermediate stages provide a hierarchy of communities which may be useful when identifying groups of users with similar sentiment, though in this study we will focus only on the level with the highest modularity.

In their paper, Blondel et al. test the algorithm's performance by running it against a small social network, a network of scientific papers and their citations, a sub-network of the internet, a network of webpages, and other datasets [5]. It performs well in all cases (having a high level of modularity with a small computation time). To verify that the Lovain Community Detection algorithm will be able to group together news discussion participants who share similar sentiment, we created a network from existing news discussions and employed a web-based study to explore the accuracy of the detected communities.

3 Methodology

We looked to the discussion systems of the websites analysed in our previous work [28] for systems which 1) have enough activity to generate a large amount of data and 2) have APIs which allow access to votes. Four of the systems provided the data needed to create a network for use with the Louvain algorithm: Facebook, Twitter, CNN, and The Telegraph.

Of these, The Telegraph and CNN use a common discussion system (Disqus), and therefore have a common API. They also, as outright news sites, have a concept of a news "story" which the other two systems do not. This makes it possible to investigate grouping users based on sentiment without needing to first separate them based on topic. For these reasons, this experiment uses data from The Telegraph and CNN websites.

3.1 Data Collection

We collected every news story which, between 22/05/2014 and 27/05/2014, featured in the "Most Popular" section of CNN or the "Most Viewed" section of The Telegraph. This resulted in 44 stories: 12 from CNN and 32 from The Telegraph. The difference can be attributed to the fact that The Telegraph's "Most Viewed" section shows 10 items at a time, whereas CNN's "Most Popular" section shows only 6. We then gathered all comments on each of these stories, including which users had voted for which comments. This resulted in 23,655 comments and 49,486 votes.

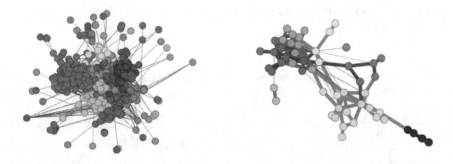

Fig. 1. Detected communities for two randomly selected stories: "Bulls take down bullfighters but still lose in Spain", and "How ITV missed the best moment of the FA Cup final"

From this data, for each story, a weighted graph was produced of relationships between the people who interacted with that story. This gave us an average of 207 nodes (s = 262) and 841 edges (s = 1621) per graph.

In each graph, nodes represent users who have interacted with the story, and edges represent the level of agreement between users (the number of times they voted the same way minus the number of times they voted differently). As the Louvain algorithm does not support negative edge weights, values less than 0 are discarded.

When used on these networks, the Louvain algorithm generates a total of 877 groups (an average of 19 groups per story, s = 24) with an average of 10 members per group (s = 29), and an average modularity of 0.516. We have calculated these numbers for each category of news on each of the news systems, which can be seen in Table 1.

Figure 1 shows the communities generated for two randomly selected stories. On the left is the CNN story "Bulls take down bullfighters but still lose in Spain" about a bullfighter being injured. The generated communities have a modularity of 0.63. There are 56 communities generated in total, though the majority of these are single-user communities which are not connected to any other nodes and are not shown in the figure. The largest community (with 23% of users) is shown in blue and consists of users supportive of bull fighting, and the second largest (19% of users) shown in red with comments typically supportive of the bulls. However the third largest community (12.4% of users), shown in light green is also supportive of the bulls and a difference in content between the two groups is not clear.

On the right is the Telegraph story "How ITV missed the best moment of the FA Cup final", about a football match between Hull City and Arsenal. This network has 9 communities and a modularity of 0.49. The largest community (27.3% of users), is shown in yellow and consists primarily of Hull supporters. The second largest (18.2% of users) is shown in green and consists primarily of supporters of Arsenal. The third largest in this case (18.2%) is shown in red and does not seem to specify a team preference, instead consisting of users criticising the article and author.

Table 1. Number of stories; average number of nodes, edges, and communities; average community size; and average modularity for each category of story, separated by news website

Category	Stories	Nodes	Edges	Communities	Community Size	Modularity
The Telegraph						
Sport	15	49.27 (s=32.15)	83.33 (s=65.59)	9.07 (s=4.46)	4.8 (s=2.14)	0.5 (s=0.09)
World	6	300.83 (s=210.76)	1562.33 (s=1451.01)	15 (s=3.74)	17.67 (s=10.27)	0.42 (s=0.06)
Domestic	5	618.4 (s=372.59)	3726.6 (s=2850.92)	23.4 (s=12.44)	25.6 (s=9.18)	0.46 (s=0.04)
Technology	2	184 (s=109)	353 (s=222)	26.5 (s=16.5)	6.5 (s=0.5)	0.59 (s=0.11)
Obituaries	1	32	54	8	4	0.46
Finance	1	367	1371	12	30	0.47
Business	1	123	193	17	7	0.61
Culture	1	118	291	16	7	0.55
CNN						
Showbusiness	4	119.5 (s=82.89)	212 (s=157.56)	22.5 (s=11.43)	4.5 (s=0.87)	0.54 (s=0.09)
Travel	3	72.67 (s=4.03)	111.33 (s=21.68)	20.67 (s=4.5)	3 (s=0.82)	0.61 (s=0.04)
World	3	527 (s=290.25)	1195 (s=863.81)	80.33 (s=55.76)	7.33 (s=4.78)	0.65 (s=0.13)
Business	1	68	110	13	5	0.47
US	1	127	257	22	5	0.67
Overall	44	207.18 (s=262.95)	841.05 (s=1621.38)	19.93 (s=23.8)	9.68 (s=9.52)	0.52 (s=0.1)

3.2   Comparison with Manual Classification

Having gathered this dataset of comments and groups, we performed an experiment to compare the results of the Louvain algorithm with human classification. This was undertaken in the form of a web-based assessment, which showed users example comments and asked them to make a judgement as to whether they expressed similar sentiment. Our hypothesis is that pairs from users in the same group will be viewed as "similar" more often than pairs from users in different groups.

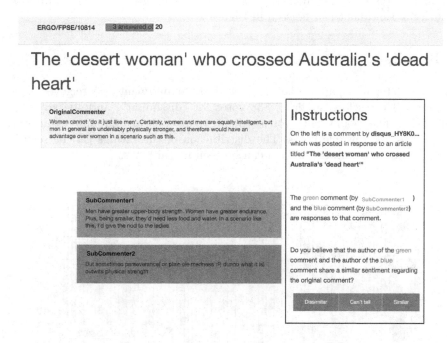

Fig. 2. The sentiment experiment interface

To generate the questions for this experiment, we identified every pair of comments posted in reply to a common parent (this was to make sure that the comments were broadly on the same topic, which makes manual classification easier for the participants, and reduces the number of times they skip the question - thus maximising the number of results received for the number of questions asked). These pairs were filtered to ensure that there was only a single question for each combination of groups and parent comment. That is, for comment i, and groups j and k there is only a single pair of comments with parent i, and one author from each of j and k. We also ensured that there were no pairs with both comments from the same author. After this filtering, 3554 pairs remained, 1658 with both authors from the same group, and 1896 with authors from different groups. For the purposes of this study, 50 pairs with same group authors and 50 pairs with different group authors were randomly selected. This gave 56 pairs from The Telegraph and 44 pairs from CNN.

Participants were then invited to view these comment pairs through an online interface. The study was shared on social media and participants were invited to share the study with their own contacts. This resulted in 154 respondents, though 20 did not complete the experiment and so the ratings they contributed were removed.

The participants were presented with 20 pairs of comments and instructed to decide for each pair of comments if their sentiments towards the parent comment were similar, dissimilar, or if they could not tell. The interface for deciding the similarity of sentiment can be seen in Figure 2.

4   Results

From the 134 participants who completed the experiment, we received 2376 ratings, 1098 votes for "similar", 838 votes for "dissimilar", and 404 votes for "can't tell". The "can't tell" votes will be treated as an inability to answer, which leaves 1936 usable votes. The distribution of the ratings can be seen in Figure 3, and an example question can be seen in Table 2.

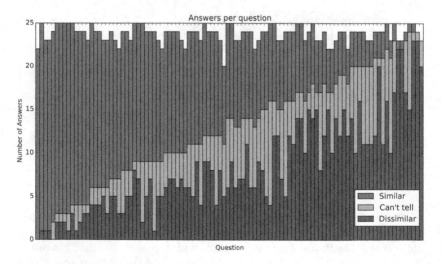

Fig. 3. The distribution of sentiment ratings

For each question, we calculated the percentage of respondents that said the sentiments expressed were "similar" (see Figure 4). As the data are not normally distributed, we used a Mann-Whitney test to compare the percentage between questions with comments from authors in the same group and questions with comments from authors in different groups. On average, comments by authors

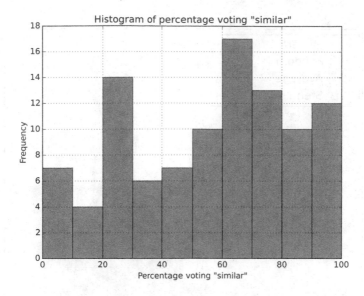

Fig. 4. The distribution of questions by percentage voting "similar"

who were allocated to the same group had a higher percentage of "similar" answers (M = 66.91, SE = 3.11) than comments by authors allocated to different groups (M = 43.6, SE = 3.91). This difference is significant (U = 653, z = -4.116, p <.001, r = -.4116).

To evaluate the algorithm's success when compared to a human, we then used the grouping status (same group or different groups) as a vote by the system (for similar and dissimilar respectively), and calculated how often the system's vote agreed with the votes of participants (62.09%). We also calculated this for each participant (M = 66.37%, s = 8.73), and generated 10,000 randomly allocated groupings to calculate how often these random groupings agreed with the votes of participants (M = 44.61%, s = 1.82). This can be seen in Figure 5.

5   Discussion

There is a significant difference between the percentage of respondents voting "similar" on questions where the authors were allocated to the same group, when compared to questions where the authors were allocated to different groups. This indicates that the Louvain Community Detection algorithm, when applied to a network produced using voting data, is able to produce groups of users who share sentiment on a given topic, in that the groups produced are more likely than random groupings to group together users judged to have similar sentiment. This is evidence that supports our original assumptions (that people will maintain consistent viewpoints within a single discussion, and that they will vote for content that they agree with).

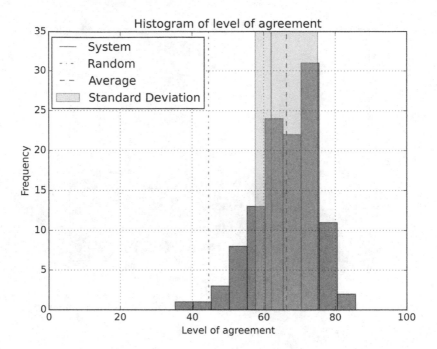

Fig. 5. The level of agreement between participants

To compare the system to the potential ability of a human to categorise the commenters by sentiment, we calculated how often the system agreed with each participant in their assessment of the similarity of comments, and also how often participants agree with each other. The average agreement of the system (62.09%) is within a single standard deviation (8.73) of the participants' average level of agreement (66.37%). This indicates that the system has comparable success to an average human when deciding if the sentiment of two comments is similar. These agreement levels are much higher than that which would be expected given random groupings (44.61%).

To further explore the quality of the sentiment groupings would require access to existing news editors who would be willing to submit their own votes on the comment pairs used in this study. This would allow a comparison between the use of community detection algorithms and best practice for other types of content (such as letters and emails to the editor), where manual filtering is used to ensure content diversity.

The relatively low levels of average agreement between participants may indicate that there is disagreement over what level of difference should be considered as "dissimilar sentiment". This should be investigated in future work and could lead to the development of systems which use the views of the user to present a broad range of viewpoints, rather than presenting a single set of content for

Table 2. Example question from sentiment experiment

Question ID	36682
Groups	Different
Parent Comment	**OriginalCommenter**: If the King did not approve the coup that kick Thaksin out, but let democracy does its job back then, Thailand would not be divided as much as today. where is the King now? It's time for him to clean up the mess he and his people created. "I believe Thai both side still listen to him". it is not too late for him to make it right. He need to say and do something now. the longer he wait the worse it will get for his country and his people. And he can only blame himself for the problem he took part in creating it.
Comment 1	**SubCommenter1**: The King and Queen are extremely old, impaired and live in seclusion in a hospital. He only makes the occasional public appearance on his birthday etc and no longer has the capacity to influence events.Their son and heir apparent is in the military so...
Comment 2	**SubCommenter2**: the scenario is simple: The king or his son ordered the general to move... they are going to twist the arm of the acting Pm to resign and appoint an interim party.. Then the army will descend on the Red and kill 100 or 200 on the excuse is that they have started something. some crazy excuse to satisfy the US three years later, they will have a controlled election.. and that is the end of that...Thailand has no democracy and never will SO the part I do not understand is that why the Japanese are still doing business there or the Germans, or the Koreans.. it is a corrupted infested state.
Similar	5
Dissimilar	15
Can't Tell	4
Percentage Similar	25%

all users, or, as in the case of many "personalised" news systems, primarily presenting viewpoints which the user identifies with.

6   Conclusions

This paper proposes using community detection algorithms to identify comments that express similar views or opinions, the long term goal being to use these groupings to reduce issues of group-think in open news systems, by presenting more diverse opinions. The success of this technique relies on the assumption that news discussion participants would maintain a consistent viewpoint within the timeframe of a single discussion, and would not strictly obey the instructions to "vote based on quality", instead allowing their biases to show through their voting behaviour.

After identifying a suitable algorithm (Louvain Community Detection), an experiment was conducted which gathered 1936 votes on the similarity of pairs of comments, and it was found that there was a significant relationship between group status (if the algorithm placed the authors in the same group or not) and the percentage of respondents who believe the pair shared similar sentiment. The system was then compared to each participant, and was found to have a level of agreement comparable to the average level of agreement between participants.

Future work may investigate the possibility of adding vote-based community detection to the promising work using explicit friend/follower relationships and textual analysis, or apply different community detection algorithms to voting networks. Using these techniques may allow for identification of more fine-grained sentiment groups.

In our own future work, we will focus on formatting and presenting full news discussions using the algorithm explored here, with a view to discovering if presenting comments in the context of like-minded groups causes people to interpret and interact with the content in a different way. The hope is that this work will lead to online news discussions which represent a wider range of viewpoints, are more welcoming of dissenting opinion, and more informative for readers.

References

1. Abu-Jbara, A., King, B., Diab, M.T., Radev, D.R.: Identifying opinion subgroups in arabic online discussions. ACL **2**, 829–835 (2013)
2. Anderson, P.J., Williams, M., Ogola, G.: The Future of Quality News Journalism: A Cross-continental Analysis, vol. 7. Routledge (2013)
3. Arguello, J., Butler, B.S., Joyce, E., Kraut, R., Ling, K.S., Rosé, C., Wang, X.: Talk to me: foundations for successful individual-group interactions in online communities. In: Proceedings of the SIGCHI Conference on Human Factors in Computing Systems, pp. 959–968. ACM (2006)
4. Bennett, W.L.: News: The Politics of Illusion. Longman, New York (1988)
5. Blondel, V.D., Guillaume, J.-L., Lambiotte, R., Lefebvre, E.: Fast unfolding of communities in large networks. Journal of Statistical Mechanics: Theory and Experiment **2008**(10) (2008)
6. Cheng, J., Danescu-Niculescu-Mizil, C., Leskovec, J.: How community feedback shapes user behavior. In: ICWSM 2014 (2014)
7. Clauset, A., Newman, M.E., Moore, C.: Finding community structure in very large networks. Physical Review E **70**(6), 066111 (2004)
8. Davies, N.: Flat Earth news: an award-winning reporter exposes falsehood, distortion and propaganda in the global media. Random House, London (2011)
9. Fortunato, S.: Community detection in graphs. Physics Reports **486**(3), 75–174 (2010)
10. Fortunato, S., Barthelemy, M.: Resolution limit in community detection. Proceedings of the National Academy of Sciences **104**(1), 36–41 (2007)
11. Giannopoulos, G., Koniaris, M., Weber, I., Jaimes, A., Sellis, T.: Algorithms and criteria for diversification of news article comments. Journal of Intelligent Information Systems **44**(1), 1–47 (2014)
12. Hampton, K., Goulet, L.S., Rainie, L., Purcell, K.: Social networking sites and our lives. Technical report, Pew Internet and American Life Project (2011)

13. Herman, E.S., Chomsky, N.: Manufacturing consent: The political economy of the mass media. Random House, London (1988)
14. Ipsos MORI: Perceptions are not reality: The top 10 we get wrong (2013). https://www.ipsos-mori.com/researchpublications/researcharchive/3188/Perceptions-are-not-reality-the-top-10-we-get-wrong.aspx
15. Jaffali, S., Jamoussi, S., Hamadou, A.B.: Grouping like-minded users based on text and sentiment analysis. In: Hwang, D., Jung, J.J., Nguyen, N.-T. (eds.) ICCCI 2014. LNCS, vol. 8733, pp. 83–93. Springer, Heidelberg (2014)
16. Kovach, B., Rosenstiel, T.: The elements of journalism: What newspeople should know and the public should expect (2007). Random House LLC
17. Kriplean, T., Toomim, M., Morgan, J., Borning, A., Ko, A.: Is this what you meant?: promoting listening on the web with reflect. In: Proceedings of the SIGCHI Conference on Human Factors in Computing Systems, pp. 1559–1568. ACM (2012)
18. McCombs, M.: A look at agenda-setting: Past, present and future. Journalism Studies 6(4), 543–557 (2005)
19. McCombs, M.E., Shaw, D.L.: The agenda-setting function of mass media. Public Opinion Quarterly 36(2), 176–187 (1972)
20. Meier, F., Elsweiler, D., Wilson, M.L.: More than liking and bookmarking? towards understanding twitter favouriting behaviour. In: ICWSM 2014 (2014)
21. Mills, R.: Researching social news - is reddit.com a mouthpiece for the 'hive mind', or a collective intelligence approach to information overload? In: Proceedings of the ETHICOMP 2011. Sheffield Hallam University (2011)
22. Mitchell, A.: State of the news media 2014: Overview. Pew Research Journalism Project (2014)
23. Muchnik, L., Aral, S., Taylor, S.J.: Social influence bias: A randomized experiment. Science 341(6146), 647–651 (2013)
24. Nardelli, A., Arnett, G.: You are probably wrong about almost everything. The Guardian, October 29, 2014. http://www.theguardian.com/news/datablog/2014/oct/29/todays-key-fact-you-are-probably-wrong-about-almost-everything
25. Papadopoulos, S., Kompatsiaris, Y., Vakali, A., Spyridonos, P.: Community detection in social media. Data Mining and Knowledge Discovery 24(3), 515–554 (2012)
26. Parau, P., Stef, A., Lemnaru, C., Dinsoreanu, M., Potolea, R.: Using community detection for sentiment analysis. In: 2013 IEEE International Conference on Intelligent Computer Communication and Processing, pp. 51–54. IEEE (2013)
27. Schultz, J.: Reviving the fourth estate: Democracy, accountability and the media. Cambridge University Press (1998)
28. Scott, J., Millard, D., Leonard, P.: Citizen participation in news: An analysis of the landscape of online journalism. Digital Journalism, (ahead-of-print), 1–22 (2014)
29. Strömbäck, J.: Four phases of mediatization: An analysis of the mediatization of politics. The International Journal of Press/Politics 13(3), 228–246 (2008)
30. Wanta, W., Golan, G., Lee, C.: Agenda setting and international news: Media influence on public perceptions of foreign nations. Journalism & Mass Communication Quarterly 81(2), 364–377 (2004)
31. Whiteley, P., Clarke, H. D., Sanders, D., Stewart, M.: The economic and electoral consequences of austerity policies in britain. In: EPSA 2013 Annual General Conference Paper, vol. 138 (2013)

Identifying Suggestions for Improvement of Product Features from Online Product Reviews

Harsh Jhamtani[1]([✉]), Niyati Chhaya[1], Shweta Karwa[2], Devesh Varshney[1],
Deepam Kedia[3], and Vineet Gupta[4]

[1] Adobe Research Big Data Intelligence Lab, Bangalore, India
{jhamtani,nchhaya,dvarshne}@adobe.com
[2] Google Inc, Mountain View, USA
shwetakarwa@gmail.com
[3] IIT Kanpur, Kanpur, India
deepamkedia@gmail.com
[4] IIM Ahmedabad, Ahmedabad, India
vineetgupta10@gmail.com

Abstract. Online forums are used to share experiences and opinions about products and services. These forums range from review sites such as Amazon (www.amazon.com) to online social networks such as Twitter (www.twitter.com). The user-generated content in these platforms capture the users' opinions and sentiments. In this work, we explore the problem of identifying suggestions from text content. The paper first defines suggestive intent and then presents a supervised learning approach to identify text that contains suggestive intent. The results show high accuracy with a $F1$ score of 0.93.

1 Introduction

Online social platforms, such as social media websites and product review forums, are extensively used to express views and share experiences. The posted content often captures the users' intent. Organizations are always looking for such information for feedback and suggestions to improve products and services. Further, marketers need to identify user segments for accurate marketing. Mining suggestions and user requirements in posted content would provide this data easily. In this paper, we present a method to extract suggestions to improve features for specific products from text posts.

In this work, we shall refer to product features as *aspects* and the corresponding product or service targeted is termed as an *entity*. For example, for an entity 'mobile phone', the aspects could be screen size and battery life. First, we define 'suggestive intent' in context of this work.

We define a text post to contain suggestive intent (SI) for a given product or service entity if it provides explicit or implicit suggestions from the author for a possible improvement of aspect(s) of the corresponding entity. Some example classes of text posts containing suggestive intent are as follows:

© Springer International Publishing Switzerland 2015
T.-Y. Liu et al. (Eds.): SocInfo 2015, LNCS 9471, pp. 112–119, 2015.
DOI: 10.1007/978-3-319-27433-1_8

- The text contains a suggestion regarding some aspect of the entity;
- The text contains a complaint regarding absence or bad quality of an aspect of the entity;
- The text contains a desire to modify or add a specific aspect to the entity;
- The text compares an aspect of two similar entities such that it can be inferred as suggestion for the entity under consideration;

Table 1 shows some example posts as either containing suggestive intent (SI post) or not containing SI (non-SI post) The rest of the paper is organized as follows. First we describe the related work. We then present an in-depth feature description followed by a supervised learning approach for classifying posts as SI post or non-SI post. We then describe the data set and experimental results.

Table 1. Examples of SI posts

Screen size of iphone 4 is too small
I wish iphone 4 had a bigger screen
Nothing could be heard on the phone
Ubuntu's GUI is not as good as that of Windows

Table 2. Examples of Non-SI posts

Iphone4 sucks!
The camera is one of the worst cameras I have used
I will never buy Nikkon camera again

2 Related Work

This section focuses on works pertaining to extracting specific types of intent from text. [3] tackled the problem of automatically identifying user's suggestions from tweets using factorization machines. However, their definition of suggestions is limited to explicit intent and proposals by the user. [6] focus on identifying user purchase intentions from social media posts, while we are focusing on identifying suggestions for product feature improvement. There has been some work in extracting wishes from text [5]. Our work focuses on identifying suggestions rather than just wishes. [14] use linguistic rules to identify suggestive wishes from a wish corpus. Their proposed method is restrictive in terms of their scope of definition especially due to the use of a rule-based approach. The generic problem of mining comparative sentences has been explored by [9], [4], [8]. However, it has not been extended to identifying specific suggestions from text.

Based on our prior art study, we realize the novelty of this work builds from the definition of suggestive intent, that not only considers complaints and wishes but also incorporates competitive analysis by considering comparative posts, and a machine learning based model to identify text posts bearing suggestive intent.

3   Feature Extraction

In this section, we describe the features used to identify posts bearing suggestive intent. We consider four categories of features, as shown in Table 3.

Table 3. Features used for Suggestive intent detection

Feature category	Corresponding features
Bag of words	- Delta TFIDF on unigrams, bigrams, and trigrams of words
Presence of Aspect	- Explicit aspect presence - Implicit aspect presence
Presence of suggestive words/ phrases	- Sentiment score of post - Presence of comparative phrases - Expression of wish - Presence of opinion phrases
Grammatical structure of the post	- Grammatical dependencies among entity, aspect, suggestive word

3.1   Delta TFIDF

Delta TFIDF is a technique to efficiently weight word scores for classification tasks [12]. It has been shown to be effective in binary classification for class imbalanced data. We extract various unigrams, bigrams, and trigrams of words from posts, which are referred to as terms. For any term t in post p, the Delta TFIDF score $V_{t,p}$ is computed as follows:

$$x = n_{t,p} * log_2(\tfrac{N_t}{P_t})$$

Here, $n_{t,p}$ is the frequency count of term t in post p. P_t and N_t are the number of posts in the SI-labeled dataset with term t and the non–SI labeled dataset with term t respectively.

3.2   Presence of an Aspect

Aspects can broadly be classified into explicit and implicit aspects [7]. SI posts contain one or more aspects, as suggestions are always directed towards some aspect of the concerned entity. We consider presence of explicit and implicit aspects in post as features.

Presence of Explicit Aspect. Work by [1] is used to find the explicit aspects present in the post. [1] uses frequent noun and noun-phrases to identify aspects, but they consider mainly those noun phrases that are in sentiment-bearing sentences or in some syntactic patterns which indicate sentiments. Several filters were applied to remove unlikely aspects, which do not have sufficient mentions along-side known sentiment words. We generated a binary valued feature which takes value 1 if there is at least one explicit aspect present in p. Otherwise it takes value 0. Examples of aspects extracted through the above mentioned method for the *phone* entity are customer service, headset, and menu options.

Presence of Implicit Aspect. Often aspect is indirectly inferred through verbs and adverbs. To check for the presence of implicit aspects, we identify occurrence of such verbs and adverbs. In order to achieve this, we leverage the work of [15], which uses association rule mining and POS filtering based approach.

We generated a binary valued feature which takes value 1 if there is at least one implicit aspect present in p. Otherwise it takes value 0. Examples of implicit aspects extracted through above method for *phone* entity are hear (refers to phone volume/ speaker quality), and die (refers to battery life).

3.3   Presence of Suggestive Phrases and Sentiment Score

Features considering sentiment score of post and the presence of suggestive phrases are described here. We define a *suggestive phrase* to be one of the following - an opinion phrase, an expression of wish, or a comparative phrase.

Sentiment Score of Post. Often complaints with negative sentiments bear suggestions for aspects of the entity. For example, 'The Raddison has poor ambiance' is a complaint and can be easily translated into a suggestion for 'The Raddison' to improve their ambience. We used *AlchemyAPI* [11] to extract the sentiment of each post. *AlchemyAPI* gives the value of sentiment in range $[-1, +1]$, -1 denoting the extreme negative polarity while $+1$ denoting extreme positive polarity. This sentiment score is used as a feature value for the classifier.

Polarity of Opinion Phrases. Sentiment score of a post alone is not sufficient as certain modifiers like longer or shorter may have different polarity for different aspects. For each explicit aspect in post p, we extract the aspect-based sentiment polarity using the approach of [1]. Moreover, the modifiers of aspects, extracted in the process, will be leveraged in grammatical dependencies based features. A binary-valued feature is created to capture the polarity of opinionated phrases. If aspect-specific sentiment polarity for at least one aspect is negative, then the binary feature value is set to 1. Otherwise, the binary feature value is set to 0. This helps identify posts which have negative polarity for at least one aspect of the concerned entity.

Expression of Wish. Many users may express suggestions through wishes. For example, the post 'I wish that IPhone 4 had a bigger screen', expresses user's suggestive intent for the entity 'Iphone 4'. These posts are characterized by presence of *wish keyword* such as *could, would, will, hope, want, wish,* or *should* [5] along with mention of aspect(s). A binary feature is created, which takes value 1 if at least one of the wish keywords as well as an aspect of the entity occurs in the post. Otherwise it is set to 0. In our dataset, 78% of posts with this feature value as 1 were SI posts, which is a significant portion of all SI posts (18% of total posts are SI posts).

Comparative Phrases. As mentioned earlier, we also consider posts comparing a given aspect of different entities. For example, the post, 'The battery life of Samsung S4 is far longer than that of IPhone 4' contains suggestive intent for IPhone but not for Samsung. The following features are considered:

Presence of Comparative or Superlative Words. We used Stanford NLP POS tagger [16] to identify the presence of comparative or superlative words. To achieve this, we look for the presence of the following POS tags by Stanford NLP - JJR, JJS, RBR, and RBS. A binary-valued feature is created which takes value 1 if atleast one word with one of the mentioned POS tags is present in the post, else the value is set to 0.

Identifying Preferences. We leverage work of [4] to identify the preferred entity in comparative sentences. A binary-valued feature is created which is set to 1 if the entity being considered by the model to identify suggestions is the non–preferred entity. Otherwise it takes the value 0.

3.4 Grammatical Dependencies

We observed that in many SI posts, *entity* is the *nominal subject* (nsubj) of the suggestive word (or phrase). Also, often the *aspect* is the *directly dependent object* (dobj) of the suggestive word (or phrase). Figure 1 shows an illustration of such dependencies for the post 'The Raddison should improve their ambiance'.

Fig. 1. An illustration of grammatical dependency based features

For any post p, we first extract the aspects and suggestive words (or phrases) through methods explained earlier. We assume that the entity corresponding to which suggestions are being mined, is already known and is mentioned in the post. The post is parsed using Stanford PCFG Parser [10], and look for above mentioned dependencies among entity, aspect, and suggestive words (or phrases). Based on the above mentioned grammatical dependencies, two features are generated. First, a binary valued feature, which takes value 1 if *entity* is the *nominal subject* (nsubj) of at least one of the suggestive words (or phrases), otherwise it takes value 0 and second, a binary valued feature, which takes value 1 if at least one *aspect* in the post is the *directly dependent object* (dobj) of the suggestive word (or phrase). Otherwise it takes the value 0.

4   Experiments and Results

This section presents the experiments and results for the classifiers used to identify suggestive intent in text.

4.1   Data Collection and Annotation

Our data set consists of 10,440 product review sentences (Reviews were split into sentences) for 14 different products ([2] and [7]). These products are various mobile phones and digital cameras. Ground truth was crowd sourced by using Amazon's Mechanical Turk. The task was to annotate the presence or absence of SI in text posts. Definitions of SI post and a non-SI post, along with some examples, were provided to all annotators. Out of 10,440 posts, 1,880 were annotated as SI and the remaining 8,560 were annotated as non-SI. Every post was annotated by 3 annotators and Cohen's kappa coefficient of inter-annotator agreement was found to be 0.91.

4.2   Results

We take a supervised learning approach. We experimented with decision trees, SVM, and Random forests classifiers, using their Scikit [13] implementations. We use 10-fold cross validation to evaluate the results. The results shown here are for experiments with Random Forest, which performed best for our dataset. As seen from the data description, there is class imbalance. Hence, AUC (Area under the curve) under the ROC (Receiver Operating Characteristic) is used as a metric for accuracy. Larger the area under the ROC curve, better is the classifier performance. We performed three experiments with different sets of features and evaluated the incremental performance.

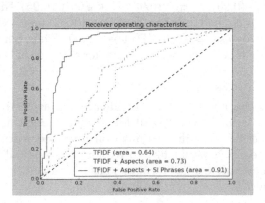

Fig. 2. ROC curves for different feature combinations

Figure 2 shows the ROC curves for different feature combinations. Following observations can be made for the obtained results:

Delta TFIDF features are important as they have significant discriminating power of distinguishing SI posts and non-SI posts as is evident from the AUC of 0.64 shown in figure 2 (*TFIDF*).

Including **Presence of aspect and sentiment score** features improve the performance as AUC of the ROC curve increased to 0.73, as shown in figure 2 (*TFIDF + Aspects*).

Including **Presence of suggestive action phrases and grammatical dependencies** features, further improved the performance as AUC of the ROC increased to 0.91 as shown in the figure 2 (*TFIDF + Aspects + SI Phrases*). The corresponding F1 score for the positive class (posts bearing suggestive intent) detection is 0.93.

Moreover, on using 'SI phrases' alone, F1 score of 0.874 is achieved, with corresponding AUC under the ROC curve to be 0.78. This signifies that SI phrases features are strong discriminators in differentiating the two classes.

Precision, recall, and F1 scores for different feature combinations, for the above mentioned experiment, are shown in Table 4

Table 4. Precision, Recall, and F1 scores for different feature combinations

Features	Precision	Recall	F1-score
TFIDF	0.803	0.868	0.834
SI Phrases	0.851	0.900	0.874
TFIDF + Aspects	0.833	0.901	0.866
TFIDF + Aspects + SI Phrases	0.898	0.971	0.933

The F1-scores for different products, when all the mentioned features are used, are as follows: ipod (0.94), MicroMP3 (0.89), Canon PowerShot SD500 (0.94), Linksys Router (0.92), Canon S100 (0.94), Diaper Champ (0.85), Nokia 6600 (0.95), Hitachi router (0.95), Norton (0.93), Nokia 6610 (0.95), Nikon coolpix 4300 (0.93), Creative Labs Nomad Jukebox Zen Xtra 40GB (0.86), Canon G3 (0.93), and Apex AD2600 Progressive-scan DVD player (0.90). Thus, our method is robust across products since it achieves high F1 scores for each one of them.

5 Conclusion

The paper studies the novel problem of identifying suggestive intent (SI) in text posts. Such a technology is useful for product design and improvement, and targeted marketing. Classifiers trained on proposed features achieve high accuracy as reported. We would like to explore the possibility of expressing such posts in a concise pre-defined format, so as to enable performing aggregate analysis of such data.

References

1. Blair-Goldensohn, S., Hannan, K., McDonald, R., Neylon, T., Reis, G.A., Reynar, J.: Building a sentiment summarizer for local service reviews. In: WWW Workshop on NLP in the Information Explosion Era, vol. 14 (2008)

2. Ding, X., Liu, B., Yu, P.S.: A holistic lexicon-based approach to opinion mining. In: Proceedings of the 2008 International Conference on Web Search and Data Mining, pp. 231–240. ACM (2008)
3. Dong, L., Wei, F., Duan, Y., Liu, X., Zhou, M., Xu, K.: The automated acquisition of suggestions from tweets. In: Twenty-Seventh AAAI Conference on Artificial Intelligence (2013)
4. Ganapathibhotla, M., Liu, B.: Mining opinions in comparative sentences. In: Proceedings of the 22nd International Conference on Computational Linguistics-Volume 1, pp. 241–248. Association for Computational Linguistics (2008)
5. Goldberg, A.B., Fillmore, N., Andrzejewski, D., Xu, Z., Gibson, B., Zhu, X.: May all your wishes come true: A study of wishes and how to recognize them. In: Proceedings of Human Language Technologies: The 2009 Annual Conference of the North American Chapter of the Association for Computational Linguistics, pp. 263–271. Association for Computational Linguistics (2009)
6. Gupta, V., Varshney, D., Jhamtani, H., Kedia, D., Karwa, S.: Identifying purchase intent from social posts. In: Eighth International AAAI Conference on Weblogs and Social Media (2014)
7. Hu, M., Liu, B.: Mining and summarizing customer reviews. In: Proceedings of the tenth ACM SIGKDD international conference on Knowledge discovery and data mining, pp. 168–177. ACM (2004)
8. Jindal, N., Liu, B.: Identifying comparative sentences in text documents. In: Proceedings of the 29th annual international ACM SIGIR conference on Research and development in information retrieval, pp. 244–251. ACM (2006)
9. Jindal, N., Liu, B.: Mining comparative sentences and relations. AAAI **22**, 1331–1336 (2006)
10. Klein, D., Manning, C.D.: Accurate unlexicalized parsing. In: Proceedings of the 41st Annual Meeting on Association for Computational Linguistics-Volume 1, pp. 423–430. Association for Computational Linguistics (2003)
11. LLC, O.: Alchemyapi (2009)
12. Martineau, J., Finin, T., Joshi, A., Patel, S.: Improving binary classification on text problems using differential word features. In: Proceedings of the 18th ACM conference on Information and knowledge management, pp. 2019–2024. ACM (2009)
13. Pedregosa, F., Varoquaux, G., Gramfort, A., Michel, V., Thirion, B., Grisel, O., Blondel, M., Prettenhofer, P., Weiss, R., Dubourg, V., et al.: Scikit-learn: Machine learning in python. The Journal of Machine Learning Research **12**, 2825–2830 (2011)
14. Ramanand, J., Bhavsar, K., Pedanekar, N.: Wishful thinking: finding suggestions and'buy'wishes from product reviews. In: Proceedings of the NAACL HLT 2010 Workshop on Computational Approaches to Analysis and Generation of Emotion in Text, pp. 54–61. Association for Computational Linguistics (2010)
15. Schouten, Kim, Frasincar, Flavius: Finding Implicit Features in Consumer Reviews for Sentiment Analysis. In: Casteleyn, Sven, Rossi, Gustavo, Winckler, Marco (eds.) ICWE 2014. LNCS, vol. 8541, pp. 130–144. Springer, Heidelberg (2014)
16. Toutanova, K., Klein, D., Manning, C.D., Singer, Y.: Feature-rich part-of-speech tagging with a cyclic dependency network. In: Proceedings of the 2003 Conference of the North American Chapter of the Association for Computational Linguistics on Human Language Technology-Volume 1, pp. 173–180. Association for Computational Linguistics (2003)

Crowdsourcing Safety Perceptions of People: Opportunities and Limitations

Martin Traunmueller^(✉), Paul Marshall, and Licia Capra

Department of Computer Science, University College London,
Gower Street, WC1E 6BT, London, UK
{martin.traunmueller.11,paul.marshall,l.capra}@ucl.ac.uk

Abstract. Online crowdsourcing has successfully been used as a paradigm to collect large amount of perceptions about our cities quickly and cheaply, enabling social scientists to quantitatively test urban theories at scale. While doing so, researchers have not focussed on getting answers from specific demographics, relying on a self–selected crowd instead. However, existing theories suggest that knowing who the respondents are is crucial for understanding safety perceptions about people (instead of, for example, about the built environment). In this case to quantitatively validate theories, it is not just the amount of data that matters, but also what demographics participate (or not). In this paper we investigate to what extent online crowdsourcing can be used for the specific case of safety perceptions about people. We built an image–based online crowdsourcing platform, collected safety perception ratings and background information from more than 700 people and used them to quantitatively evaluate established theories based on qualitative research. On one hand, we show in this paper that online, image-based crowdsourcing can be used to gather perceptions about people too, not just architecture, confirming established theories based on qualitative work. Furthermore, we are able to uncover detailed interactions that would be challenging to grasp using qualitative methods. On the other hand, we show limitations of using crowdsourcing as a method. By not controlling who makes up the crowd, we were not able to investigate all theories as we did not reach all user groups that have been discussed in qualitatitve research.

Keywords: Crowdsourcing · Perception · Social studies · Crime studies · Theory validation

1 Introduction

The perception of safety has great impact on quality of life for people all over the world [19,31]. Theories from social and criminological sciences have explored the relationship between perceptions of safety and people, both in terms of *who we are*, and *who we see*. Depending on background and demographic properties, such as age [34], gender [5] and ethnicity [2,20] people perceive safety differently.

© Springer International Publishing Switzerland 2015
T.-Y. Liu et al. (Eds.): SocInfo 2015, LNCS 9471, pp. 120–135, 2015.
DOI: 10.1007/978-3-319-27433-1_9

For instance, research has found that the most fearful groups are women and the elderly, who are surprisingly least at risk of being victimized [14,21,22,28], whereas young men, who are most at risk, show the least fear of crime [10,30]. Based on qualitative methods, these works offer semantically rich and detailed insights; however, the cost associated with using these methods makes it difficult to replicate such studies at scale, across different cultures, and over time. To reach a larger number of people quickly and cheaply, recent research in urban studies and computer science suggests online crowdsourcing as a complementary method to gather perceptions of happiness and safety, amongst others, in relation to the built environment. For example, Salesses et al. [26] and Quercia et al. [24] present two randomly selected Google Street View images of a city on a webpage to the user who is asked to select the one appearing safer or happier. In this way, a large amount of data is collected that allows researchers to test previously defined theories about the built environment.

In this paper, we investigate to what extent online crowdsourcing can be used by social scientists to validate theories of safety about *people* instead. While works such as Salesses et al. [26] and Quercia et al. [24] did not need to differentiate who provided opinions to validate theories about the built environment, we discuss theories where it matters *who* gives the opinions. As we do not control who the self–selected volunteers are, we do not know to what extent crowdsourcing is viable. While online market–places, such as Amazon Mechanical Turk (AMT) aim to offer such control of crowd selection, they actually show homogeneity of their population as well, and hence would not offer a satisfying approach for our purposes [12,27]. To understand the extent to which online crowdsourcing could be used, we built an online crowdsourcing platform similar to Salesses et al. [26] and Quercia et al. [24], presenting images of pre–selected types of people on white background to participants, who were asked to rate them in terms of safety perception. Before taking part, participants were asked to provide some information about themselves, such as age, gender, ethnicity and if they are London residents (the city where the study was conducted and that was expected to attract most participants). We were then able to correlate safety perceptions about people with respondent characteristics. We ran the study from January until April 2015 and collected 13,560 answers from over 700 participants.

Our analysis shows that online, image–based crowdsourcing can be used to gather perceptions not just of architecture, but about people too, confirming established theories that were based on qualitative work. We were also able to uncover detailed interactions that would be challenging to grasp using qualitative methods. We also describe limitations of using crowdsourcing as a method. By not controlling who makes up the crowd, we were not able to investigate all theories as we did not reach to all user groups that have been discussed in qualitatitve research.

The remainder of the paper is structured as follows: first we give an overview of studies on safety perception, including recent work on quantitative methods. Then we outline the development and deployment of *Streetsmart*, an online crowdsourcing platform developed to gather safety perceptions about people,

define hypotheses and describe our analysis steps. We then present the results of our study, discuss its limitations, and outline its future directions.

2   Related Work

2.1   People Demographics and Safety Perception

Previous work in social science suggests that people's demographic background has a significant impact on how they perceive the environment and how others perceive them. For instance, Eiland et al. [3] found that ethnicity, age and gender of people affect the judgement of facial expressions shown on photographs. Furthermore, findings of the 'Implicit Association Test' [7] show that there are differences of implicit cognition when combining specific attributes, such as 'pleasant' and 'unpleasant', with different ethnicities, such as 'Black' and 'White'. These differences are also found when it comes to safety perception.

Matei et al. [17] explored the role of the media in relation to people's fear perception in Los Angeles. The study used Geographical Information System (GIS) technology to process hand–drawn mental maps [16] drawn by 215 study participants of seven neighbourhoods all over the city. Analysis revealed that it is not actual crime activity, but the concentration of certain ethnicities, that has the major impact on participants fear perception. Another study [9] uses mental maps to explore relationships between usage of public space and the violence perceived by adolescents in South Africa, differentiated by gender, age and residential background (urban / rural). Findings suggest that teenage girls show most significant movement restrictions in public areas, compared to other participant groups, indicating that this group perceives more dangers in their communities than others.

Studies like these show how people's demographic background, such as age [34], gender [5] and ethnicity [2, 20] impact our safety perceptions towards others: on the one hand these properties define our own personality and how we perceive other people. On the other, they also define our appearance and hence how other people perceive us. Previous work suggests that, for instance, women and the elderly are among the most fearful groups, who surprisingly are least at risk of being victimized by others [14, 21, 22, 28]. Young men show the least fear of crime [10, 30], while being most at risk of being victimized and most feared by other people. Furthermore, Matei et al. [17] found that people's ethnical background has a major impact on how others perceive them, with non–Caucasian and non–Asian background being the most feared among residents in Los Angeles.

However, most of these findings have been derived from qualitative research methods, such as victimization surveys, public perception questionnaires and semi–structured interviews [4, 13]. While such methods offer rich insight, they are also costly and time consuming, and thus these studies have been conducted on a relatively small scale. Using this methodology, it is difficult to test whether the findings still hold with a broader participant base, in different areas, and as time passes and society changes.

2.2   Crowdsourcing Perceptions

Current research in computer science proposes crowdsourcing as a different method to perform certain types of perception studies at scale. For instance, *Urbanopticon*[1] presents 360 degree Google Street View images to users who are asked to guess their geographic location on a map. As 'happy' places are found to be easily recognized by people [16], a collective mental map is drawn with the aim of detecting 'happy' places in the city. *Urbangems* [24] takes a similar approach to crowdsource perceptions of 'happiness', 'beauty' or 'calmness' of a city based on visual cues [32]. The online platform[2] shows two random images of Greater London to the user, who is asked to rate images on these attributes. In this way, *Urbangems* aims to identify visual cues of the built environment and the perceived attributes people attach to them. These were used in a follow–up study to generate 'happy' and 'beautiful' routes through the city [25]. Following the same methodology, *PlacePulse*[3] extends the research into the perception of safety, among other attributes, and not only for one European city, but for different cultures using images of both American and European cities [26]. Results show two main differences, as perceptions in American cities were found to be more clustered, compared to the European ones. Furthermore, the research explored the relationship between perceived safety and actual crime activity for New York City and found significant correlations between them: areas with a higher rate of 'class' and 'uniqueness' were perceived as safer and showed indeed a lower crime rate than others. Based on these findings, researchers developed *Streetscore*,[4] an algorithm that identifies visual cues in Google Street View images using computer vision technology to automate the process [18].

Online crowdsourcing using images shows promising results for urban studies researching effects of architecture on human perceptions. However, while these works did not need to differentiate between demographic properties of the crowd, we consider theories about safety perceptions where demographic differences do matter and explore to what extent online crowdsourcing can be used in these cases too.

3   Streetsmart

We built an online crowdsourcing platform called *Streetsmart* to gather safety perception data about different demographics of people. Our goal was to understand the suitability of this form of collaborative work to validate and further develop theories in this context. In this section, we detail how the platform was built.

Fig. 1. Image examples, as used in the pilot study, showing the selected people on white background with black perspective lines, indicating a street.

3.1   Selecting Features

First, we decided what features of people the study should explore. From the literature we know that age, gender and ethnicity matter to safety perception, both on *who we are* (e.g., a man and a woman may perceive the same person differently), and *who we see* (e.g., the same person may perceive a young or elderly person differently). Many other factors about who we see may impact safety perceptions, such as clothing, posture, facial expression, etc. To explore what other factors we could include, we conducted an exploratory pilot study, where images of people were shown to participants, who were then asked to speak aloud about their feeling of safety and reasons for their decision. We interviewed 21 people living in London, between the age of 21-64 years, including 13 female and 8 male. Their ethnicities included Caucasian, Black, Indian, Asian and Arab, covering all main ethnical groups in London.[5] Each session lasted for 30 minutes, answers were audio recorded and transcribed for analysis.

The images used in the pilot study were selected from online repositories under a creative commons license. As well as including different age groups, gender and ethnicities, we included people's appearance in terms of signifiers of religious, sexual or sub–cultural orientation, the person's facing–direction, differences in fashion, posture and gestures, differences in activity (walking / standing), the number of people in the picture (single / group), mix of people within a group and if the person was concealed or not, as for instance wearing a hood. 138 images of people were subtracted from the image background and placed on a white canvas showing only perspective lines indicating a street, as shown in Figure 1. This step was necessary to focus participants on the person only without any distraction from the backdrop, which is known to have an impact on our safety perception [6]. People shown in the picture were scaled to appear at the same distance of 6-7 meters to the participant, as in real situations this distance enables us to see enough visual detail in other people to assess whether to be fearful or not [8].

[1] http://urbanopticon.org/ann-arbor/ May, 2015.
[2] http://urbangems.org March, 2015.
[3] http://pulse.media.mit.edu May, 2015.
[4] http://streetscore.media.mit.edu May, 2015.
[5] http://data.london.gov.uk/census/ May, 2015.

Each participant was shown all 138 images and was asked to elaborate about their safety perception towards each person and the reasons for it. After transcribing audio recordings, we used thematic analysis [1] to process the output. We generated initial codes for potential themes represented in the majority of answers and detected patterns emerging over the data. Then we iterated the process and finally were able to describe two main themes to be included in our study beside the already known variables of *age, gender* and *ethnicity*. These were: people's *facing–direction* and if the person's face is *concealed or not*, for example by a hood. Based on these properties, we then selected the images to be used in the study.

3.2   Selecting Images

According to the literature and our findings from the pilot study, we defined 5 variables to take into account when selecting the images: *age, gender, ethnicity, facing–direction* and *concealed*, as shown in Table 1. Using images of concealed people brought up limitations in categorizing them on the remaining variables, as these properties were hidden. This lead us to use this variable as 'special variable' represented as an extra set of images in addition to the others. The remaining variables were defined as 'main variables' (as found in the literature, including *age, gender* and *ethnicity*) and as 'sub variable' (*facing–direction*), found in the pilot. We broke the main and sub variables down by the properties, presented in Table 1 describing 72 different categories including 3× age (0–20, 21–40, 41+),[6] 2× gender (male, female), 4× ethnicity (Caucasian, Black, Asian, Arab) and 3× facing–direction (towards me, away from me, not aware of me).

Table 1. Table showing breakdown of selected variables.

Variables	Properties	Type
Age	Teenager (14 – 28 years) Grown-up (29 – 55 years) Elder (56+ years)	main
Gender	Male Female	main
Ethnicity	Caucasian Black Asian Arab	main
Facing–Direction	towards me away from me not aware of me	sub
Concealed		special

We then selected pictures of people covering this range from which we subtracted backgrounds, using online marketplace *Fiverr*[7]. To have them look like "normal" people on the street, we avoided images taken by professionals (from

---

[6] http://www.ons.gov.uk/ons/index.html May, 2015.
[7] http://fiverr.com May, 2015.

magazines for instance) or people posing for the camera and focused on snapshots with creative commons license instead. From every category we chose three different images, resulting in a total of 216 images (in this work we focus on individuals; groups of people will be part of a future study).

3.3   Online Crowdsourcing

Fig. 2. Web–based user–interface of *Streetsmart*, showing one image of a person on white background at a time. The user is asked to rate his/her perception of safety according to the image on the slider. By pressing the 'Next' button a new image is presented; by pressing 'Exit' the study can be ended at any time.

Next we built the *Streetsmart* website. The entry page to the website asked crowdworkers to provide basic background information on who they are. Following the literature, we asked their *gender, age* and *ethnicity*. Furthermore, to be able to understand possible relationships with their geographical / cultural background, we asked which *continent* they come from and if they are a London resident. Then the safety perception survey followed (see Figure 2). We asked crowdworkers to complete a run of 25 safety evaluations using a slider built as a 70 point Likert–scale. We randomized the pictures to use from a pool of 216, so to avoid the same crowdworker seeing the same image twice, and to obtain roughly the same number of answers for each image in each category.

4   Method

4.1   Data Collection

We obtained ethical approval to conduct the study in September 2014. We then deployed *Streetsmart* from January 2015 until April 2015 and advertised on various social media platforms (Facebook (advertisment, page and group postings), LinkedIn, Twitter and Reddit). Users could share the webpage link via embedded Facebook–Like and Tweet buttons. Throughout these three and a half months, we were able to collect 13,560 votes from 716 users. As it is necessary to know the background information about our participants, we removed votes from users who did not want to reveal their background, resulting in 8,292 votes from 537 users, most completing their run (82%). The ratio between male (52%) and female (48%) participants was almost even, with most participants aged between 21 – 40 years (68%), followed by 41+ year olds (25%) and 0 – 20 year olds (7%). For ethnicity we found a high majority of Caucasian (79%), followed by Asian (7%) and mixed/multiple ethnic groups (5%). 1/3 of our participants were London residents (31%). Overall most participants came from Europe (79%), followed by North America (9%), South America (4%), and Asia (4%).

4.2   Hypotheses

Based on the literature and our pilot study, we formulated the following four hypotheses:

- images showing men would have lower safety ratings then those showing women (Gi);
- ethnicities other than Caucasian and Asian in the image would have lower safety ratings (Ei);
- people in the image looking at the participant would have lower safety ratings than those who were looking away or were not aware of him/her (Fi);
- concealed people would have lower safety ratings than those who were not concealed (Ci).

4.3   Analysis

With the collected data, we were now able to draw relationships between participant's safety perception towards the person in the image. We used a factorial repeated Analysis of Variance (ANOVA) with planned contrasts. Effects that were not covered by the hypotheses were explored with post hoc analysis, using the Tukey correction for multiple comparisons. To do so, we first aggregated our collected data to combine the participant information (*gender, age, ethnicity, from London / not from London, where in London, from which continent*), with the properties of the person in the image(*gender, age, ethnicity, facing–direction, concealed / not concealed*) and the safety ratings. As the variable *concealed* does not allow us to define other variables such as the *age, gender* and *ethnicity* of the person in the picture as they stay in the dark, it prevented a polynomial comparison between the variables so was tested in a separate ANOVA.

Looking at the frequency distribution of ratings we observe that the data is highly skewed showing a long tail distribution: most people feel very safe in relation to other people, whereas only some feel unsafe. This is a common situation in multi–factorial experiments in human–computer interaction (HCI) that work with Likert responses. To transform our data so that we could build an ANOVA model, we used the *Aligned Rank Transform* [33] for non–parametric factorial data analysis, which aligns the data in a pre–processing step before applying averaged ranks. With transformed data, we were then able to conduct our analysis.

5   Results

5.1   Who We See

Table 2. Table showing breakdown of our discussed hypotheses, with each contained variable and their means of voting scores. Voting ranged from 0 (indicating not safe) to 70 (indicating very safe), p<0.001 *** p<0.01 **p<0.05 *.

Hypothesis	Var Name	contained Properties	Mean M	
Gender image	Gi	male	55.5	**
		female	59.57	**
Ethnicity image	Ei	Black	56.95	*
		Asian	59.5,	**
		Caucasian	58.18	**
		Arab	55.57	*
Facing image	Fi	away from me	57.08	*
		towards me	57.09	***
		not aware of me	58.46	**
Concealed image	Ci	concealed	46.13	***
		not concealed	57.54	***

We started with the theory testing by analysing main effects. We took the safety score as the dependent variable, and all properties of the user and the person in the image as independent variables. In Table 2 we present the mean safety scores M for each variable of the model ($M \in [0, 70]$ with 0 indicating not safe and 70 indicating very safe). Next we summarize the findings.

Gi – *Gender of person in the image*: according to Warr [30] men are perceived as less safe than women. We can support this hypothesis as there was a main effect of gender ($F(1, 7885) = 128.38, p < 0.001$) with men being perceived as less safe than women.

Ei – *Ethnicity of the person in the image*: according to Matei et al. [17] Caucasian and Asian people are perceived as more safe than other ethnicities. We can support this hypothesis with our findings. We found a significant main effect of ethnicity of the person in the image ($F(3, 7885) = 21.99, p < 0.001$). Contrasts revealed that Black and Arab people were rated less safe than Asian and Caucasian people ($p < 0.001$). Post hoc analysis showed significant differences between Asian and Black people ($p < 0.001$) (indicating that Asian people are perceived safer than Black people), Arab and Asian people ($p < 0.01$) (indicating that Asian people are perceived safer than Arab people), Arab and Caucasian people ($p < 0.01$) (indicating that Caucasian people are perceived safer than Arab people) and Black and Asian people ($p < 0.001$) (indicating that Black people are perceived as less safe compared to Asian ethnicities).

Fi – *Facing direction of the person in the image*: from our pilot study, we know that people looking at the participant were perceived as less safe as people looking away or are not aware. There was significant main effect of the facing–direction of the person in the image on safety rating by the users ($F(2, 7885) = 6.50, p < 0.001$). The contrast revealed that people facing towards the participant in the image were rated less safe compared to people facing away or who are not

Table 3. Table showing breakdown of our statistically significant interactions with mean voting scores for each variable. Voting ranged from 0 (indicating not safe) to 70 (indicating very safe), p<0.001 *** p<0.01 ** p<0.05 *.

Interaction	Var Name	contained Properties	Mean M	
Age : Ethnicity	$Ai : Ei$	young Asian	60.2	*
		young Black	56.1	*
		young Caucasian	57.3	**
		young Arab	56.8	*
Facing : Ethnicity	$Fi : Ei$	Asian, away	59.1	**
		Black, away	55.9	**
		Caucasian, towards user	57.4	**
		Arab, towards user	54.6	*
Gender : Age	$Gi : Ai$	young male	55.5	***
		grown–up male	54.5	**
		elder male	56.5	***
		young female	59.7	**
		grown–up female	59.7	***
		elder female	59.3	**

aware of the user ($p < 0.001$). Post hoc analysis showed significant differences in ratings of people who are not aware of ($p < 0.01$), or who are facing away from the participant ($p < 0.05$). This indicates that people turning their back completely to the user are perceived as less safe than those not aware.

Ci – *Concealment of the person in the image*: we found in our pilot study that concealed people were perceived as less safe than those who are not concealed. The outcome of our study ($F(1, 8290) = 156.92, p < 0,001$) supports this hypothesis; concealed people in the image received lower rates than not concealed ones.

Interactions. Our results for the main effects show to what extent online crowdsourcing can be used to validate theories on urban safety perceptions. However, as in reality such factors are not isolated but happen in interaction with each other, we exploreds some possible interactions too. In Table 3 we present three examples of significant interactions that show clear differences within each group when observed in more detail.

$Ai : Ei$ – There was a significant interaction of *age* with *ethnicity* of the person in the image ($F(6, 7885) = 2.31, p < 0.01$). Simple effects analysis revealed while teenagers were found to be perceived as least safe for Black ($p < 0.05$) and Caucasian ethnic groups ($p < 0.01$), they were perceived as most safe with the groups of Asian ($p < 0.05$) and Arab population ($p < 0.05$).
In Figure 3 we visualize this interactions on the left interaction plot. While safety perception towards elder (red) and grown–ups (green) follow a similar curve, teenagers (blue) – generally perceived as least safe – show higher scores for the ethnic groups of Asian and Arab people.

$Fi : Ei$ – There was also a significant interaction between *facing–direction* and the *ethnicity* of the person in the image ($F(6, 7885) = 1.84, p < 0.01$). While Caucasian ($p < 0.01$) and Arab people ($p < 0.05$) were perceived as least safe when they looked towards the user, Black ($p < 0.01$) and Asian people ($p < 0.01$) were perceived as least safe when facing away from the user. In

Figure 3 we visualize this interactions on the middle interaction plot. We observe that generally people that are not aware of the user (green) have been rated as safest. People looking towards the user (blue) – generally rated as least safe – show a higher score, compared to people looking away (red) when they are Black or Asian only.

$Ai : Gi$ – We found also a significant interaction between *age* and the *gender* of the person in the image ($F(2, 7885) = 3.77, p < 0.01$). While all age groups of men are perceived as less safe compared to all age groups of women, results show that, within the group of men only, grown–ups ($p < 0.01$) were perceived as less safe than elder men ($p < 0.001$). In Figure 3 we visualize this interaction on the right interaction plot. We observe that generally all age groups of women received higher safety scores compared to men. Overall age matters more in the case of men compared to women, where scores were more similar over all ages. Elder people (red) are perceived as most safe and grown–ups (green) as least safe in the case of men.

These examples show clearly how differences within demographic groups of people we see matters to our safety perception. We now turn our attention to demographics characteristics of crowdworkers.

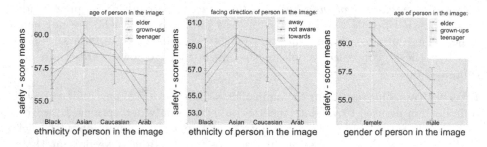

Fig. 3. Interaction plots: age and ethnicity (left), facing–direction and ethnicity (middle), age and gender (right) of the person in the image. y–axis shows means of perceived safety scores, x–axis shows the variable 1 and colours indicate variable 2 of each interaction.

5.2   Who We Are

Safety perceptions not only depend on *who we see* but also on *who we are*, the demographics and geographical background of the crowdworker. To analyse safety perceptions based on who the crowdworkers are, we explored our dataset in a next step according to the background information provided by the user: *gender, age, ethnicity, from London, where in London, continent*. As mentioned above, we received very diverse sample sizes for each of the properties, leading to limitations in terms of further evaluations due to lack of statistical power for smaller groups. From our data, we were able to reach statistical power for *gender, age* and *from London*.

Table 4. Table showing breakdown of significant effects on different user groups and their means of voting scores, ranging from 0 (indicating not safe) to 70 (indicating very safe), p<0.001 *** p<0.01 ** p<0.05 *.

User Groups	Effect	Person in the Image	Mean M
0 – 20	Main effect Mi ***		50.36
	Interaction Ai **	teenager	51.2
		grown-up	49.9
		elder	50.0
	Interaction Ei **	Arab	46.6
		Caucasian	51.4
21 – 40	Main effect Mi ***		57.69
	Interaction Ai **	teenager	57.7
		grown-up	57.3
		elder	58.0
	Interaction Ei **	Arab	55.5
		Caucasian	58.3
41+	Main effect Mi ***		58.46
	Interaction Ai **	teenager	58.5
		grown-up	58.0
		elder	58.9
	Interaction Ei **	Arab	57.6
		Caucasian	59.0
from London	Main effect Mi ***		58.4
	Interaction Ei **	Arab	56.5
		Caucasian	59.2
not from London	Main effect Mi ***		57.04
	Interaction Ei **	Arab	55.0
		Caucasian	57.6

Table 4 shows significant effects related to the different crowdworker groups, defined by their age and location, whether he/she is from London of not. Next we will summarize these findings. Compared to the overall findings presented in Table 2, where responses from all crowdworkers were aggregated, we now see the differences in perceptions depending on crowdworker's background.

Main effect Mi – We found a main effect Mi of participant's age on perceptions of safety ($F(2, 7529) = 37.23, p < 0.001$). Contrasts revealed that participants between 0–20 rated images as less safe than participants of 21–40 ($p < 0.05$) and 41+ ($p < 0.05$). Participants in the 21–40 age group rated images as less safe than those in the age group of 40+ ($p < 0.01$). We also found a main effect Mi on where participants came from, if from London or outside London ($F(1, 7529) = 13.58, p < 0.001$). Contrasts revealed that Londoners felt safer compared to participants coming from outside London ($p < 0.001$). Overall these findings suggest that elder people and people from a city such as London feel safer in relation to the appearance of other people, defined by age, gender and ethnicity, compared to others.

Interaction Ai – We also found a significant interaction Ai between the age of the user and the age of the person shown in the image ($F(4, 7529) = 3.13, p < 0.01$). Contrasts revealed that participants aged between 0–20 rated images lower than participants of 21–40 ($p < 0.05$) and 41+ ($p < 0.05$), and participants of 21–40 rated images lower than participants of 41+ ($p < 0.05$). Looking at mean scores, we observe while grown–ups are perceived as least safe by all three age groups, people perceived as most safe change with age of the participant: participants in the age of 0–20 perceived teenagers as most safe, while the other two age groups (21–40 and 41+) felt most safe with elder people.

Interaction Ei – There was a significant interaction Ei between the age of the user and the ethnicity of the person in the image, in particular in the cases of Arab and Caucasian backgrounds ($F(6, 7529) = 3.92, p < 0.01$). Contrasts revealed that participants aged between 0–20 rated images with both Arab and Caucasian people lower than people aged 21–40 ($p < 0.05$) and 40+ ($p < 0.05$). Participants in the age group of 21–40 rated both ethnicities lower than participants in the 41+ age group ($p < 0.05$). Furthermore we found significant interaction between participants' geographical background – if coming from or outside London – and the ethnic background of the person in the picture ($F(3, 7529) = 2.07, p < 0.05$). Contrasts show that participants from outside London rated images with Arab and Caucasian people lower than participants coming from London ($p < 0.05$). From these findings we can assume that people living in a multi–cultural city like London are more used to different ethnicities compared to people coming from outside London, and hence feel less intimidated by them.

In summary, we see how crowdsourced safety perceptions differ between each discussed group of crowdworkers, as different variables matter to different people. We found an increase in safety perception based on other people's age and ethnic background with increasing age of crowdworkers, from younger (0–20) to elder groups (40+). When looking at the age of the person in the image, crowdworkers felt mostly safe with people of the same age group, as participants in the age group 0–20 perceive teenagers as safest, while participants of 21–40 and 40+ perceive elder people as safest. Furthermore, we found that crowdworkers coming from London feel safer in relation to other ethnicities, such as Arab and Caucasian ethnicities, compared to crowdworkers from outside London.

6   Discussion and Future Work

Key Findings and Limitations. In this paper we have explored the viability of using online crowdsourcing to gather safety perceptions about people, and using the collected data to test theories quantitatively at near zero cost. In doing so, we were able to confirm a number of established theories on a large scale, previously only elaborated qualitatively. We have explored new features, such as the *facing–direction* and *concealment* of a person. Furthermore, interactions between these features can now be easily analysed, as we have shown in three examples, thus giving a new approach to researchers in social and urban studies to identify more features and in more depth. However, in terms of the representativeness of answers, we hit a barrier by not being able to control who makes up the crowd: certain demographics of crowdworkers were not reached. This is an important limitation that researchers aiming to use crowdwork to test theories need to be aware of. If theories require specific demographics or compositions, then we should either detect who is under–represented and use alternative methods, such as interviews, or we find ways of reaching out to them.

Future Work. Our future work spans two orthogonal directions. As people often appear in groups, we want to investigate safety perceptions about groups of people. Matei et al. [17] suggests that different group sizes and compositions, defined by age, gender and ethnicity for instance, affect our safety perceptions. With the aim to answer these findings quantitatively, we will develop *Streetsmart* further and crowdsource perception data focusing on differences related to the number and mix of people in a scene. Furthermore, to advance theories of urban studies, we are interested in how the presence of people affects our safety perception when combined with information about the built environment [24, 26]. To do so, we will run a similar study as presented in this paper, including people *and* the built environment in the image.

Besides the expansion of this study, we want to further investigate the limitations of crowdsourcing as a method. Crowdsourcing has shown to suffer from self–selection bias due to lack of crowd–control. For instance, Quattrone et al. [23] discuss differences found in spatial crowdsourcing datasets in the case of Open Street Map (OSM) and found that there is significant geographic bias. Kazai et al. [15] showed that lack of crowd–control has an effect on the quality of task outcome with significant quality differences for a number of experimental tasks between Asian and American crowdworkers. Online market–places, such as Amazon Mechanical Turk (AMT), aim to offer control of crowd selection, but in fact show homogenity of their population as well, such as education level and nationality in worker demographics. For instance, Ross et al. [12] showed that demographics of AMT workers have simply been shifting from moderate–income U.S. citizens to young educated people from India. Ipeirotis et al. [11] found that turkers were younger, mainly female, with low income and live in smaller families compared to U.S. internet users. More recent work [27] discusses the effect that culture has on the so–called "gold standards" obtained through online market-places such as AMT, specifically in the context of understanding semantic relatedness. By comparing task performance of AMT-turkers, scholars and scholar-experts, this work shows how communities matter and that AMT–derieved data is often not representative.

There exist different approaches to engage and motivate specific crowds, based on gamification for instance [29]. Still, it is not clear *who* they are attracting. As a second research direction, we want to explore different strategies, such as gamification, social cause and rewards, and analyse what type of crowd each strategy is most successful in engaging, and towards what type of task.

Acknowledgements. The research leading to these results is part of a PhD that is being funded by the Intel Collaborative Research Institute: Cities.

References

1. Braun, V., Clarke, V.: Using thematic analysis in psychology. Qualitative Research in Psychology **3**(3), 77–101 (2006)
2. Day, K.: Embassies and sanctuaries: womens experiences of race and fear in public space. Environment and Planning D **17**, 307–328 (1999)
3. Eiland, R., Richardson, D.: The influence of race, sex and age on judgments of emotion portrayed in photographs **43** 167–175 (2009)
4. Farrall, S., Bannister, J., Ditton, J., Gilchrist, E.: Questioning the measurement of the 'fear of crime'. British Journal of Criminology **37**(4), 658–679 (1997)
5. Felson, M., Clarke, R.: Opportunity Makes the Thief: Practical theory of crime prevention. Home Office (1998)
6. Gehl, J.: Cities for People. Island Press (2010)
7. Greenwald, A., McGhee, D., Schwartz, J.: Measuring individual differences in implicit cognition: The implicit association test. **74**, 1464–1480 (1998)
8. Hall, E.: The hidden Dimension. Anchor Books (1966)
9. Hallman, K.K., Kenworthy, N.J., Diers, J., Swan, N., Devnarain, B.: The contracting world of girls at puberty: Violence and gender-divergent access to the public sphere among adolescents in south africa. Poverty, Gender, and Youth Working Paper **25** (2013)
10. Hollway, W., Jefferson, T.: The risk society in an age of anxiety: situating fear of crime. The British Journal of Sociology **48**(2), 255 (1997)
11. Ipeirotis, P.: Demographics of mechanical turk (working paper) (2010)
12. Ross, J., Irani, I., Silberman, M., Zaldivar, A., Tomlinson, B.: Who are the Crowdworkers? Shifting Demographics in Mechanical Turk. In: Proceedings of CHI2010 (2010)
13. Jackson, J.: Validating new measures of the fear of crime. Int. J. Social Research Methodology **8**(4), 297–315 (2005)
14. Katz, C.M., Webb, V.J.: Fear of gangs: a test of alternative theoretical models. Justice Quarterly **20**(1), 95–130 (2003)
15. Kazai, G., Kamps, J., Milic-Frayling, N.: The face of quality in crowdsourcing relevance labels: demographics, personality and labeling accuracy. In Proc of CIKM 2012 (2012)
16. Lynch, K.: The Image of the City. MIT Press (1960)
17. Matei, S., Ball-Rokeach, S., Linchuan, J.: Qiu. Fear and misperception of los angeles urban space - a spatial-statistical study of communication-shaped mental maps. Communication Research **28**(4), 429–463 (2001)
18. Naik, N., Philipoom, J., Raskar, R., Hidalgo, C.: Streetscore - predicting the perceived safety of one million streetscapes. In: CVPR Workshop on Web-scale Vision and Social Media (2014)
19. Nasar, J., Fisher, B.: Proximate physical cues to fear of crime. Landscape and Urban Planning **26**, 161–178 (1993)
20. Pain, R.: Gender, race, age and fear in the city. Urban Studies **38**(5–6), 899–913 (2001)
21. Painter, K.: The influence of street lighting improvements on crime, fear and pedestrian street use, after dark. Landscape and Urban Planning **35**(2–3), 193–201 (1996)
22. Pantazis, C.: fear of crime, vulnerability and poverty. The British Journal of Criminology **40**(3), 414–436 (2000)

23. Quattrone, G., Capra, L., De Meo, P.: There's no such thing as the perfect map: quantifying bias in spatial crowd-sourcing datasets. In: Proc of CSCW 2015 (2015)
24. Quercia, D., O'Hare, N., Cramer, H.: Aesthetic capital: What makes london look beautiful, quiet, and happy. In: CSCW 2014 (2014)
25. Quercia, D., Schifanella, R., Aiello, L.: The shortest path to happiness: Recommending beautiful, quiet, and happy routes in the city. In: Proc of HT 2014 (2014)
26. Salesses, P., Schechtner, K., Hidalgo, C.: Image of The City: Mapping the Inequality of Urban Perception. PLoS ONE **8**(7), e68400 (2013). doi:10.1371/journal.pone.0068400
27. Sen, S., Giesel, M., Gold, R., Hillmann, B., Lesicko, M., Naden, S., Russell, J., Wang, Z., Hecht, B.: Turkers, scholars, arafat and peace: Cultural communities and algorithmic gold standards. In: Proc of CSCW 2015 (2015)
28. Taylor, R.B., Hale, M.: Testing alternative models of fear of crime. The Journal of Criminal Law and Criminology **77**(1), 151–189 (1986)
29. von Ahl, L., Dabbish, L.: Designing games with a purpose. Communications of the ACM **51**, 58–67 (2008)
30. Warr, M.: Fear of victimization: why are women and the elderly more afraid. Social Science Quarterly **65**(3), 681–702 (1984)
31. Warr, M., Ellison, C.: Rethinking social reactions to crime: personal and altruistic fear in family households. American Journal of Sociology **106**(3), 551–578 (2000)
32. Wilson, J.Q., Kelling, G.L.: broken windows: The police and neighborhood safety. Atlantic Monthly **249**(3), 29–38 (1982)
33. Wobbrock, J., Findlater, L., Gergle, D., Higgins, J.: The aligned rank transform for nonparametric factorial analyses using only anova procedures. In: Proc of CHI 2011 (2011)
34. Zako, R.: Young people's gatherings. Urban Public Realm **66** (2009)

A Real-Time Crowd-Powered Testbed for Content Assessment of Potential Social Media Posts

Himel Dev[1]([✉]), Mohammed Eunus Ali[1], Jalal Mahmud[2], Tanmoy Sen[1],
Madhusudan Basak[1], and Rajshakhar Paul[1]

[1] Department of Computer Science and Engineering,
Bangladesh University of Engineering and Technology, Dhaka, Bangladesh
{himeldev,sen.buet,madhusudan.buet,raaz.cse08}@gmail.com,
eunus@cse.buet.ac.bd
[2] IBM Research - Almaden, San Jose, CA, USA
jumahmud@us.ibm.com

Abstract. Increasing eminence of online reputation of individuals along with the tendency to avoid unpleasant real-life and/or virtual events such as cyber-bullying, social awkwardness, unintentional false news propagation etc., have made many social users concerned about their social media posts. To avoid the miscellaneous unpleasant events and to ameliorate online reputation, rigorous assessment of proposed posts before broadcasting them in actual social media has become crucial for these users. We observe that, such pre-screening of a proposed post requires human evaluation or feedback regarding different aspects of the post, which in turn assists the associated user in deciding whether or not he/she should broadcast the post in actual social media. In this paper, we address this issue and propose a crowd-powered testbed that allows a social media user to get a real-time evaluation of his/her proposed post. This assessment of a proposed post includes a positive/negative recommendation indicating whether or not the post should be broadcasted in actual social media.

Keywords: Social media · Crowdsourcing

1 Introduction

Social media has become the mainstream platform for sharing feelings and opinions among friends, colleagues, acquaintances, fans, and activists. Everyday, hundreds of millions of users express their opinions regarding various issues or events in the form of social media posts. These posts have remarkable impacts on different aspects such as socializing, marketing, economy, psychology, politics, and etc., of our day-to-day life. Due to such remarkable impacts in our everyday life, social media posts have turned into consequential issue, having both positive and negative consequences in the life of associated users. In this paper, we propose a

© Springer International Publishing Switzerland 2015
T.-Y. Liu et al. (Eds.): SocInfo 2015, LNCS 9471, pp. 136–152, 2015.
DOI: 10.1007/978-3-319-27433-1_10

crowd-powered testbed for a social media user to get a real time evaluation of a potential social media post, which ultimately helps the user to avoid *unintended consequences* of a social media action.

Though social media posts are initially thought of as private activities of individuals, due to their remarkable impacts on society and others, activities on social media are no longer considered as private. Each individual has to take the responsibility of his/her own actions and may face serious consequences of ill-judged social media post(s). For example, if an individual posts disparaging remarks about his/her boss on a Facebook page, he/she may face with repercussions. A plethora of real-life social media post related incidents recently emerged in the newspapers [1–4]. Let us have a look at some of these examples: "H.S. Teacher loses job over Facebook posting"- reported by Boston News on August 18, 2010 [1], "Ohio prison worker fired after posting Facebook comments about John Kasich"- reported by Huffington Post on January 19, 2012 [2], "Demotion as a consequence of controversial Facebook post"- reported by NC Jolt on February 2014 [3]. Though the aforementioned examples are related to employee-employer relations over social media issue, there are reported cases [4], where an individual faced legal actions over a social media post.

In all of the above cases, we observe that a careless social media post can bring much difficulty in the user's life in the form social awkwardness to job-loss. Things can go much worse if a user posts something that can be used to file a lawsuit against him/her [5]. Moreover, a social media post conveying false information or news can trigger a rumor leading to a social unrest, which had severe consequences observed in the recent past [4,6]. In addition, a user may get affected by social media anxiety disorder, by being constantly on alert for a social media post's response [2,7]. Though the above consequences of social media posts involve many other parties, just like any other physical action, an individual is also responsible for what he or she posts online. Hence, it is crucial for users to carefully assess the content of a post before broadcasting it in social media.

Apart from avoiding negative consequences of social media actions, users are also interested in building reputations through social media posts due to the astounding popularity of social media. While a good post can help building the reputation of the user by influencing the perceptions of other people about him/her [8,9], an incautious social media post with any inappropriate content can damage a user's reputation that will leave a lasting impression within his/her audience. Such a negative post can put a user in awkward situations, create havoc on his/her professional and personal life, and the image created through hard work and determination can be destroyed in seconds. Therefore, content assessment of a post before broadcasting it in social media has become a critical need for many social media users such as virtual celebrities and individuals who use social media reputation to build their career or business.

A major strategy to avoid the negative consequences and actuate the positive consequences via social media posts involves diligent content management of these posts. The focal operation of this content management procedure requires rigorous assessment of proposed posts before broadcasting them in actual social media. We observe that, such pre-screening of a proposed post requires evaluation of different aspects of the post such as whether the post contains any inappropriate content, whether the post will receive appreciation from the audience, whether the post prompts any legal issue, etc. More importantly, we observe that, evaluating these aspects require both domain knowledge and commonsense knowledge, which in turn imply human involvement in the form of crowd feedback. Based on the above observations, we hypothesize that, the crowd-powered evaluation of a proposed post based on human feedback can assist a user in deciding whether or not he/she should broadcast the post in a social media.

We propose a crowd-powered testbed, Cassandra, that allows a social media user to get a real-time evaluation of his/her proposed social media post (which we call the *potential post* in this paper). Our proposed solution consists of five major phases. In the first phase, our system automatically extracts *topics* from the *potential social media post*. In the second phase, the system identifies specialists, who are familiar with or have interest in these topics, by analyzing their past social media posts. After identifying a list of specialists on the subject matter, in the third phase, the system selects the final list of reviewers from the list of specialists based on diversified personality. Feedback from reviewers with diversified personality ensure un-biased assessment of a post, and more importantly reduce the chance of false positive assessment that has much worse consequences than other forms of misclassification. In the fourth phase, the post is forwarded to the selected specialist reviewers, who rate the post based on the definite guidelines provided by the system. Finally, in the fifth phase, our system applies a two-layer artificial neural network classifier on the ratings of the reviewers to recommend a post as either positive or negative indicating whether or not the post should be disseminated in actual social media.

To evaluate our system, we have developed a web-based prototype, which allows a user to connect through his/her social media account. A user can post anonymously, and receive positive/negative feedback based on the specialists' ratings. The experimental study shows that our crowd-powered approach results in a highly accurate feedback system for pre-screening potential social media posts.

2   System Design

We have designed a comprehensive crowed-powered social media post assessment system, Cassandra, which facilitates a user to get a real-time evaluation of his/her potential social media post. Cassandra takes a user post as an input and gives a positive/negative recommendation for the post based on the ratings of expert reviewers. Figure 1 depicts the high-level diagram of the architecture of our proposed system, Cassandra. The proposed system has five major phases, (i) topic extraction, (ii) specialist identification, (iii) representative selection, (iv) feedback collection, and (v) report generation, which we describe in the following subsections.

2.1 Topic Extraction

To extract topics from the potential post, first, our algorithm extracts tokens of form ⟨*word, part-of-speech*⟩ or ⟨*word group, phrase*⟩ from the post. During this token extraction, the algorithm uses the concept of fuzzy string matching to convert misspelled and/or short-handed word/word-groups into proper tokens (e.g., '*abt*'may refer to '*about*', '*chk*'may refer to '*check*', etc.). This is specially required for parsing Tweets, which often contain short-handed words due to 140 character limit. After token extraction, the algorithm filters the extracted token stream by eliminating the trivial tokens such as pronouns, adverbs, prepositions, auxiliary verbs, question words, etc. Then, the algorithm matches the remaining tokens with our relevance dictionary to determine the correlation of the post with different topics. The relevance dictionary contains the tokens that form the system's knowledge base and maps these tokens to different topics. The dictionary has initially been populated by applying a modified version of *topical keyphrase extraction* method proposed by Zhao et al. [10] on a dataset[1] containing more than 460,000 social media posts. Currently, the dictionary consists of more than 52,000 tokens classified into 290 topics. Note that, the tokens in relevance dictionary can map to more than one topic label (e.g., the word/token '*play*'can be used with all sports and music). Such tokens with multiple topic labels are called *homonyms*. While considering a homonym during correlation determination, our algorithm considers the degree of its relevance to the corresponding topic by using frequency information from prior posts by the user(s). For example, if the word '*play*'has been used 5 times by a user, 2 times for music related posts and 3 times for sports related posts, the relevance score of the token

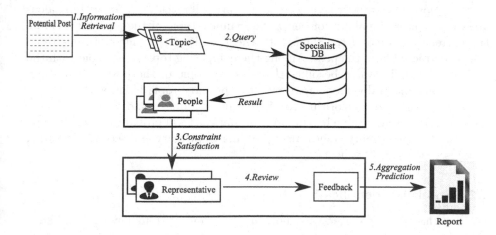

Fig. 1. Cassandra System Architecture

---

[1] The dataset has been collected over a span of eight weeks by accumulating sample posts from a social media post stream using a stream API.

Table 1. Topic Extraction Example

Potential Post	It's a green tea in bed kind of morning
Non-trivial Tokens	$T =\{$(green, noun), (tea, noun), (bed, noun), (morning, noun)$\}$
Relevance Scores	$\{$(green, [Life = 0.3, Food = 0.7]), (tea, [Food = 1]), (bed, [Life = 1]), (morning, [Life = 1])$\}$
Correlation	$c_{Life} = 0.3 + 1 + 1 = 2.3$, $c_{Food} = 0.7 + 1 = 1.7$
Probability	$p_{Life} = \frac{c_{Life}}{|T|} = \frac{2.3}{4} = 0.575$, $p_{Food} = \frac{c_{Food}}{|T|} = \frac{1.7}{4} = 0.425$
Primary Topic	Life ($\arg\max_x p_x$)
Threshold	0.8
Secondary Topic(s)	Food ($p_{Life} + p_{Food} >= 0.8$)

'*play*'corresponding to the topics music and sports for the particular user will be 0.4 and 0.6, respectively. If the user never used the particular token before, our algorithm considers the general frequency information of the token (number of times the token appeared under the particular topic vs the number of times it appeared in total). For tokens with a single topic label, relevance score (i.e., the degree of relevance) is 1. Using the relevance scores described above, our algorithm determines the correlation of a post with different topics. In particular, it determines the correlation of a post and a particular topic by summing the relevance scores corresponding to each of the extracted tokens of the post and the topic. Based on the determined correlations, the algorithm picks a primary topic, and optionally a couple of secondary topics for the potential post. Note that, secondary topics are only required when the primary topic does not predominate the subject matter of the post. More specifically, for each topic having a correlation value greater than zero for a particular post, our algorithm calculates the topic's probability of being the primary topic of the post by dividing the correlation value of the topic with the post by the number of non-trivial tokens extracted from the post. Among the competing topics, the topic with the highest probability value becomes the primary topic of the post. In addition, if the probability value associated with the primary topic is lower than a predefined threshold, our algorithm considers one or more secondary topics. These secondary topics are chosen in a manner so that the total probability, achieved by summing the probability values corresponding to primary and secondary topics, exceeds the threshold value.

2.2 Specialist Identification

We use the term *specialist* to denote users of our system having knowledge regarding one or more topics in social media. We have designed a database containing information regarding users of our system and their expertise in different topics. When a user joins our system via his/her social media account, our system creates a portfolio of the user, focusing on the expertise of the user in different topics. To identify the expertise of a user in different topics, our system (i) fetches the user's prior posts in social media, (ii) extracts topics from these posts as per the first phase, and (iii) determines his/her expertise in these topics based on the popularity of the associated posts in social media.

To identify the degree of expertise of a particular user in a particular topic, we extend the scholastic concept of h-index [11] to define topic-wise h-index for social media as follows:

Definition 1. *A social media user has index* h *in topic* T, *if* h *of his/her* N_p *posts regarding topic* T *have at least* h *shares each, and the other* $(N_p - h)$ *posts have no more than* $(h - 1)$ *shares each. In other words, a user with an index of* h *in a topic* T *has* h *social media posts on topic* T, *each of which has been shared at least* h *times by other social media users.*

Note that, our system considers the social media posts no longer than six months old for calculating the h-index of users, and updates these scores in a period of six months. Currently, to return the specialists associated with a topic, we return the set of users having a h-index of 5 or higher in that topic. In case of posts with a primary topic and one or more secondary topics, we return the set of users having a h-index of 5 or higher in the primary topic and 3 or higher in each of the secondary topics of the post.

2.3 Representative Selection

Our system selects a group of representatives from the set of specialists identified in the second phase. This selection is primarily done by diagnosing the personality traits of the identified specialists, and picking individuals who have dissimilar personality traits from one another. Picking individuals of dissimilar personality traits allows our system to diversify the psychological group of representatives. Since individuals with dissimilar personality traits often see a problem or content from distinct perspectives, the suggested psychological diversity of representatives is beneficial for our decision making problem where different aspects of the problem or content need to be explored. This is evident from our experimental evaluation, where we find that the psychological diversity of representatives reduce the amount of false negative classifications, i.e., it helps to prevent faulty posts from getting positive recommendation.

In addition to diversifying the personality traits of the representatives, we impose certain constraints on selecting the representatives. Some of these constraints are compulsory constraints which must be fulfilled, whereas others are preference constraints which should be fulfilled for a fair review process. The compulsory constraint of the representative not being a friend or follower of the user submitting the potential post serves the purpose of preventing spoiler (i.e., seeing a post before it appears in social media). It also mitigates some privacy concerns of users. The optional constraint of selecting representatives from the spatial region of the user submitting the potential post allows the user to get feedback from people of similar cultural sphere. This is especially required in case of posts that express opinions or ideas regarding issues or events confined within a spatial region. The compulsory constraint of not passing new review requests to users having several pending review requests allows the review process to be cosy to the reviewers.

Personality Identification: In recent times, social media has emerged as a massive source of diverse insightful written information where authors reveal themselves to the world, revealing personal details and insights into their lives in the form of social media posts. As an immediate outcome, personality identification of users from their social media posts has gained a huge momentum in recent years [12][13]. The state-of-the-art personality identification from social media primarily relies on the big five model of personality, where an individual is associated with five scores that correspond to the five main personality traits of the individual (i.e., Openness, Conscientiousness, Extraversion, Agreeableness, Neuroticism) [12]. In our system, we use this personality model to identify the personality traits of our users by using an approach similar to Golbeck et al. [13]. Note that, we identify the personality traits of a user during portfolio creation (i.e., when the user joins our system).

Personality Diversification: The goal of the DIVERSITY(S, k) algorithm is to find k diverse (in terms of personality traits) representatives from the set of specialists, S. The set of representatives is denoted by R. Here, each element of S or R refers to a point in five dimensional space corresponding to the five scores in big five model.

Algorithm 1 shows the pseudocode for our representative diversification algorithm. The algorithm starts with the selection of a random representative from

---

Algorithm 1. DIVERSITY(S, k)

Input   : A set of specialists $S = \{s_1, ..., s_n\}$, number of representatives k
Output: A set of k representatives $R = \{r_1, ..., r_k\}$
1.1 $R \leftarrow \{rand\_get(S)\}$;
1.2 $Init(D, P, CP)$;
1.3 **while** $|R| < k$ **do**
1.4 $sum \leftarrow 0$;
1.5   **for** *each* $s \in S$ **do**
1.6 $d_{new} \leftarrow dist(s, r_{new})$;
1.7   **if** $d_{new} < d$ **then** $d \leftarrow d_{new}$;
1.8 $sum \leftarrow sum + d$;
1.9 $cp \leftarrow 0$;
1.10   **for** *each* $s \in S$ **do**
1.11 $p \leftarrow \frac{d}{sum}$; $cp_{new} \leftarrow p + cp$; $cp \leftarrow cp_{new}$;
1.12 $u \leftarrow U(0, 1)$;
1.13   **for** *each* $s \in S$ **do**
1.14   **if** $u < cp$ **then**
1.15 $R \leftarrow R \cup \{s\}$; $S \leftarrow S \setminus \{s\}$; *break*;
1.16

---

set of specialists S, adding it to R and deleting it from S (Line 1.1). This is done using the $rand\_get(S)$ function which removes a random element from S and returns it. Then, for each element s in S, the algorithm incrementally calculates its minimum distance from the elements in R (Lines 1.4-1.8). Here, for each element s in S, D stores the minimum distance d corresponding to the element and the elements of R, P stores the probability p of the element to be added to R, CP stores the corresponding cumulative probability cp. The algorithm calculates the probability p for each element s in S based on its distance d. The greater the distance d corresponding to the element and the elements of R, the higher the probability p of the element to be added to R (Lines 1.9-1.11). This strategy ensures that the points in R are well separated from each other and thus leads to diversity. Later, the algorithm selects a random element from S based on the determined probability values, adds it to R and deletes it from S (Lines 1.12-1.15). The process continues until the set R contains k elements (Line 1.3, Line 1.16).

2.4 Feedback Collection

Our system asks the selected representatives to review the potential post. The review is double-blind, i.e., the identity of the user submitting the potential post is concealed from the reviewers, and vice versa. This is done to protect the privacy of users. We consider feedback of two types. The first feedback is a compulsory one that involves rating the post on different factors or issues. Currently, we consider rating a post in terms of four issues: *'popularity'*, *'controversy'*, *'cyber-bullying'*, *'legal-issue'*, and a scale of five for each rating. The second feedback is an optional one that involves writing a short review of 140 characters. We use support vector machine (SVM) with a polynomial kernel to categorize written reviews into neutral, positive and negative based on their semantic orientation. The SVM classifier is a feasible one considering accuracy and training time.

Note that, there is a time constraint involved in reviewing a post. Currently, this time constraint is imposed by the user submitting the potential post.

Four Factors: We observe that, the four factors, *'popularity'*, *'controversy'*, *'cyber-bullying'* and *'legal-issue'*, can potentially identify the effects of different types of social media posts. In particular, whereas the *'popularity'* factor deals with whether a post is mundane or not, the other three factors specifically determine whether a post can lead to any negative consequence. By investigating these factors, one can determine whether the post is worthy of being disseminated in social media. Therefore, we have considered these factors in our evaluation process. Note that, the list of hazards-to-be-avoided is not closed, so there is potential for the crowd to incorporate new types of hazards in our system and avoid those as well.

Rating the Four Factors: Our system provides a guideline for rating each of the four factors. We observe that, evaluating these factors require commonsense knowledge along with domain (topics of the post) knowledge. We hypothesize that, a domain expert with adequate commonsense knowledge can evaluate the

Table 2. Sample Negative Posts and Corresponding Issues

Sample Post	Issue
I think save Kader Molla is your Islam. Right?	Controversial
Not sure if you are pregnant or really fat	Bullying
When someone loves you, the way they talk about you is different	Monotonous
I have a bomb on one of your planes, but forgot which one	Legal Issue

factors effectively. This is evident from the experimental evaluation, where the system has identified good/bad posts with high accuracy, based on the ratings given by the reviewers.

2.5   Report Generation

Our system generates a report evaluating different aspects of the potential post based on the feedback given by the representatives. Currently, the report contains aggregation of ratings given by the representatives, and a positive/negative recommendation indicating whether or not the potential post should be broadcasted in actual social media.

Aggregate Ratings: The aggregate ratings for each issue (e.g., popularity) is a weighted combination of the ratings provided the representatives and their h-index corresponding to the primary topic of the potential post. More specifically, if n representatives having h-index $h_1, h_2, ..., h_n$ in primary topic T of a potential post have rated the post $r_1, r_2, ..., r_n$ in terms of a particular issue j, then, the aggregate ratings, R_j, corresponding to that particular issue of the potential post is determined as: $R_j = \frac{\sum_{i=1}^{n} h_i * r_i}{\sum_{i=1}^{n} h_i}$.

Recommendation System: The recommendation system we use is a two layer neural network that classifies a post as positive/negative indicating whether or not it should be broadcasted in actual social media. For each post, our proposed neural network considers the aggregate ratings provided by the representatives along with the SVM outputs corresponding to the written reviews as input attributes, and predicts the class (i.e.,positive/negative) of the post based on the value of these attributes. The input attributes can have either discrete or real value, whereas the output is discrete. We observe that, the heterogeneous properties of neural networks are well suited to serve these purposes. The input of a neural network can be real values which can be either highly correlated or independent of one another, and the output can be discrete. Further, neural networks allow anomalous examples, which is appropriate for any crowdsourced data. For these reasons, we use neural network to serve our purpose of classifying posts based on crowd feedback.

The initialization of parameters (i.e., learning rate and momentum) of the neural network is of critical importance. The proposed neural network unfolds with high learning rate and low momentum, which allows more exploration in the beginning of the learning process. As the number of example increases, learning rate of the classifier declines. On the other hand, the momentum of the neural network gradually increases to a moderate value which ensures that the target function will avoid local minima without making the system unstable.

Note that, the neural network has initially been trained on a dataset[2] containing reviews of more than 12,000 posts. Each post has been reviewed by at least three reviewers and later on, entirely based on these reviews (not based on the content of the posts), five moderators labeled the posts as positive/negative.

3 Real-Time Crowdsourcing with Cassandra

One major limitation of crowd-powered systems involves latency. Immediate response from a crowd-powered system can hardly be expected as human workers are often reluctant to stay connected for a long period of time. To resolve this issue, the concept of *real time crowdsourcing* has emerged [14]. Real-time continuous crowdsourcing explores [14] using the crowd to complete tasks that would usually require workers to remain connected for a long period of time. In our system, the concept of real-time continuous crowdsourcing can be efficacious in collecting feedback from the specialists. A tweet or a facebook status may have its own temporal significance, which may be deceased due to late posting. Therefore, a user may ask for immediate feedback, before his/her post becomes insignificant in social media. In such cases, we anticipate either some specialists to remain inclined to our system for an appreciable amount of time or the review to be provided by others in the quickest possible time.

One approach we consider is to maintain a queue of at least a minimum number of reviewers for each topic category. This can be considered as a recruiting pool of reviewers or specialists and keep them standby until a post of their expertise appears [15]. This queuing theory is designed in such a way that it can work out on how to optimize the process of recruitment according to how often the posts come up, how much time is required for reviewing a post etc. Note that, the reputation of a reviewer (i.e.; h-index) is considered during candidate selection in queue.

Our system is designed to be cosy to the reviewers, which allows the reviewers to provide feedback within a short period. Although an optional feedback choice of Cassandra involves writing a short review, reviewers are often indifferent to writing reviews. On the other hand, the compulsory feedback in the form of rating allows the reviewers to provide quick response [16].

A multilayer rewarding scheme has been modeled with a view to encouraging the reviewers to submit effective responses quickly. More like Chorus [17], Cassandra pays a reviewer a small reward for each review, a medium reward if the review is compatible with reviews from other expert reviewers of the same topic, and a large reward if ultimately the user agrees with the review by either posting or not posting the potential post in actual social media. The difference

---

[2] The posts in this dataset have been sampled from the dataset used for populating the dictionary. Later, these posts have been reviewed and labeled by subjects recruited by the authors.

among the reward values corresponding to different tiers have been set as such that the reviewers are induced to give effective reviews. For example, if the large reward is 10 times more attractive than the medium one, reviewers tend to show much more commitment to provide responses using more of their intellect and expertise. However, a constraint is imposed on the number of reviews by a reviewer within a particular time span to prevent a reviewer from giving too many responses for the participation reward. Note that, the final reward value is discounted with response time as the system is time critical. In particular, the reward value is exponentially discounted by a factor of $\gamma^{i-\epsilon}$, where $0 <= \gamma <= 1$ is a constant that determines the relative value of reward, i is response time, and ϵ is a threshold on response time.

Like many popular QA communities (e.g., Stack Overflow, Stack Exchange etc.), we are currently offering virtual rewards in the form of different membership privileges. For receiving feedback on each post, users need to earn a fixed reward value. Users earn rewards by rating other posts. This mechanism capitalizes on anticipated reciprocity: users are performing an action in the hopes of receiving something in return.

4 System Prototyping

We have developed a working prototype of Cassandra, which is currently a web-based application having Twitter integration. A Twitter user can join our system by signing up via his/her Twitter account. When a user joins our system, our system analyzes the user's Twitter profile and creates a portfolio of the user incorporating connection with other users, expertise in different topics and personality traits.

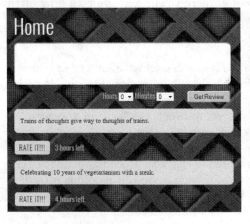

(a) Home Screen   (b) Feedback Screen

Fig. 2. Application Screenshot

In Figure 2(a), we show a screenshot of the home screen associated with a user. The home screen includes a provision to ask for reviews, and displays the pending review requests. In Figure 2(b), we show a screenshot of the feedback screen corresponding to a positively recommended post. The feedback screen displays the aggregate ratings by the reviewers and the corresponding recommendation sign.

5 Performance Evaluation

We have performed a small-scale evaluation of our proposed system, Cassandra. Our evaluation of Cassandra focuses on the following questions: (i) how reliable is Cassandra in classifying the potential social media posts, (ii) what is the benefit of personality diversification during representative selection, (iii) how much latency is involved in the response from the crowd. We explore these questions in the following subsections.

5.1 Reliability

We measure the reliability of our proposed testbed by monitoring the forecast of various posts in our testbed and the real performance of these posts in actual social media. We investigate whether different posts receive similar or dissimilar feedback in our testbed and actual social media platform. In addition, we compare these results with the results obtained by a multinomial naive Bayes classifier. We perform the investigation in two phases.

In the first phase of exploration, we have used past social media posts by subjects from four countries. More specifically, we have collected public posts by random social media users from user streams (using stream API) and inquired the corresponding users about the consequences of these posts. Based on user response, we have determined the consequence of 17,000 posts. We know whether these posts received appreciation from the audience or not. We also know whether or not these posts led to unfavourable events (e.g., conflict among the audience). We have labeled these posts based on their outcomes. As per our labeling, 63% (10,710 posts) of the posts are positive posts which received much appreciation from the audience and the rest 37% (6,290 posts) are negative posts which either did not receive much appreciation or led to unfavourable events. Later, we have used these past social media posts as trial posts in our testbed. Note that, all these posts are time invariant, i.e., they do not express opinions or ideas about past issues or events. Otherwise, the posts may seem obsolete to the reviewers in current epoch. Based on the reviews by the specialist reviewers, our recommendation system has classified each post as either positive or negative. We have compared these labels with the real labels of the posts based on their performance in actual social media. If a negatively recommended post either received negative feedback from the audience or led to unfavourable events, we have considered it as true negative. If such a post did not receive negative feedback, we have considered it as false negative. Similarly, if a positively recommended post received appreciation from the audience, we have considered it as true positive. If such a post did not receive appreciation from the audience, we have considered it as false positive. Based on the number of true positive (6,993), true negative (4,951), false positive (1,339) and false negative (3,717) classifications, we have calculated accuracy, precision, recall and F-score for our recommendation system. The results are summarized in the third row of Table 3. We can see that, in case of past social media posts, the precision and recall values are well balanced and thus lead to a high F-score value demonstrating the effectiveness of our system in identifying a good/bad post.

In this phase of exploration, we have also trained a multinomial naive Bayes classifier using 80% of the 17,000 labeled posts and later used the classifier to classify the rest 20% posts. We have repeated this operation 5 times, following the procedures associated with a standard 5-fold cross validation technique. The results are summarized in the fourth row of Table 3. We can see that, in this case, the naive Bayes classifier is unreliable in identifying a good/bad post.

Table 3. Performance Comparison among Cassandra, Multinomial Naive Bayes Classifier and Cassandra without Personality Diversification (Cassandra**)

System	Old Posts				New Posts			
	Accuracy	Precision	Recall	F-score	Accuracy	Precision	Recall	F-score
Cassandra	0.70	0.84	0.65	0.73	0.79	×	×	×
Naive Bayes	0.47	0.61	0.55	0.58	0.60	×	×	×
Cassandra**	0.65	0.76	0.67	0.71	0.76	×	×	×

In the second phase of exploration, the trial posts have been provided by 47 users recruited by the authors. In total, 2,795 posts have been contributed by these users. We have tested the performance of these posts in our testbed as per the first phase and later on, the corresponding users have tried only the positively recommended posts (1,561 posts) in actual social media. Note that, negative posts have not been disseminated in social media to prevent unpleasant consequences. Based on the performance of the positively recommended posts in actual social media, we have tagged them with their real labels and compared these real labels with the label (+) provided by our system. Then, as per the first phase, we have calculated accuracy for our recommendation system. Note that, we have not calculated precision, recall and F-score for this phase as we do not know the real label of the negatively recommended posts. The results are summarized in the third row of Table 3. We can see that, in case of user contributed posts, our testbed has achieved a high level of accuracy in identifying a good post.

In this phase, we have trained the multinomial naive Bayes classifier using the 17,000 old posts and later used the classifier to classify the new posts. The results are summarized in the fourth row of Table 3. We can see that, in this case, the naive Bayes classifier is unreliable in identifying a good/bad post as well.

User Recruitment: The 47 users have been recruited using two strategies. The first 29 users have been recruited using the well known non-probabilistic sampling approach, *snowball sampling*. We have initially recruited 4 users from our known community. Then the preliminary users have been asked to choose and motivate future users among their acquaintances. They have chosen another 9 participants, who in turn have chosen 16 other participants. Thus, the potential community recruited by *snowball sampling* has extended to 29 users. The next 18 users of our system have been recruited via *random sampling*. More specifically, we have invited random social media users to try our system and 18 of them have agreed to use it. Note that, all (47) these users regularly use social media.

Reviewer Recruitment: 31 of the 47 recruited users have served as reviewers to assess the posts. In addition to these 31 users, 19 individuals who are highly reputed (high *h*-index in various topics) in social media have been recruited by the authors to assess the posts. All these reviewers have reviewed the posts based on their expertise inferred by their past social media posts.

5.2 Benefit of Personality Diversification

One of the key design concerns of Cassandra has been whether the personality diversification of reviewers plays any role in improving the performance of the system. To investigate this, each post has been evaluated by at least six reviewers belonging to two groups. The first group consists of reviewers with diverse personality traits (as per the standard system design) and the second group consists of reviewers with random personality traits. Note that, the two groups may have overlapping members. The results corresponding to the first group have already been presented in the third row of Table 3. The results corresponding to the second group are summarized in the fifth row of Table 3. We can see that, for the first set of experiment (involving past social media posts), the precision value has steeply declined whereas the recall value has slightly improved. This implies that the number of false positives (i.e., negative posts that have been positively recommended) has increased whereas the number of false negatives has decreased. This is an alarming issue considering the fact that false positives have much worse consequences compared to the false negatives. Hence, a post that can have negative consequences must be identified, even at the cost of some positive posts to be misidentified as negative posts. Therefore, the system is expected to have a high precision, even at the cost of sacrificing recall. Based on the experimental results, we can hypothesize that the personality diversification of users plays a significant role in this regard.

5.3 Latency

We have monitored the response time of reviewers to measure the latency involved in getting a post evaluated by our system. To do so, we invited 23 new users (via *random sampling*) to try our system. After a period of two weeks (when these users started using our system in a regular manner), we started tracking the review time. We have found that, 75% posts have been reviewed within a period of five and a half hours. In extreme cases, the reviewers have taken less than nine hours to review a post. We believe, this is a reasonable time considering the reviewers have not been persuaded by means of incentives.

6 Related Work

To the best of our knowledge, Cassandra is the first system to deal with content assessment of potential social media posts. However, it is motivated by earlier works on self-censorship and social media content management.

Sleeper et al. [18] proposed the scheme of self censorship in Facebook by analyzing the users' tendency of not sharing a content or sharing it within a desired group. In their work, they answered questions that are related to the type of content that are not worth sharing, the associated causes, the subset of not sharable contents which can be shared among restricted groups, and the attributes which classify these groups.

André et al. [19] designed a website with a view to collecting a large corpus of follower ratings in order to understand the broad continuum of reactions in-between, which are not shared publicly. They categorized the tweets with different labels using the Naaman's manual tagging scheme [20]. They asked the users to give feedback whether the tweets were worth reading, ok, not worth reading. As an extended scheme, they provided the users options like funny, exciting, useful, informative, arrogant, boring, depressing etc., so that the reason behind their feedback could be deciphered.

7 Conclusion

In this work, we have proposed a crowdsourced testbed, Cassandra, which takes a potential social media post from a user as an input and gives the user a positive/negative recommendation indicating whether or not the post should be broadcasted in actual social media. We have built a prototype of Cassandra and evaluated our proposed feedback system, which shows the efficacy and deficiency of our feedback system. With the help of our system, a social media user can avoid unpleasant consequences such as cyber-bullying, social awkwardness, unintentional false news propagation, etc., due to incautious social media posts. Our system also helps a user to build an online reputation through careful social media posts. To the best of our knowledge, this is the first approach to deal with the problem of content assessment of posts before broadcasting them in actual social media.

References

1. News, B.: H.s. teacher loses job over facebook posting. http://www.wcvb.com/H-S-Teacher-Loses-Job-Over-Facebook-Posting/11284946
2. Huffingtonpost: Ohio prison worker fired after posting facebook comments about john kasich. http://www.huffingtonpost.com/2012/01/19/ohio-prison-fired-facebook-john-kasich_n_1217125.html
3. Wheaton, C.: Demotion as a consequence of controversial facebook post not in violation of the first amendment. http://ncjolt.org/demotion-as-a-consequence-of-controversial-facebook-post-not-in-violation-of-the-first-amendment/
4. Maharaj, S.: Sensitive facebook posts may carry legal consequences. http://globalnews.ca/news/849720/sensitive-facebook-posts-may-carry-legal-consequences/
5. NewAge: Death threat to pm on facebook: Former buet teacher gets 7yrs in jail. http://www.newagebd.com/detail.php?date=2013-06-28&nid=54778#.Uyg1CVfgOf0
6. terminalx: Bangladeshi intelligence intercepting social sites. http://www.terminalx.org/2012/10/bangladesh-intelligence-intercepting-social-sites.html
7. DepAnxD: How social media causes depression anxiety. http://www.depressionanxietydiet.com/how-social-media-causes-depression-anxiety/#sthash.XzTpaAeF.dpuf
8. Rosenberg, J., Egbert, N.: Online impression management: Personality traits and concerns for secondary goals as predictors of self-presentation tactics on facebook. J. Computer-Mediated Communication 17(1), 1–18 (2011)
9. Krämer, N.C., Winter, S.: Impression management 2.0. Journal of Media Psychology: Theories, Methods, and Applications 20(3), 106–116 (2010)
10. Zhao, W.X., Jiang, J., He, J., Song, Y., Achananuparp, P., Lim, E.P., Li, X.: Topical keyphrase extraction from twitter. In: Proceedings of the 49th Annual Meeting of the Association for Computational Linguistics: Human Language Technologies - Volume 1, HLT 2011, pp. 379–388. Association for Computational Linguistics, Stroudsburg (2011)
11. Hirsch, J.E.: An index to quantify an individual's scientific research output that takes into account the effect of multiple coauthorship. Scientometrics 85(3), 741–754 (2010)
12. Quercia, D., Kosinski, M., Stillwell, D., Crowcroft, J.: Our twitter profiles, our selves: predicting personality with twitter. In: 2011 IEEE Third International Conference on Social Computing (socialcom) Privacy, Security, Risk and Trust (Passat), pp. 180–185 (2011)
13. Golbeck, J., Robles, C., Edmondson, M., Turner, K.: Predicting personality from twitter. In: 2011 IEEE Third International Conference on Social Computing (Socialcom) Privacy, Security, Risk and Trust (Passat), pp. 149–156 (2011)
14. Lasecki, W.S., Murray, K.I., White, S., Miller, R.C., Bigham, J.P.: Real-time crowd control of existing interfaces. In: Proceedings of the 24th Annual ACM Symposium on User Interface Software and Technology, pp. 23–32 (2011)
15. Bigham, J.P., Jayant, C., Ji, H., Little, G., Miller, A., Miller, R.C., Miller, R., Tatarowicz, A., White, B., White, S., Yeh, T.: Vizwiz: nearly real-time answers to visual questions. In: Proceedings of the 23nd Annual ACM Symposium on User Interface Software and Technology, pp. 333–342 (2010)

16. Lasecki, W.S.: Real-time conversational crowd assistants. In: CHI Extended Abstracts, pp. 2725–2730 (2013)
17. Lasecki, W.S., Wesley, R., Nichols, J., Kulkarni, A., Allen, J.F., Bigham, J.P.: Chorus: a crowd-powered conversational assistant. In: Proceedings of the 26th Annual ACM Symposium on User Interface Software and Technology, pp. 151–162 (2013)
18. Sleeper, M., Balebako, R., Das, S., McConahy, A.L., Wiese, J., Cranor, L.F.: The post that wasn't: exploring self-censorship on facebook. In: Proceedings of the 2013 Conference on Computer Supported Cooperative Work, CSCW 2013, pp. 793–802. ACM, New York (2013)
19. André, P., Bernstein, M., Luther, K.: Who gives a tweet?: evaluating microblog content value. In: Proceedings of the ACM 2012 Conference on Computer Supported Cooperative Work, CSCW 2012, pp. 471–474. ACM, New York (2012)
20. Naaman, M., Boase, J., Lai, C.H.: Is it really about me?: message content in social awareness streams. In: Proceedings of the 2010 ACM Conference on Computer Supported Cooperative Work, CSCW 2010, pp. 189–192. ACM, New York (2010)

Adaptive Survey Design Using Structural Characteristics of the Social Network

Jarosław Jankowski[1]([✉]), Radosław Michalski[1], Piotr Bródka[1],
Przemysław Kazienko[1], and Sonja Utz[2]

[1] Department of Computational Intelligence,
Wrocław University of Technology, Wrocław, Poland
{jaroslaw.jankowski,radoslaw.michalski,piotr.brodka,kazienko}@pwr.edu.pl
[2] Leibniz-Institut für Wissensmedien, Tübingen, Germany
s.utz@iwm-tuebingen.de

Abstract. The implementation of new methods that increase the quality and effectiveness of research processes became an unique advantage to online social networking sites. Conducting accurate and meaningful surveys is one of the most important facets for research, wherein the representativeness of selected online samples is often a challenge and the results are hardly generalizable. This study presents a proposal and analysis based on surveys with representativeness targeted at network characteristics. Hence, the main goal of this study is to follow the measures' computed for the main network during survey and focusing on acquiring similar distributions for sample.

Keywords: Social network analysis · Network sampling · Adaptive surveys

1 Introduction

The rapid development in the field of information technology, particularly the Internet, has changed the way humans do things and pervades all aspects of life. The popularity and accelerated growth following the popularity of social networking sites opened a wide array of opportunities to study, understand, and leverage their unique capabilities. These sites not only provide an alternative avenue for research conduction, but also allow for an in-depth analysis of their impact in terms of human behavior and online activities [1,7]. Recent studies in this field focused primarily on new algorithms and their application in various areas [13,16,19]. Due to the complexity of the network structures, the analyses are usually performed using some samples to find structures that are smaller, but share similar properties and distributions. The information gathered from social network analysis can be enhanced by using either typical surveys or new approaches based on adaptive surveying that optimizes quality and response rates, still being cost efficient [17]. Research in this area is still in the early stages and adaptive methods are rarely implemented. Furthermore, another motivation

© Springer International Publishing Switzerland 2015
T.-Y. Liu et al. (Eds.): SocInfo 2015, LNCS 9471, pp. 153–163, 2015.
DOI: 10.1007/978-3-319-27433-1_11

for research on the development of sampling methods is to improve the selection of elements during the remainder of the sampling and, as a result, it yields an increase in the representativeness of the data. The majority of studies in the social media field utilize online surveys and focus greatly on social networking sites, such as Facebook [3,24]. Students or self-selected individuals are the most common respondents in this study. Although it is possible to extract behavioural data from social media and use it as the basis for analysis, psychologists and sociologists are often interested in the subjective experience of social media users and surveys are still the most suitable tool [20]. In this study, a new method for validating and enhancing the representativeness of online samples is presented. This paper moves the outcomes of an earlier work in new field of applications and proposes additional sampling algorithms [10]. Performed research assumes that it might be useful to utilize network measures such as centrality or degree as a basis in determining the representativeness of an online survey within the entire population. The presented approach is based on selecting an adequate set of candidates in each step of the multistage process to improve the representativeness of the sample in terms of network measures. To identify opinion leaders, centrality measures are usually crucial, as different network characteristics may also be considered depending on the research goal and its area of application. The proposed method can be adapted to different research goals by using weighted sampling. The obtained sample derived from the use of online surveys, which are usually based on voluntary participation, may have other characteristics than the random sampling due to the probability of having low response rates. Hence, the proposed method makes it possible to direct the selection process towards expected characteristics of the sample.

2   Related Work

Research based on surveys uses several techniques and surveys can either be identified as static and adaptive [21]. Static surveys are not dependent on collected observations, while adaptive surveys are partially based on data from observations [17]. Adaptive surveys are a means of increasing both responses and the quality of the research by selecting samples characterized by the lowest mean square error on the sample values [5]. Aside from sampling direction, other adaptive components can include offering different incentives and using responsive survey designs or questionnaire structures [8,18]. The design-based approach to survey sampling uses variables of interest as fixed values, while model-based variables of interest are defined as random variables with joint distribution [22]. During surveys, interventions can be made to decrease variances of selected variables in the respondent pool by targeting sampling to key subgroups [7]. Most of the presented methods on which target surveys are focused are the sample representativeness in terms of attributes. However, problems may arise in the representativeness of the surveyed users with network distributions. The majority of studies in this area focus more on the representativeness of the population in terms of demographic values or other attributes and more often than not the parameters of the analysed social networks are overlooked [3,4]. Standard methodology does not ensure that all possible knowledge is used while building the representative data sample. A sample built using only attribute distributions may not be representative in terms of network measures and their characteristics. Additionally, there is the question of what a representative sample means. This study raises questions whether focus only on the attributes or consider checking network parameters and their distributions as well. Moreover, researchers need to decide which network measures ought to be considered. This area relates to sampling methods which deliver information about the network, evolution of data structures and make possible further enhancement of survey design [5]. Sampling is treated as conventional when it uses free-standing sampling frames and does not use the acquired data in the sampling process. The first group of methods in the conventional survey is a two-stage self-weighted probability based on random-node selection focused on uniform or proportional to node degree [14]. The other group is based on graph sampling, which includes snowball sampling [6]. Apart from theoretical work, some studies were conducted using real online social systems, such as Facebook and Twitter [2,7]. In contrast to static designs, adaptive sampling can be applied after the results of the earlier stages are collected, and it is used for direct sampling [23]. Adaptive sampling is a technique wherein its purpose is to improve the selection of elements during the remaining stages of the sampling collection, thereby improving the representativeness of the data that the entire sample yields. In marked contrast to conventional sampling designs, which have problems with sampling hidden populations, the adaptive sampling technique can take advantage of spatial patterns in the population to obtain more precise measures of population abundance. Hence, by using the adaptive approach, the sampling intensity may be increased when necessary. There are approaches targeted at adaptive cluster sampling based on the selection of neighbours in the network only if a given condition related to cluster location is satisfied [22]. Respondent-driven sampling, which was introduced by Heckathorn, allows researchers to make asymptotically unbiased estimates about the social network connecting the hidden population [9]. It is an extension of snowball sampling and patterns of recruitment are used to calculate inclusion probabilities for different types of nodes that are based on the recruitment of members of the population by other sampled members. It collects information about ties from each participant, but its limitation is that it can be inaccurate in clustered networks due to homophily and separated communities. The proposed adaptive approach is based on the collection of network data from respondents, and adaptive sampling is based on moving to other regions of the network after obtaining enough samples from the identified cluster. Sampling delivers information about the network evolution of data collection methods and with the continuous development of technology, further enhancement of survey design is possible.

The main goal of this research is to demonstrate a new approach for generating a network sample by selecting nodes for surveying in such a manner that they will follow several distributions. The objective is to minimize the distance between the surveyed sample and the whole network profile by using the Kullback-Leibler measure, which is an important concept in statistical mechanics. A network sampling approach should result in conformity with network measure distributions to follow. In a natural way, the available sampling methods are targeted to nodes with a high number of representatives. Certain nodes will not be selected in the samples and some concentrations of the samples can be observed. The method proposed in this paper makes it possible to perform long-tail sampling in order to get representatives from areas that are hidden and not easily accessed by random sampling and other methods. The method using the approach from [10] is extended with new algorithms what eases performing long tail sampling in order to get representatives from areas that are hardly reachable by random sampling and other typical approaches. Furthermore, this study also seeks to define a separate weight for each network measure and to force sampling to focus on a selected network measure or set of measures, which makes it possible to obtain a sample specific to the area of application that the researcher needs.

3   Methodological Background

Based on a set of network measure distribution, a balanced adaptive distribution fitting approach can be used. The main goal of this approach is to create representative survey responses based on a selected set of participants in terms of distance from the whole network distribution. A function minimizing the distance from the vector of network distributions is proposed, and the network members are selected to fit the reference distributions for the whole network, which are known in advance. In Figure 1, the sample process of conducting a survey in the social network with the proposed approach is presented. The survey system with the conventional static approach, is responsible for conducting surveys within social network. Using the static survey module users fill out a survey on a non-incentivised basis. When using the conventional approach, the process ends at this stage and analyses are carried out on the obtained data sample, even if the sample is not representative. During an example surveying process the initial sample is obtained from the set of users denoted as U. Afterwards, analysis and decision is being made whether a representative sample should be built from the obtained data by weighting, or whether the data collection process should be continued. The final sample S_i should be localized between Upper and Lower boundaries in terms of the distance from $n = 1, ..., M$ network measures. If the results yield that it is not possible to collect additional data, the obtained sample is analysed in terms of network distributions. Some of the selected nodes from the sample with the highest negative impact on overall sample evaluation are being excluded from further analysis.

The consequence of such an approach, however, would be that only a subset of the obtained data is actually used. If the sampling is continued, a multistage survey control system uses a vector of reference distributions D^* based on the whole social network and a vector of nodes' properties distributions D_i computed

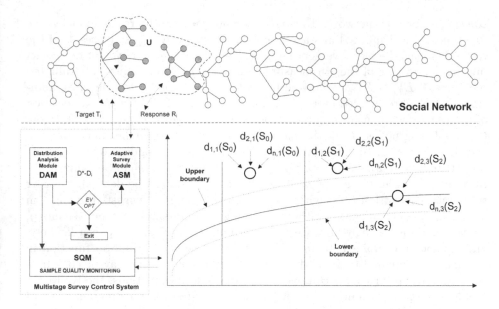

Fig. 1. Sample surveying process controlled by a multistage survey control system

for sample S_i at the ith stage. The main goal of the system is to build a sample S_i that is characterized by a small distance from the whole network distribution D^*. If D_i differs from D^* significantly, the sample is treated as not representative and additional participants have to be drawn. The distribution analysis module DAM and the adaptive survey module ASM are included within the systems. The DAM is responsible for the distribution analysis of the vector D_t. At each stage, the distance from the whole social network distribution is measured by the evaluation function EV_i as a part of sample quality monitoring module (SQM). If the value of this function is at an acceptable level (EV_{opt}), the process terminates at this stage and the sample S_i can be used for analysis. If not, the DAM system selects the group of candidates in target set T_i. These users are requested and those who respond from set R_i are passed to the next stage. The proposed approach may be represented as a series of steps. In stage $t = 0$ for static initial survey and for every measure, should be found the best distribution function approximation for the whole network. The resulting vector will consist of functions and their parameters describing the particular reference and expected measure distribution - D^*. Next, initiate the set of target users T_0 (selected in any way) and obtain the initial set of those who responded and were surveyed R_0, $S_0 = \varnothing$. In the next phase stage i is initiated with validation and extension. Sample set S_i, which is composed of the users surveyed so far ($S_i = S_{i-1} \cup R_{i-1}$) updated and the approximation functions as above D_i calculated. Then the difference is evaluated, i.e. the Kullback-Leibler divergence, $KL_{i,j}$ [12], between two functions: measure the distributions from the tested (D_i) and reference set (D^*), and separately measure j out of all M measures considered for each network. To generalize and evaluate the result globally, the function EV_i is computed in stage i, as follows: where $KL_{i,j}$ is the result of the recent Kullback-Leibler divergence for the jth measure, w_j is a weight representing the importance of the jth measure, and M is the total number of evaluated measures. If $EV_i > EV_{opt}$ (the goal has not been obtained yet), collect an additional set R_i of responded users (for expanding the set S_{i+1}) by: (i) the selection of the new target T_i using e.g. the proposed K-bins algorithm (see below), where the number of nodes $|T_i|$ is the parameter set by the researcher depending on the time allowed t, (ii) requesting users from T_i the responding ones form R_i; repeat steps 2-5 until $EV_i <= EV_{opt}$. Preliminary decisions shall be made at the beginning, wherein the set of M network measures should be defined. In order to calculate network metrics within a reasonable time, it is suggested to focus on local measures (e.g. degree), rather than the global measures (e.g. betweenness), as various measures may have various distributions, i.e. normal, power law, etc. and cannot be combined. In that case, every measure distribution function in D_t shall be compared to D^* and the result of the comparison may be combined later. In order to provide a comparison for distribution functions, the Kullback-Leibler divergence was utilized, as it has the ability to compare different types of distributions. Different domains of various measures could be an issue that may arise in this study. To be able to make $KL_{i,j}$ values comparable, all the measures should be normalized into a similar domain, e.g. range $[0, 1]$. As described above, the adaptive nature of the algorithm lies in the method of choosing nodes, if the value of EV_i is not satisfying. During research three algorithms were evaluated and apart from one that presents the most promising results, other algorithm with important properties are presented. Algorithms for node sampling proposed by authors are presented below.

Alg. 1 Introduce normalized measure M which aggregates all evaluated measures into a single one by summing all values of those measures. Sort M in ascending order and choose additional nodes starting from the node with lowest M value towards the one with the highest value.

Alg. 2 with four variants: For each element calculate chosen measure values, order those in ascending order and choose nodes with lowest values of that particular measure. Each variant chooses different measure as the base one: 1 - in-degree centrality, 2 - out-degree centrality, 3 - node total degree, 4 - clustering coefficient.

Alg. 3 Introduce normalized measure M that combines all evaluated measures into a single one by summing all normalized values of those values. Calculate M for all nodes in the network and create the histogram of it with twenty bins. If the number of total nodes in the network is n and the number of currently evaluated users should be i, for every bin choose i/n representatives starting from borders of the bin towards the middle of it. Then evaluate the resulting set of representatives.

The first two algorithms over-represent the users most often met in the network, because most local measures are following the power law distribution. That enables the sampling of the most typical users. The third one proposes to model the referencing distribution adequately. However, for Algorithms 1 and 3 that combined selected measures it might be the case that if two metrics are contradictory, the result may not be satisfying.

4   Results

All the procedures and data gathered are perceived in the same manner by all respondents. In this case, it is demonstrated be presenting the results of a survey performed in a social networking site which is based on a graphical virtual world for adolescents which was the subject of earlier research related to diffusion of information in social networks [11] and multilayer community detection [15] done by the authors of this work. An online survey was conducted with the experimental survey system covering motivations, self-disclosure and self-presentation among portal users and was filled out by 373 of the respondents while 9,631 respondents logged into the system during the examination period and were identified basically by their unique identifier. Observed network had 294,935 edges and identified 74.97% of reciprocal links. Average node in-degree within the network was 30.62 while among surveyed users it was 99.26. Similar difference was observed for out-degree for whole network 30.62 and 89.02 for surveyed users. An average clustering coefficient for whole network was 0.07 while 0.08 was observed for surveyed users. Results show that the structural measures computed for the full network and for the surveyed users showed significant differences between the whole and the surveyed network profile. A preliminary analysis of the data shows that survey derived from voluntary participation did not necessarily follow random sampling. Rather, obtained dataset would be closer to the random sample with the increasing response rate value. However, due to low response rate, the obtained sample has a profile other than random and it can be additionally improved using the proposed method. In the next step, this dataset was used to empirically test the approach based on the multistage survey control system. Authors analysed the results from this survey in terms of survey participant representativeness in comparison to the whole network measure distribution. Proposed algorithms dedicated to the usage within the Distribution Analysis Module (DAM) were evaluated with random sampling results as a baseline. Firstly, algorithms were evaluated for chosen measures: in-degree, out-degree, total degree centrality and clustering coefficient. Figure 2 shows that combining anticorrelated measures, like degree centrality and clustering coefficient is not effective in terms of finding adequate representation of network, yet when they follow together like in Figure 3, the goal is achieved.

In the next stage was verified how the EV_i values within Sample Quality Monitoring module (SQM) will look like when analysing only similar measures: total degree, in-degree and out-degree centralities. Comparing the results of the different analyses shows that combining measures is possible, but with some caveats. Firstly, it is nearly impossible to combine measures that are divergent in terms of their distributions. Secondly, the last algorithm shows that it is possible to "catch" the distribution function parameters while using lower number of

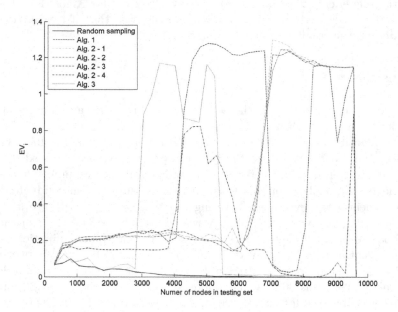

Fig. 2. Random sampling vs. proposed sampling algorithms for all measures combined

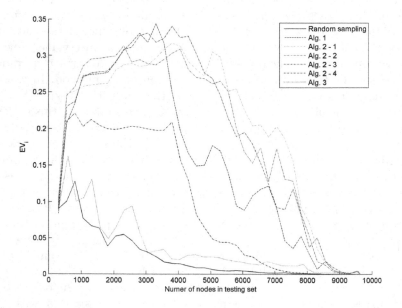

Fig. 3. Random sampling vs. proposed sampling algorithms for measures: in-degree, out-degree, total degree centrality

nodes even while combining measures. With the increase of S_i the EV_i value does not decrease linearly. It leads to conclusion that increasing the number of surveyed users is effective only to some point, after which there will be little benefit while gaining additional surveys. The results regarding random samples after sampling 10% of the network EV_i was reduced for 30% and sampling 20% of the network gives 72% reduction.

5   Summary

The continuous engagement with social network systems, along with the innovation of web technology, allows the movement from a traditional to an online-based environment. That being said, it provides an avenue to consider both theoretical and empirical studies. As technology advances, the demand for new methods also increases, allowing the development of research processes that are more beneficial, effective, and of above par quality. Although adaptive methodologies were the subject of prior research studies, they are not frequently applied to online research. Hence, an alternative solution was presented in this paper as iterative and adaptive in getting people to be surveyed that can be implemented within almost any web survey-capable system. The results of the experiment lead to several interesting conclusions. The study reveals that random sampling does not always deliver an adequate sample to the network measure distributions, specifically when combining multiple performance measures and if the sample is relatively small (up to 5% of nodes). Moreover, this study further reveals that after reaching a specified number of users, the effectiveness of further sampling decreases drastically. The implication of this is that by taking into consideration the entire population, there is a greater probability of getting faster results with smaller distance from the compound network distributions through the application of the proposed method, along with the K-bins algorithm, rather than through the use of typical random sampling. The main limitation of the proposed approach, however, is the availability of measure distributions for the whole network in order for the method to be implemented easily by social network operators. Proposed approach can be an alternative technique to increase the representativeness of online samples while reducing effort and costs by focusing on the most representative users. Although this experimental study demonstrated a new approach in the methodological level, i.e. iterative, getting new respondents independently for distribution regions of network measures, additional algorithms, and methods can be further developed to support the proposed idea. Therefore, future research could investigate other networks and answer the question how to produce a better iterative sampling algorithm leading to faster error reduction using non-equal sized bins or various measures of distance given as examples.

Acknowledgments. The work was partially supported by European Union's Seventh Framework Programme for research, technological development and demonstration under grant agreement no 316097 [ENGINE], by the National Science Centre, the decision no. DEC-2013/09/B/ST6/02317, as well as by the funds for statutory activities of the Faculty of Computer Science and Management of Wrocław University of Technology.

References

1. Abbasi, M.-A., Chai, S.-K., Liu, H., Sagoo, K.: Real-world behavior analysis through a social media lens. In: Yang, S.J., Greenberg, A.M., Endsley, M. (eds.) SBP 2012. LNCS, vol. 7227, pp. 18–26. Springer, Heidelberg (2012)
2. Ahn, Y.Y., Han, S., Kwak, H., Moon, S., Jeong, H.: Analysis of topological characteristics of huge online social networking services. In: Proceedings of the 16th International Conference on World Wide Web, pp. 835–844. ACM (2007)
3. Back, M.D., Stopfer, J.M., Vazire, S., Gaddis, S., Schmukle, S.C., Egloff, B., Gosling, S.D.: Facebook profiles reflect actual personality, not self-idealization. Psychological Science (2010)
4. Couper, M.P., Groves, R.: Moving from prespecified to adaptive survey design. In: Couper, M.P., Traugott, M.W., Lamias, M.J. (eds.) Modernisation of Statistics Production Conference, Stockholm, Sweden (2001). Web Survey Design and Administration, vol. 65(2), pp. 230–253. Public Opinion Quarterly (2009)
5. Deville, J.C., Tillé, Y.: Variance approximation under balanced sampling. Journal of Statistical Planning and Inference **128**(2), 569–591 (2005)
6. Frank, O., Snijders, T.: Estimating the size of hidden populations using snowball sampling. Journal of Official Statistics-Stockholm **10**, 53–53 (1994)
7. Gjoka, M., Kurant, M., Butts, C.T., Markopoulou, A.: A walk in facebook: Uniform sampling of users in online social networks (2009). arXiv preprint arXiv:0906.0060
8. Groves, R.M., Heeringa, S.G.: Responsive design for household surveys: tools for actively controlling survey errors and costs. Journal of the Royal Statistical Society: Series A (Statistics in Society) **169**(3), 439–457 (2006)
9. Heckathorn, D.D.: Extensions of respondent-driven sampling: analyzing continuous variables and controlling for differential recruitment. Sociological Methodology **37**(1), 151–207 (2007)
10. Jankowski, J., Michalski, R., Bródka, P., Kazienko, P., Utz, S.: Knowledge acquisition from social platforms based on network distributions fitting. Computers in Human Behavior **51**(B), 685–693 (2015)
11. Jankowski, J., Michalski, R., Kazienko, P.: The multidimensional study of viral campaigns as branching processes. In: Aberer, K., Flache, A., Jager, W., Liu, L., Tang, J., Guéret, C. (eds.) SocInfo 2012. LNCS, vol. 7710, pp. 462–474. Springer, Heidelberg (2012)
12. Kullback, S., Leibler, R.A.: On information and sufficiency. The Annals of Mathematical Statistics, 79–86 (1951)
13. Lee, S.H., Kim, P.J., Jeong, H.: Statistical properties of sampled networks. Physical Review E **73**(1), 016102 (2006)
14. Maiya, A.S., Berger-Wolf, T.Y.: Online sampling of high centrality individuals in social networks. In: Zaki, M.J., Yu, J.X., Ravindran, B., Pudi, V. (eds.) PAKDD 2010, Part I. LNCS, vol. 6118, pp. 91–98. Springer, Heidelberg (2010)

15. Michalski, R., Jankowski, J., Brodka, P., Kazienko, P.: The same network-different communities? the multidimensional study of groups in the cyberspace. In: 2014 IEEE/ACM International Conference on Advances in Social Networks Analysis and Mining (ASONAM), pp. 864–869. IEEE (2014)
16. Rusmevichientong, P., Pennock, D.M., Lawrence, S., Giles, C.L.: Methods for sampling pages uniformly from the world wide web. In: AAAI Fall Symposium on Using Uncertainty Within Computation, pp. 121–128 (2001)
17. Schouten, B., Calinescu, M., Luiten, A., et al.: Optimizing quality of response through adaptive survey designs. Survey Methodology **39**(1), 29–58 (2013)
18. Singh, J., Howell, R.D., Rhoads, G.K.: Adaptive designs for likert-type data: An approach for implementing marketing surveys. Journal of Marketing Research, 304–321 (1990)
19. Stumpf, M.P., Wiuf, C., May, R.M.: Subnets of scale-free networks are not scale-free: sampling properties of networks. Proceedings of the National Academy of Sciences of the United States of America **102**(12), 4221–4224 (2005)
20. Thelwall, M.: Social networks, gender, and friending: An analysis of myspace member profiles. Journal of the American Society for Information Science and Technology **59**(8), 1321–1330 (2008)
21. Thompson, S., Seber, G.: Adaptive sampling. Wiley series in probability and statistics Show all parts in this series (1996)
22. Thompson, S.K.: Adaptive sampling in graphs. In: Proceedings of the Section on Survey Methods Research, American Statistical Association, pp. 13–22 (1998)
23. Thompson, S.K.: Adaptive network and spatial sampling. Survey Methodology **37**(2), 183–196 (2011)
24. Utz, S., Krämer, N.: The privacy paradox on social network sites revisited: The role of individual characteristics and group norms. Cyberpsychology: Journal of Psychosocial Research on Cyberspace **3**(2), 2 (2009)

Digital Stylometry: Linking Profiles Across Social Networks

Soroush Vosoughi[(✉)], Helen Zhou, and Deb Roy

MIT Media Lab, Cambridge, MA, USA
{soroush,hlzhou}@mit.edu, dkroy@media.mit.edu

Abstract. There is an ever growing number of users with accounts on multiple social media and networking sites. Consequently, there is increasing interest in matching user accounts and profiles across different social networks in order to create aggregate profiles of users. In this paper, we present models for *Digital Stylometry*, which is a method for matching users through *stylometry* inspired techniques. We experimented with linguistic, temporal, and combined temporal-linguistic models for matching user accounts, using standard and novel techniques. Using publicly available data, our best model, a combined temporal-linguistic one, was able to correctly match the accounts of 31% of 5,612 distinct users across Twitter and Facebook.

Keywords: Stylometry · Profile matching · Social networks · Linguistic · Temporal

1 Introduction

Stylometry is defined as, "the statistical analysis of variations in literary style between one writer or genre and another". It is a centuries-old practice, dating back the early Renaissance. It is most often used to attribute authorship to disputed or anonymous documents. Stylometry techniques have also successfully been applied to other, non-linguistic fields, such as paintings and music. The main principles of stylometry were compiled and laid out by the philosopher Wincenty Lutosawski in 1890 in his work "Principes de stylomtrie" [10].

Today, there are millions of users with accounts and profiles on many different social media and networking sites. It is not uncommon for users to have multiple accounts on different social media and networking sites. With so many networking, emailing, and photo sharing sites on the Web, a user often accumulates an abundance of account profiles. There is an increasing focus from the academic and business worlds on aggregating user information across different sites, allowing for the development of more complete user profiles. There currently exist several businesses that focus on this task [13,19,20]. These businesses use the aggregate profiles for advertising, background checks or customer service related tasks. Moreover, profile matching across social networks, can assist the

T.-Y. Liu et al. (Eds.): SocInfo 2015, LNCS 9471, pp. 164–177, 2015.
DOI: 10.1007/978-3-319-27433-1_12

growing field of social media rumor detection [15,21,23,24], since many malicious rumors are spread on different social media platforms by the same people, using different accounts and usernames.

Motivated by traditional stylometry and the growing interest in matching user accounts across Internet services, we created models for *Digital Stylometry*, which fuses traditional stylometry techniques with big-data driven social informatics methods used commonly in analyzing social networks. Our models use linguistic and temporal activity patterns of users on different accounts to match accounts belonging to the same person. We evaluated our models on 11,224 accounts belonging to 5,612 distinct users on two of the largest social media networks, Twitter and Facebook. The only information that was used in our models were the time and the linguistic content of posts by the users. We intentionally did not use any other information, especially the potentially personally identifiable information that was explicitly provided by the user, such as the screen name, birthday or location. This is in accordance with traditional stylometry techniques, since people could misstate, omit, or lie about this information. Also, we wanted to show that there are implicit clues about the identities of users in the content (language) and context (time) of the users' interactions with social networks that can be used to link their accounts across different services.

Other than the obvious technical goal, the purpose of this paper is to shed light on the relative ease with which seemingly innocuous information can be used to track users across social networks, even when signing up on different services using completely different account and profile information (such as name and birthday). This paper is as much of a technical contribution, as it is a warning to users who increasingly share a large part of their private lives on these services.

The rest of this paper is structured as follows. In the next sections we will review related work on linking profiles, followed by a description of our data collection and annotation efforts. After that, we discuss the linguistic, temporal and combined temporal-linguistic models developed for linking user profiles. Finally, we discuss and summarize our findings and contributions and discuss possible paths for future work.

2   Related Work

There are several recent works that attempt to match profiles across different Internet services. Some of these works utilize private user data, while some, like ours, use publicly available data. An example of a work that uses private data is Balduzzi et al. [2]. They use data from the *Friend Finder* system (which includes some private data) provided by various social networks to link users across services. Though one can achieve a relatively high level of success by using private data to link user accounts, we are interested in using only publicly available data for this task. In fact, as mentioned earlier, we do not even consider publicly available information that could explicitly identify a user, such as names, birthdays and locations.

Several methods have been proposed for matching user profiles using public data [7,9,11,14,16,22,26,27]. These works differ from ours in two main aspects. First, in some of these works, the ground truth data is collected by assuming that all profiles that have the same screen name are from the same users [7,9]. This is not a valid assumption. In fact, it has been suggested that close to 20% of accounts with the same screen name in Twitter and Facebook are not matching [6]. Second, almost all of these works use features extracted from the user profiles [9,11,14,16,22,26,27]. Our work, on the other hand, is blind to the profile information and only utilizes users' activity patterns (linguistic and temporal) to match their accounts across different social networks. Using profile information to match accounts is contrary to the best practices of stylometry since it assumes and relies on the honesty, consistency and willingness of the users to explicitly share identifiable information about themselves (such as location).

3   Data Collection and Datasets

For the purposes of this paper, we focused on matching accounts between two of the largest social networks: Twitter and Facebook. In order to proceed with our study, we needed a sizeable (few thousand) number of English speaking users with accounts on both Twitter and Facebook. We also needed to know the precise matching between the Twitter and Facebook accounts for our ground truth.

To that end, we crawled publicly available, English-language, Google Plus accounts using the Google Plus API[1] and scraped links to the users' other social media profiles. (Note that one of the reasons why we used Twitter and Facebook is that they were two of the most common sites linked to on Google Plus). We used a third party social media site (i.e., Google Plus), one that was not used in our analysis to compile our ground truth in order to limit selection bias in our data collection.

We discarded all users who did not link to an account for both Twitter and Facebook and those whose accounts on either of these sites were not public. We then used the APIs of Twitter[2] and Facebook[3] to collect posts made by the users on these sites. We only collected the linguistic content and the date and time at the which the posts were made. For technical and privacy reasons, we did not collect any information from the profile of the users, such as the location, screen name, or birthday.

Our analysis focused on activity of users for one whole year, from February 1st, 2014 to February 1st, 2015. Since we can not reliably model the behaviour patterns of users with scarce data, users with less than 20 posts in that time period on either site were discarded. Overall, we collected a dataset of 5,612 users with each having a Facebook and Twitter account for a total of 11,224 accounts.

---

[1] https://developers.google.com/+/web/api/rest/
[2] https://dev.twitter.com/rest/public
[3] https://developers.facebook.com/docs/public_feed

Figure 1 shows the distribution of the number of posts per user for Twitter and Facebook for our collected dataset. In the figure, the data for the number of posts has been divided into 500 bins. For the Twitter data, each bin corresponds to 80 tweets, while for the Facebook data, it corresponds to 10 posts. Table 1 shows some statistics about the data collected, including the average number of posts per user for each of the sites.

Table 1. Statistics about the number of posts by the users of the 5,612 accounts collected from Twitter and Facebook.

	Twitter	Facebook
Mean	1,535	155
Median	352	54
Maximum	39,891	4,907
Minimum	20	20

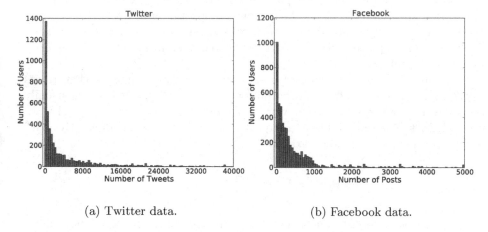

(a) Twitter data.   (b) Facebook data.

Fig. 1. Distribution of number of posts per user for Twitter and Facebook, from our collected dataset.

4   Models

We developed several linguistic, temporal and combined temporal-linguistic models for our task. These models take as input a user, u, from one of the sites (i.e., Twitter or Facebook) and a list of N users from the other service, where one of the N users, $u\prime$, is the same as u. The models then provide a ranking among candidate matches between u and each of the N users. We used two criteria to evaluate our models:

- Accuracy: percentage of cases when a model's top ranked candidate is $u\prime$.
- Average Rank: the average rank of $u\prime$ within the ranked list of candidates generated by a model.

A baseline random choice ranker would have an accuracy of $1/N$, and an average rank of $N/2$ (since $u\prime$ may appear anywhere in the list of N items).

4.1 Linguistic Models

A valuable source of information in matching user accounts, one used in traditional stylometry tasks, is the way in which people use language. A speaker or writer's choice of words depends on many factors, including the rules of grammar, message content and stylistic considerations. There is a great variety of possible ways to compare the language patterns of two people. However, first we need a method for modelling the language of a given user. Below we explain how this is done.

Language Models. Most statistical language models do not attempt to explicitly model the complete language generation process, but rather seek a compact model that adequately explains the observed linguistic data. Probabilistic models of language assign probabilities to word sequences $w_1...w_\ell$, and as such the likelihood of a corpus can be used to fit model parameters as well as characterize model performance.

N-gram language modelling [4,8,12] is an effective technique that treats words as samples drawn from a distribution conditioned on other words, usually the immediately preceding $n-1$ words, in order to capture strong local word dependencies. The probability of a sequence of ℓ words, written compactly as w_1^ℓ is $\Pr(w_1^\ell)$ and can be factored exactly as

$$\Pr(w_1^\ell) = \Pr(w_1) \prod_{i=2}^{\ell} \Pr(w_i|w_1^{i-1})$$

However, parameter estimation in this full model is intractable, as the number of possible word combinations grows exponentially with sequence length. N-gram models address this with the approximation $\tilde{\Pr}(w_i|w_{i-n+1}^{i-1}) \approx \Pr(w_i|w_1^{i-1})$ using only the preceding $n-1$ words for context. A bigram model ($n=2$) uses the preceding word for context, while a unigram model ($n=1$) does not use any context.

For this work, we used unigram models in Python, utilizing some components from NLTK [3]. Probability distributions were calculated using Witten-Bell smoothing [8]. Rather than assigning word w_i the maximum likelihood probability estimate $p_i = \frac{c_i}{N}$, where c_i is the number of observations of word w_i and N is the total number of observed tokens, Witten-Bell smoothing discounts the probability of observed words to $p_i^* = \frac{c_i}{N+T}$ where T is the total number of observed word types. The remaining Z words in the vocabulary that are unobserved (i.e. where $c_i = 0$) are given by $p_i^* = \frac{T}{Z(N+T)}$.

We experimented with two methods for measuring the similarity between n-gram language models. In particular, we tried approaches based on *KL-divergence* and *perplexity* [5]. We also tried two methods that do not rely on n-gram models, *cosine similarity of TF-IDF vectors* [17], as well as our own novel method, called the *confusion model*.

The performance of each method is shown in Table 2. Note that all methods outperform the random baseline in both accuracy and average rank by a great margin. Below we explain each of these metrics.

Table 2. Performance of different linguistic models, tested on 5,612 users (11,224 accounts), sorted by accuracy. Best results are shown bold.

Model	Performance	
	Accuracy	AverageRank
Random Baseline	0.0002	2,806
Perplexity	0.06	1,498
KL-divergence	0.08	2,029
TF-IDF	0.21	999
Confusion	**0.27**	**859**

KL-Divergence. The first metric used for measuring the distance between the language of two user accounts is the Kullback-Leibler (KL) divergence [5] between the unigram probability distribution of the corpus corresponding to the two accounts. The KL-divergence provides an asymmetric measure of dissimilarity between two probability distribution functions p and q and is given by:

$$KL(p||q) = \int p(x) ln \frac{p(x)}{q(x)}$$

We can modify the equation to prove a symmetric distance between distributions:

$$KL_2(p||q) = KL(p||q) + KL(q||p)$$

Perplexity. For this method, the similarity metric is the perplexity [5] of the unigram language model generated from one account, p and evaluated on another account, q. Perplexity is given as:

$$PP(p,q) = 2^{H(p,q)}$$

where $H(p,q)$ is the cross-entropy [5] between distributions of the two accounts p and q. More similar models lead to smaller perplexity. As with KL-divergence, we can make perplexity symmetric:

$$PP_2(p,q) = PP(p,q) + PP(q,p)$$

This method outperformed the *KL-divergence* method in terms of average rank but not accuracy (see Table 2).

TF-IDF. Perhaps the relatively low accuracies of perplexity and KL-divergence measures should not be too surprising. These measures are most sensitive to the variations in frequencies of most common words. For instance, in its most straightforward implementation, the KL-divergence measure would be highly sensitive to the frequency of the word "the". Although this problem might be mitigated by the removal of stop words and applying topic modelling to the texts, we believe that this issue is more nuanced than that.

Different social media (such as Twitter and Facebook) are used by people for different purposes, and thus Twitter and Facebook entries by the same person are likely to be thematically different. So it is likely that straightforward comparison of language models would be inefficient for this task.

One possible solution for this problem is to look at users' language models not in isolation, but in comparison to the languages models of everyone else. In other words, identify features of a particular language model that are characteristic to its corresponding user, and then use these features to estimate similarity between different accounts. This is a task that *Term Frequency-Inverse Document Frequency*, or TF-IDF, combined with *cosine similarity*, can manage.

TF-IDF is a method of converting text into numbers so that it can be represented meaningfully by a vector [17]. TF-IDF is the product of two statistics, *TF* or Term Frequency and *IDF* or Inverse Document Frequency. Term Frequency measures the number of times a term (word) occurs in a document. Since each document will be of different size, we need to normalize the document based on its size. We do this by dividing the Term Frequency by the total number of terms.

TF considers all terms as equally important, however, certain terms that occur too frequently should have little effect (for example, the term *"the"*). And conversely, terms that occur less in a document can be more relevant. Therefore, in order to weigh down the effects of the terms that occur too frequently and weigh up the effects of less frequently occurring terms, an Inverse Document Frequency factor is incorporated which diminishes the weight of terms that occur very frequently in the document set and increases the weight of terms that occur rarely. Generally speaking, the Inverse Document Frequency is a measure of how much information a word provides, that is, whether the term is common or rare across all documents.

Using TF-IDF, we derive a vector from the corpus of each account. We measure the similarity between two accounts using cosine similarity:

$$Similarity(d1, d2) = \frac{d1 \cdot d2}{||d1|| \times ||d2||}$$

Here, $d1 \cdot d2$ is the dot product of two documents, and $||d1|| \times ||d2||$ is the product of the magnitude of the two documents. Using TD-IDF and cosine similarity, we achieved significantly better results than the last two methods, with an accuracy of 0.21 and average rank of 999.

Confusion Model. TF-IDF can be thought of as a heuristic measure of the extent to which different words are characteristic of a user. We came up with a new, theoretically motivated measure of "being characteristic" for words. We considered the following setup:

1. The whole corpus of the $11,224$ Twitter and Facebook accounts was treated as one long string;
2. For each token in the string, we know the user who produced it. Imagine that we removed this information and are now making a guess as to who the user was. This will give us a probability distribution over all users;
3. Now imagine that we are making a number of the following samples: randomly selecting a word from the string, taking the *true user*, TU for this word and a guessed user, GU from correspondent probability distribution. Intuitively, the more often a particular pair, $TU = U_1, GU = U_2$ appear together, the stronger is the similarity between U_1 and U_2;
4. We then use mutual information to measure the strength of association. In this case, it will be the mutual information [5] between random variables, $TU = U_1$ and $GU = U_2$. This mutual information turns out to be proportional to the probabilities of U_1 and U_2 in the dataset, which is undesirable for a similarity measure. To correct for this, we divide it by the probabilities of U_1 and U_2;

We call this model the *confusion model*, as it evaluated the probability that U_1 will be confused for U_2 on the basis of a single word. The expression for the similarity value according to the model is $S \times log(S)$, where S is:

$$S = \sum_w p(w)p(U_1|w)p(U_2|w)$$

Note that if $U_1 = U_2$, the words contributing most to the sum will be ordered by their "degree of being characteristic". The values, $p(w)$ and $p(u|w)$ have to be estimated from the corpus. To do that, we assumed that the corpus was produced using the following auxiliary model:

1. For each token, a user is selected from a set of users by multinomial distribution;
2. A word is selected from a multinomial distribution of words for this user to produce the token.

We used Dirichlet distributions [1] as priors over multinomials. This method outperforms all other methods with an accuracy of 0.27 and average rank of 859.

4.2 Temporal Models

Another valuable source of information in matching user accounts, is the activity patterns of users. A measure of activity is the time and the intensity at which users utilize a social network or media site. All public social networks, including publicly available Twitter and Facebook data, make this information available. Previous research has shown temporal information (and other contextual information, such as spatial information) to be correlated with the linguistic activities of people [18, 25].

We extracted the following discrete temporal features from our corpus: month (12 bins), day of month (31 bins), day of week (7 bins) and hour (24 bins). We chose these features to capture fine to coarse-level temporal patterns of user activity. For example, commuting to work is a recurring pattern linked to a time of day, while paying bills is more closely tied to the day of the month, and vacations are more closely tied to the month.

We treated each of these bins as a word, so that we could use the same methods used in the last section to measure the similarity between the temporal activity patterns of pairs of accounts (this will also help greatly for creating the combined model, explained in the next section). In other word, the 12 bins in month were set to $w_1 \ldots w_{12}$, the 31 bins in day of month to $w_{13} \ldots w_{43}$, the 7 bins in day of week to $w_{44} \ldots w_{50}$, and the 24 bins in time were set to $w_{51} \ldots w_{74}$. Thus, we had a corpus of 74 words.

For example, a post on *Friday, August 5th at 2 AM* would be translated to $\{w_8, w_{17}, w_{48}, w_{53}\}$, corresponding to August, 5th, Friday, 2 AM respectively. Since we are only using unigram models, the order of words does not matter. As with the language models described in the last section, all of the probability distributions were calculated using Witten-Bell smoothing. We used the same four methods as in the last section to create our temporal models.

Table 3 shows the performance of each of these models. Although the performance of the temporal models were not as strong as the linguistic ones, they all vastly outperformed the baseline. Also, note that here as with the linguistic models, the *confusion model* greatly outperformed the other models.

Table 3. Performance of different temporal models, tested on 5,612 users (11,224 accounts), sorted by accuracy. Best results are shown bold.

| | Performance | |
Model	Accuracy	AverageRank
Random Baseline	0.0002	2,806
KL-divergence	0.02	2,491
Perplexity	0.03	2,083
TF-IDF	0.07	1,503
Confusion	**0.10**	**1,458**

4.3 Combined Models

Finally, we created a combined temporal-linguistic model. Since both the linguistic and the temporal models were built using the same framework, it was fairly simple to combine the two models. The combined model was created by merging the linguistic and temporal corpora and vocabularies. (Recall that we treated temporal features as words). We then experimented with the same four methods as in the last two sections to create our combined models.

Table 4 shows the performance of each of these models. Across the board, the combined models outperformed their corresponding linguistic and temporal models, though the difference with the linguistic models were not as great. These results suggest that at some level the temporal and the linguistic "styles" of users provide non-overlapping cues about the identity of said users. Also, note that as with the linguistic and temporal models, our combined *confusion model* outperformed the other combined models.

Another way to evaluate the performance of the different combined models is through the rank-statistics plot. This is shown in Figure 2. The figure shows the distribution of the ranks of the 5, 612 users for different combined models. The x-axis is the rank percentile (divided into bins of 5%), the y-axis is the percentage of the users that fall in each bin. For example, for the *confusion model*, 69% (3880) of the 5, 612 users were correctly linked between Twitter and Facebook when looking at the top 5% (281) of the predictions by the model. From the figure, you can clearly see that the *confusion model* is superior to the other models, with *TF-IDF* a close second. You can also see from the figure that the rank plot for the random baseline is a horizontal line, with each rank percentile bin having 5% of the users (5% because the rank percentiles were divided into bins of 5%).

Table 4. Performance of different combined models, tested on 5,612 users (11,224 accounts), sorted by accuracy. Best results are shown bold.

Model	Performance	
	Accuracy	AverageRank
Random Baseline	0.0002	2,806
KL-divergence	0.11	1,741
Perplexity	0.11	1,303
TF-IDF	0.26	902
Confusion	**0.31**	**745**

5 Evaluation Against Humans

Matching profiles across social networks is a hard task for humans. It is a task on par with detecting plagiarism, something a non-trained person (or sometimes

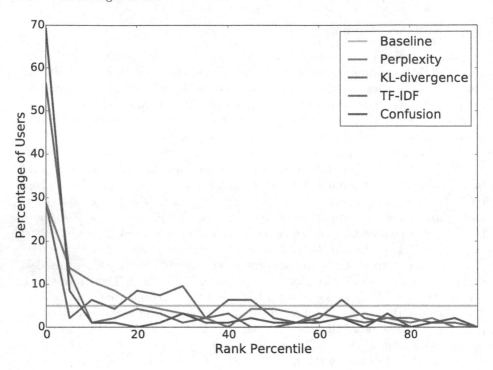

Fig. 2. Rank percentiles of different combined temporal-linguistic models.

even a trained person) cannot easily accomplish. (Hence the need for the development of the field of stylometry in early Renaissance.) Be that as it may, we wanted to evaluate our model against humans to make sure that it is indeed outperforming them.

We designed an experiment to compare the performance of human judges to our best model, the temporal-linguistic confusion model. The task had to be simple enough so that human judges could attempt it with ease. For example, it would have been ludicrous to ask the judges to sort $11,224$ accounts into $5,612$ matching pairs.

Thus, we randomly selected 100 accounts from distinct users from our collection of $11,224$ accounts. A unique list of 10 candidate accounts was created for each of the 100 accounts. Each list contained the correct matching account mixed in with 9 other randomly selected accounts. The judges were then presented with the 100 accounts one at a time and asked to pick the correct matching account from the list of 10 candidate accounts. For simplicity, we did not ask the judges to do any ranking other than picking the one account that they thought matched the original account. We then measured the accuracy of the judges based on how many of the 100 accounts they correctly matched. We had our model do the exact same task with the same dataset. A random baseline model would have a one in ten chance of getting the correct answer, giving it an accuracy of 0.10.

We had a total of 3 English speaking human judges from Amazon Mechanical Turk (which is an tool for crowd-sourcing of human annotation tasks)[4]. For each task, the judges were shown the link to one of the 100 account, and its 10 corresponding candidate account links. The judges were allowed to explore each of the accounts as much as they wanted to make their decision (since all these accounts were public, there were no privacy concerns).

Table 5 shows the performance of each of the three human judges, our model and the random baseline. Since the task is so much simpler than pairing 11, 224 accounts, our combined confusion model had a much greater accuracy than reported in the last section. With an accuracy of 0.86, our model vastly outperformed even the best human judge, at 0.69. Overall, our model beat the average human performance by 0.26 (0.86 to 0.60 respectively) which is a 43% relative (and 26% absolute) improvement.

Table 5. Performance of the three human judges and our best model, the temporal-linguistic confusion model.

Model	Accuracy
Random Baseline	0.10
Human C	0.49
Human B	0.63
Human A	0.69
Average Human	0.60
Combined Confusion	**0.86**

6   Discussion and Conclusions

Motivated by the growing interest in matching user account across different social media and networking sites, in this paper we presented models for *Digital Stylometry*, which is a method for matching users through stylometry inspired techniques. We used temporal and linguistic patterns of users to do the matching.

We experimented with linguistic, temporal, and combined temporal-linguistic models using standard and novel techniques. The methods based on our novel *confusion model* outperformed the more standard ones in all cases. We showed that both temporal and linguistic information are useful for matching users, with the best temporal model performing with an accuracy of .10 and the best linguistic model performing with an accuracy of 0.27. Even though the linguistic models vastly outperformed the temporal models, when combined the temporal-linguistic models outperformed both with an accuracy of 0.31. The improvement in the performance of the combined models suggests that although temporal information is dwarfed by linguistic information, in terms of its contribution to digital stylometry, it nonetheless provides non-overlapping information with the linguistic data.

---

[4] https://www.mturk.com/

Our models were evaluated on 5, 612 users with a total of 11, 224 accounts on Twitter and Facebook combined. In contrast to other works in this area, we did not use any profile information in our matching models. The only information that was used in our models were the time and the linguistic content of posts by the users. This is in accordance with traditional stylometry techniques (since people could lie or misstate this information). Also, we wanted to show that there are implicit clues about the identity of users in the content (language) and context (time) of the users' interactions with social networks that can be used to link their accounts across different services.

In addition to the technical contributions (such as our *confusion model*), we hope that this paper is able to shed light on the relative ease with which seemingly innocuous information can be used to track users across social networks, even when signing up on different services using completely different account and profile information. In the future, we hope to extend this work to other social network sites, and to incorporate more sophisticated techniques, such as topic modelling and opinion mining, into our models.

Acknowledgments. We would like to thank Ivan Sysoev for his help with developing the confusion model. We would also like to thank William Powers for sharing his insights on user privacy in the age of social networks. Finally, thanks to all the human annotators for their help with the evaluation of our models.

References

1. Balakrishnan, N., Nevzorov, V.B.: A primer on statistical distributions. John Wiley & Sons (2004)
2. Balduzzi, M., Platzer, C., Holz, T., Kirda, E., Balzarotti, D., Kruegel, C.: Abusing social networks for automated user profiling. In: Jha, S., Sommer, R., Kreibich, C. (eds.) RAID 2010. LNCS, vol. 6307, pp. 422–441. Springer, Heidelberg (2010)
3. Bird, S., Klein, E., Loper, E.: Natural language processing with Python. O'Reilly Media, Inc. (2009)
4. Charniak, E.: Statistical language learning. MIT press (1996)
5. Cover, T.M., Thomas, J.A.: Elements of information theory. John Wiley & Sons (2012)
6. Goga, O., Loiseau, P., Sommer, R., Teixeira, R., Gummadi, K.P.: On the reliability of profile matching across large online social networks. In: Proceedings of the 21th ACM SIGKDD International Conference on Knowledge Discovery and Data Mining, pp. 1799–1808. ACM (2015)
7. Iofciu, T., Fankhauser, P., Abel, F., Bischoff, K.: Identifying users across social tagging systems. In: ICWSM (2011)
8. Jurafsky, D., Martin, J.H.: Speech and language processing: An introduction to natural language processing, computational linguistics, and speech recognition, 2nd edn. Prentice Hall (2008)
9. Labitzke, S., Taranu, I., Hartenstein, H.: What your friends tell others about you: low cost linkability of social network profiles. In: Proc. 5th International ACM Workshop on Social Network Mining and Analysis, San Diego, CA, USA (2011)

10. Lutosawski, W.: Principes de stylomtrie (1890)
11. Malhotra, A., Totti, L., Meira Jr., W., Kumaraguru, P., Almeida, V.: Studying user footprints in different online social networks. In: Proceedings of the 2012 International Conference on Advances in Social Networks Analysis and Mining (ASONAM 2012), pp. 1065–1070. IEEE Computer Society (2012)
12. Manning, C.D., Schütze, H.: Foundations of statistical natural language processing. MIT press (1999)
13. Peekyou: http://www.peekyou.com/
14. Peled, O., Fire, M., Rokach, L., Elovici, Y.: Entity matching in online social networks. In: 2013 International Conference on Social Computing (SocialCom), pp. 339–344. IEEE (2013)
15. Qazvinian, V., Rosengren, E., Radev, D.R., Mei, Q.: Rumor has it: identifying misinformation in microblogs. In: Proceedings of the Conference on Empirical Methods in Natural Language Processing, pp. 1589–1599. Association for Computational Linguistics (2011)
16. Raad, E., Chbeir, R., Dipanda, A.: User profile matching in social networks. In: 2010 13th International Conference on Network-Based Information Systems (NBiS), pp. 297–304. IEEE (2010)
17. Rajaraman, A., Ullman, J.D.: Mining of massive datasets. Cambridge University Press (2011)
18. Roy, B.C., Vosoughi, S., Roy, D.: Grounding language models in spatiotemporal context. In: Fifteenth Annual Conference of the International Speech Communication Association (2014)
19. Social Intelligence Corp: http://www.socialintel.com/
20. Spokeo: http://www.spokeo.com/
21. Takahashi, T., Igata, N.: Rumor detection on twitter. In: 2012 Joint 6th International Conference on Soft Computing and Intelligent Systems (SCIS) and 13th International Symposium on Advanced Intelligent Systems (ISIS), pp. 452–457. IEEE (2012)
22. Vosecky, J., Hong, D., Shen, V.Y.: User identification across multiple social networks. In: First International Conference on Networked Digital Technologies, NDT 2009, pp. 360–365. IEEE (2009)
23. Vosoughi, S.: Automatic detection and verification of rumors on Twitter. Ph.D. thesis, Massachusetts Institute of Technology (2015)
24. Vosoughi, S., Roy, D.: A human-machine collaborative system for identifying rumors on twitter. In: ICDM Workshop on Event Analytics using Social Media Data (2015)
25. Vosoughi, S., Zhou, H., Roy, D.: Enhanced twitter sentiment classification using contextual information. In: Proceedings of the 6th Workshop on Computational Approaches to Subjectivity, Sentiment and Social Media Analysis, pp. 16–24. Association for Computational Linguistics, Lisboa, September 2015. http://aclweb.org/anthology/W15-2904
26. You, G.W., Hwang, S.W., Nie, Z., Wen, J.R.: Socialsearch: enhancing entity search with social network matching. In: Proceedings of the 14th International Conference on Extending Database Technology, pp. 515–519. ACM (2011)
27. Zafarani, R., Liu, H.: Connecting corresponding identities across communities. In: ICWSM (2009)

Autoregressive Model for Users' Retweeting Profiles

Soniya Rangnani[✉], V. Susheela Devi, and M. Narasimha Murty

Computer Science and Automation Department, Indian Institute of Science,
Bangalore 560012, Karnataka, India
{soniya.rangnani,susheela,mnm}@csa.iisc.ernet.in

Abstract. Social media has become an important means of everyday communication. It is a mechanism for "sharing" and "resharing" of information. While social network platforms provide the means to users for resharing (aka retweeting), it remains unclear what motivates users to retweet. Previous studies have shown that history of users' interaction and properties of the message are good attributes to understand the retweet behaviour of users. They however, do not consider the fact that users do not read all the blogs on their site. This results in shortcomings in the models used. We realised that simple feature engineering is also not enough to address this problem. To mitigate this, we propose an incremental model called Influence Time Content (ITC) model for predicting retweeting behavior by considering the fact that users do not read all their tweets. We have tested the effectiveness of this model by using real data from Twitter. In addition, we also investigate the parameters of the model for different classes of users. We found some interesting distinguishing patterns in retweeting behavior of users. Less active users are more topically motivated for retweeting a message than active users, who on the other hand, are social in nature.

1 Introduction

Microblogging services like "Twitter"[1] have given people a platform to obtain, share, and spread ideas, as and when they wish. Users can post tweets (name given to a message or post on Twitter) through either a web service, or some third-party applications available at all times. Users often give credit for a message to other users (who posted the same message before the former), through a mechanism known as *retweeting*. Retweeting a tweet amounts to sharing it directly with followers (neighbors in a Twitter network), with acknowledgement to its source (*i.e.* the original tweet-er). This retweeting mechanism strongly indicates the direction of information flow. The task of predicting retweet behavior has been used in various applications, such as business intelligence, micro-blog retrieval, etc.

An active user can easily receive more than a thousand tweets a day, making it nearly impossible for a user to keep track of his complete *timeline* (name given

---

[1] http://www.twitter.com/

© Springer International Publishing Switzerland 2015
T.-Y. Liu et al. (Eds.): SocInfo 2015, LNCS 9471, pp. 178–193, 2015.
DOI: 10.1007/978-3-319-27433-1_13

to a user's Twitter page). To help users do so, a starting point is to understand how they "consume" messages and interact with other users in their network. Our approach to answering these questions begins with a characterization of the behavior of Twitter users. Users are found to have different tweeting behavior. We categorize users into two classes – *active* and *less active/ ordinary*. Active users periodically (also, frequently) log into the Twitter system, thus are less likely to skip the tweets appearing on their timelines. A retweet action may not be observed due to two reasons – either the user saw the tweet but didn't retweet, or he didn't see it at all. It is challenging to address this issue, especially for ordinary users. We realised that a simple classification method will not be able to model such user centric time dependent decisions. A model should incorporate timely attentiveness of users. Therefore, the temporal aspect plays an important role in analyzing retweet behavior of users. Considering these challenges, we first study the temporal behavior of users' activities, which directly reflects their availability pertaining to the upcoming post. We propose a model for predicting retweet behavior of users, and following are the key points of our contribution.

1. We design a user-centric, and temporally localized incremental classification model, exploiting time as well as content of tweets.
2. Experiments shown on actual Twitter dataset demonstrates effectiveness of our approach for ordinary as well as active users.
3. We further investigate retweeting behavior of users based on their activity.

The paper is organized as follows. We formally state the problem in Section 2, after which, the related work follows (Section 3). We study the temporal behavior of users in Section 4. The detailed description of our solution is presented in Section 5. Features and setting parameters for the model is explained in Sections 6 and 7 respectively. The form and volume of the Twitter data we use, have been discussed in Section 8 along with the experimental setup. Section 9 presents the results. We finally conclude along with a note on future directions in Section 10.

2   Problem Statement

To model a microblogging social network (e.g., Twitter), we adopt the following terminology. Let $G = (U, E)$ be the social graph with $U = \{u_1, u_2, \ldots, u_N\}$ being the set of N users (nodes) and $E \subseteq U \times U$, the set of directed edges, where the direction is in accordance with the information flow[2]. Thus, $(u, v) \in E$ denotes that user u is followed by v, indicating that information flows from user u to user v.

We introduce functions *follower* $(F_r : U \rightarrow 2^U)$ and *followee* $(F_e : U \rightarrow 2^U)$, defined as follows:

$$F_r(u) = \{v \in U \mid (u, v) \in E\}, \text{ and}$$
$$F_e(v) = \{u \in U \mid (u, v) \in E\}.$$

---

[2] Note that the direction is not based on a usual follower-followee network.

Thus, for a given user u, $F_r(u)$ denotes the set of users who *follow* u, which means a message posted by u will reach all members in $F_r(u)$. Similarly, $F_e(u)$ denotes those whom u follows, that is, a message posted by any member of $F_e(u)$ will reach u.

Let M be a universal pool (a set) of messages that were tweeted. And let T denote a universal timeline. We define functions $tt : M \to T$ and $tu : M \to U$, which give the timestamp $tt(m)$ and tweeter $tu(m)$, respectively, of message $m \in M$ when it was tweeted (*i.e.*, first posted).

For a user v, let $M_v = \{m_1, m_2, \ldots, m_{|M_v|}\}$ denote the messages that have reached v, denoted, *w.l.o.g.*, in order of tweet times, *i.e.*, $tt(m_1) \leq tt(m_2) \leq \ldots \leq tt(m_{|M_v|})$. Formally,

$$M_v = tu^{-1}(F_e(v)) := \bigcup_{u \in F_e(v)} tu^{-1}(u).$$

Let us now define the retweet decision function, δ as follows:

$$\delta : U \times M \to \{0, 1\}$$
$$(u, m) \mapsto \begin{cases} 1, \text{if } u \text{ retweets } m, \\ 0, \text{ otherwise.} \end{cases}$$

We assume here that a user u retweets only messages in M_u, *i.e.*, $\delta(u, m) = 0$ for all $m \notin M_u$. Let the set of messages M be partitioned into train and test sets M_{tr} and M_{te} respectively. Finally, we can define two decision functions $\delta_{tr} := \delta|_{U \times M_{tr}}$ and $\delta_{te} := \delta|_{U \times M_{te}}$. Predict future retweet decisions δ_{te}, given past retweet decisions δ_{tr}.

3   Related Work

Many existing works estimate the number of nodes adopting a piece of information [3,4,14], its lifespan [5], and its popularity [6]. However, very few prediction models in the literature aim to predict whether a node will participate in the diffusion of a particular information, depending on the time at which the tweet was posted. Luo *et al.* [7] consider the contents of tweets and predict the activated nodes (retweeters). However, they have not discussed explicitly about influence among the users. They showed that retweeting behavior mostly depends on diffusion history and tweet properties. Similar observations were seen in the models by Uysal, *et al.* [8] and Chen, *et al.* [9]. These works were not able to make any remarkable observations on the temporal and topical factors affecting retweetability, and were left for further investigation. Their models predict well for active participators, and those who read almost all the tweets posted on their timeline. Another work by Zhang *et al.* [10] combines a function called the influence locality function, and additional features including personal attributes, instantaneity, topic similarity, etc. as predictors of retweeting behavior. One assumption that they make is that, for a given user, retweet decisions by his neighbors is already known, and is used in predicting the user's retweet decision. The assumption is valid but not useful in real world situation unless the retweet decisions of neighbors are predicted first. Also, their analysis is not user-centric. This may lead to poor prediction results for less active users. We compare our results with the work of Luo *et al.* [7] and Zhang, *et al.* [10] as these are the closest to our work.

Recent works have not addressed the problem of incomplete reading behavior of users and online setting of the problem. Analyzing Twitter data, we realized that retweeting is a time-dependent user centric process. We propose a novel way of quantifying important features of information diffusion process, *viz.*, influence, content and time. The conventional way to include *time* as a feature to a classification model [7–10] is not enough to address the problem. We propose a novel online method to model the retweet decisons of users using ensemble of classifiers. Recent works have shown that *content* is a poor predictor of retweet decisions [7,8]. We validate our model on real world Twitter users of all kinds of activity patterns so as to explain retweet decisions of users. We analyzed *content* and found some interesting observations on prediction of retweet actions of users which are not explained in recent works.

4   Activity-Based User Profiling

Bild *et al.*[11] study the aggregate user behavior on Twitter. Retweet rate (retweet counts per user) follows power law distribution. It can be seen that the number of users who send more than ten retweets weekly are much less than those who send one retweet. The whole retweet rate probability mass is distributed around users found less active.

We take three classes of users: *active*, *ordinary* and *passive* users. Active users, on an average, send at least ten tweets in a week. Ordinary users send at least one and at most nine tweets per week. We chose such characterization of users as per the distribution of tweet rates explained in the work of Bild *et al.*[11]. Passive users participate in the tweeting activity very rarely. Active and ordinary users are the only ones found to take part in the information spread in the system. To study and understand retweeting behavior for these different classes of users, it is important to understand how users in Twitter interact with their timeline. A user logs into the system and actively engages in reading, tweeting and retweeting activities. Let us now define two states: ON and OFF, which correspond to the user being online (logged in) and offline respectively. We cannot capture the delimiters of the given states due to unavailability of data.

We analyzed the retweeting activity of a user for 142 days. We define **Retweet fraction** for a time interval as the ratio of the number of retweets by a user to the number of tweets posted by a subset of his followees during that time interval.

If we solve our problem as a binary classification problem with a set of features of tweets and activity as predictors, and the labels as retweet (positive class) or no retweet (negative class). There could be two reasons for not retweeting a tweet–either the user has not seen the tweet because of being in the OFF state, or he has read the tweet, but did not retweet it on purpose. The latter case is an active decision not to retweet, and hence belongs to the no retweet class, while the former case is an incidental and involuntary no-retweet action, and hence constitutes noise. If we use such noisy instances, a highly imbalanced dataset gets generated. In such cases, for ordinary users, classification performance may result in high recall but low precision values.

Retweet fraction values of some users are shown in Fig. 1. Temporal fluctuations can be extremely drastic, as shown in Figure 2. It can be observed that the patterns of retweeting behavior between two consecutive months differ significantly. Lesser variance is observed in the average retweet fraction values for smaller intervals. We call it **temporal locality**. We can estimate the activity level of the current day using the activity level of the previous days. It can be clearly seen that, generally, variation in retweet fraction on consecutive days is just 0.2 provided the user is in the same (ON/OFF) state.

Hourly State Variation of a User. We assume a hourly time resolution for modeling a user. It is important to consider at a given hour, how receptive (to retweet) a user generally is. We assume that an individual's activity level (total number of posts) in a particular hour follows a Poisson distribution. Let $x_{dh}^{(u)} \in \mathbb{N}$ be the number of tweets (including retweets) posted by user u on day d and during hour $h \in \{1, 2, \ldots, 24\}$. For an hourly analysis over multiple days, we consider, for each hour h, the sequence $x_{1h}^{(u)}, x_{2h}^{(u)}, \ldots, x_{nh}^{(u)}$ of number of tweets posted by user u for n consecutive days. The maximum likelihood estimate of the activity level of the user u at hour h may be represented as

$$\lambda_{uh} = \frac{1}{n} \sum_{d=1}^{n} x_{dh}^{(u)} \tag{1}$$

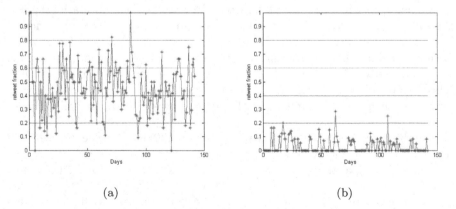

(a)   (b)

Fig. 1. (a) a specific user who has at least moderate daywise retweet activity (active user), (b) a specific user has both zero as well as moderate daywise activity (ordinary user)

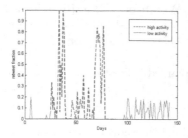

Fig. 2. Daywise retweet behavior showing large temporal fluctuation

It is observed that the activity level of the users is much less during night hours, as compared to day time. For an ordinary user, it is observed to have peak values for one or two hours on a particular day. It depicts the ON/OFF state of the user. When a user logs into the system, he sees only latest tweets posted by his followees. The tweets are arranged chronologically, with the newly posted tweets appearing before older ones. A user takes action on the newest tweets first. He may or may not read all previously unread tweets. We assume that users log in during their respective peak hours, remain active for some time, and then their activity dies off thereafter.

5   Retweet Decision Model

Suppose user u follows user v, and v posts a tweet m through a microblogging service, we want to model the retweeting behavior of user u *i.e.*, predicting whether u will retweet the tweet m or not.

We assume that retweeting behavior of a user is determined by the following three factors:

- **Pairwise Influence:** It can be seen as the transmission rate of information from a user to another user.
- **Content:** Two different pieces of information will propagate differently depending on the respective fields of interest of different users in the network.
- **Time:** It is important to consider how receptive a user is, to retweet at a particular time instant.

Due to the three components–*influence, time,* and *content*–we call our model, the ITC model. The three main components (I, T, and C) are the predictors of the model with binary class labels {retweet, no retweet}. The classification model is expected to be:

- **Incremental:** Retweet behavior involves a pattern of actions extending over time. Due to the structure of decision making problems, an incremental approach is meaningful.

- **Temporally Local Weighted Classification:** Decision of retweeting depends significantly on temporal behavior of the user. It is observed that decisions of a user can be explained better by his activity in the recent past; we call it "temporally local".
- **Noise Sensitive** (explained in Section 4)**:** The work should also model the behavior of a user who is not logged in, and his obvious inaction–which is mistaken as deliberate–should be detected as noise.

We have already explained the shortcomings of using the conventional classification model for solving the problem (Section 4). As a consequence, with all the training examples and prior knowledge, a unique best hypothesis may exist, but will be very complex. We design an ensemble of classifiers to model the problem.

5.1   Learning the Model (Training)

Let $M_u^{(d)} \subseteq M_u$ denote the set of messages sent by followees of user u during day d. Let $y_u^{(d)}(m) := \delta(u, m)$ be the binary label corresponding to a message $m \in M_u^{(d)}$ posted on day d.

We classify $M_u^{(d)}$ into two classes, where the positive class corresponds to the message being retweeted by user u. A message is classified as a negative instance if it is not retweeted. We have extracted three features $\mathbf{x}(m) := (x_i(m), x_t(m), x_c(m))$ of message m corresponding to the retweet action of a user, including *influence, time* and *content*. We train a classification model h_{ud} on the tweet instances posted by user u on day d. We have used logistic regression for the day-wise classification model.

5.2   Testing the Model (Prediction)

For predicting the tweets generated on day d, we design an ensemble of the day-wise classifiers h_{ud}. Let $m \in M_u^{(d)}$ and $\mathbf{x}(m) := (x_i(m), x_t(m), x_c(m))$ be the feature vector of message m. We predict the class for $\mathbf{x}(m)$ (written $\mathbf{x}$ for brevity) as follows:

$$g_{ud}(\mathbf{x}) := \sum_{d'=d-k}^{d-1} \alpha_{d-d'} \cdot h_{ud'}(\mathbf{x})$$

$$\delta(u, m) = \begin{cases} 0, & \text{if } g_{ud}(\mathbf{x}) \leq 0.5, \\ 1, & \text{if } g_{ud}(\mathbf{x}) > 0.5. \end{cases}$$

where $1 \leq k < d$, and $\alpha_1, \alpha_2, \ldots, \alpha_k \in [0, 1]$ are the weights of the classifiers. The choice of parameters k and $\alpha_1, \alpha_2, \ldots, \alpha_k$ are discussed later in this section.

5.3   Why Daywise Classifiers?

We predict the spread of a message one day before it is posted. It sufficiently serves real world practicality such as implementation of various applications like personalised tweet recommendation, trend analysis, and so on. Users have different interests at different times. We have evidence of temporal locality in retweet behavior of users (See Section 4). Increasing the duration from daywise prediction to more than one day, may lead to poor performance in classification. On the other hand, decreasing the duration to less than a day, may effect the implementation of applications in practical sense.

6   Features

6.1   Pairwise Influence

Influence can be seen as transmission rate of information on that edge. In practice, we only observe log of actions of users or cascade data originating from users. Influence among them is not available. We use three sets of features: *structural*, *user-profile based* and *user-pair based*. These are explained in detail below.

Structural Features: Many topological properties of nodes of a social graph explain the information spread. These include *eigenvector centrality*, *in-degree*, *out-degree*, and *clustering coefficient* of user nodes.

User-Profile Based Features: These features are collected from a user's profile information and observing tweet/retweet activity in the training period. *Number of tweets, retweet ratio, social pressure* and *number of tweets which either mention or retweet the user* comprise this set of features.

Let's define $R_{tr}(u) := \{m \in M_u \mid \delta(u, m) = 1\}$ as the set of retweets by user u during the training period, and $rt(u, m)$ represents the timestamp of a retweet $m \in R_{tr}(u)$.

Retweet ratio $\rho(u)$ of user u is the ratio of number of retweets by the user and total number of his original tweets in the training period.

$$\rho(u) := \frac{|R_{tr}(u)|}{|tu^{-1}(u) \cap M^{(tr)}|}$$

Social pressure [12] $\psi(u)$ for user u can be estimated as the expected number of active neighbors of u right before u himself gets active. It can be formulated as follows:

$$\psi(u) = \frac{\sum\limits_{m \in R_{tr}(u)} |\{v \in F_e(u) : rt(v, m) < rt(u, m)\}|}{|R_{tr}(u)|}$$

User-Pair Based Features: *Homophily* and *social influence* are two aspects being captured in this set of features. The phenomenon of people tending to communicate with those similar to them in socially significant ways is called homophily. It is found to have an important implication in information flow in social networks. Homophily is captured by the number of common URLs and common hashtags found in tweets of pairs of users, and by the number of common followees. For an edge from user u to user v, *number of tweets of user u retweeted by user v* and *number of times user v has mentioned user u* are the features characterizing social influence.

An edge (u, v) denotes the spread of message from user u to user v. We classify edges of the social graph in two classes. Positive class corresponds to the edges which have participated in at least one diffusion sequence in the training data. An edge is classified as a negative instance, if it has not participated in any of the diffusion. We have extracted twenty one features (first two sets of features for the users u and v respectively, and the third set for the pair (u,v)), and Logistic Regression was applied to the data. Influence can be seen as the confidence score of an edge to estimate participation in diffusion. We define the supervised classification task as follows.

Let $PLR_{v,u}$ be the estimate of Logistic regression for a pair of users (u, v) whose feature vector is X.

$$PLR_{v,u} = P(Y = 1 \mid X).$$

6.2   Content

Content of tweets shared by a user highly signifies the topics he follows. It is intuitive to assume that a user tends to share those tweets which are similar in content to the past ones. A content-based similarity score $K_\sigma(u, m)$ between the words in the history of user u and words of tweet m can be formulated to capture u's willingness to retweet (diffuse the content). Let $\kappa(m)$ denote the content (set of distinct words) of tweet m and $vocab_u$ denote distinct words in the posts of user u.

$$K_\sigma(u, m) = \frac{|\kappa(m) \cap vocab_u|}{|\kappa(m)|}$$

To model thematic interest among a connected pair of users (u, v) and tweet m, we formulate a combined content factor $K_f(u, v, m)$ as

$$K_f(u, v, m) = K_\sigma(u, m) \cdot K_\sigma(v, m)$$

6.3   Time

As discussed before, we assume an hourly time resolution. We analyze the activeness of a user u per hour by assuming a Poisson distribution on the number of tweets sent in the hour h and fitting the parameter λ_{uh} using Equation 1. The receptivity of a user for a piece of information is calculated as *average activation time* $(\overline{\tau})$ for user u. It is the average number of hours a user takes to retweet a tweet. Let h_m be the hour of the day at which tweet m is posted. The temporal aspect is captured for the tweet m and the user u using $\lambda_{u(h_m + \overline{\tau})}$.

7   Parameters k and α_i

Cross-Validation: The choice of k can be empirically fixed using cross-validation. The parameters α_i incorporate the temporal locality in the model (Section 4). Intuitively, the decision for a tweet on day d depends on the activity in the past: dependence on recent past being more than that on the far past. The weight of the classifiers could be accordingly chosen, i.e., weights decreasing as we move further back in time. This entails the choice of using a decreasing function for the values of α_i, i.e., $\alpha_t \leq \alpha_{t+1}$ for $t = 1, 2, \ldots, k-1$.

In accordance with the above discussion, we choose the following two functions (with parameter k fixed using cross-validation) for analysis:

$$\alpha_t = \frac{1}{k} \qquad \text{for } t = 1, 2, \ldots, k. \tag{2}$$

$$\alpha_t = 1 - \frac{1}{1 + e^{-t}} \qquad \text{for } t = 1, 2, \ldots, k. \tag{3}$$

Autoregressive Model: We have introduced the term *retweet fraction* previously in Section 4. We can call *retweet fraction* as 'activity level' of a user as it specifies a level of tweeting activity of a user on that day. Let $r_1, r_2, \ldots, r_n$ be the series of active levels of a user over n days. It is a time-varying process. Current activity level can be predicted by past observations in user's activity. A linear time-series model for representing this response process r_t can be of the form

$$r_t = c + \sum_{t'=1}^{k} (\gamma_{t'} * r_{t-t'}) + e_t \tag{4}$$

where parameters γ_i are the parameters of the model, c is a constant and e_t is white noise (uncorrelated process with mean zero).

A model with a similar nature is the autoregressive (AR) [13] model. It is found to be flexible in representing a wide range of time series pattern. Here, the observation r_t is found to exhibit linear association between lagged observations. An AR process that depends on k past observations is called an AR model of degree k, denoted by AR(k). We apply maximum likelihood estimation to find parameters of the observed univariate time series. Let $\gamma_1', \gamma_1', \ldots, \gamma_p'$ be the estimated parameters for the model AR(p) for $r_1, r_2, \ldots, r_n$ series. We regenerate the series r_i' with initial values as $r_1, r_2 \ldots, r_p$ and estimated paramters γ_i' using Equation 4. We calculate root mean square error e_p the estimated series r_i' and actual series r_i as

$$e_p = \frac{1}{n} \sqrt{\sum_{i=1}^{n} (r_i' - r_i)^2}$$

We fit AR(p) for a range of values of p and calculate calculate RMS errors e_p. We choose k such that it minimizes over the RMS values.

Data: v, $M_v^{(tr)}$, $Y_v^{(tr)}$

Result: $(PLR_{vu})_{u \in F_e(v)}$, $(\lambda_{vh})_{h=1}^{24}$, $(k,\ (\alpha_i)_{i=1}^k)$

Step 1: Calculate Influence component for v and $F_e(v)$ using $M_v^{(tr)}$ and $Y_v^{(tr)}$ as $(PLR_{vu})_{u \in F_e(v)}$.

Step 2: Estimate hourly parameters for v as $(\lambda_{vh})_{h=1}^{24}$

Step 3: Fit AR model parameters k and $(\alpha_i)_{i=1}^k$

Algorithm 1: Training for user v

$$k = \underset{p}{\operatorname{argmin}}\ e_p^2 \tag{5}$$

The normalized absolute values indicate the importance of $(t-i)^{\text{th}}$ element in the estimation of the tth element. We normalize the values of parameters γ_i of the model $AR(k)$ which is obtained by Equation 5. These parameters appear to be the appropriate values for modeling α_i as

$$\alpha_t = \gamma_t \text{ for } t = 1, 2, .., k \tag{6}$$

Algorithms: Algorithm 1 shows training of the model for user v. Algorithm 2 shows steps for predicting messages posted during d^{th} day for a user.

8   Experimental Setup

8.1   Dataset

To analyze the retweeting behavior, we collect data from a Twitter graph using web service calls of the Twitter REST API[3]. To ensure availability of diffusion sequences in the data, we select five seed nodes (*BarackObama, Surgeon_General, youtube, kohl_nick, iamsrk*), collect hundred recent tweets by them, and their retweeters (users who have retweeted the tweets). Then, 2500 users from each seed nodes' retweeters are selected. These users, called *central users*, are studied and used for further experiments.

The seed nodes post tweets about a variety of topics. Leaders like Barack Obama post their opinions on the current events around the globe. Surgeon General account writes about the news about White House and govenment. The account *kohl_nick* is a ordinary user. Youtube is content centric account. The account *iamsrk* is personal account of Indian actor Shahrukh Khan. He posts his daily activities and communicates with peer accounts on Twitter. A wide range of audience users constitute the followers of these seed nodes. This may help us

---

[3] https://dev.twitter.com/rest/

Data: v, $M_v^{(d)}$, $(PLR_{vu})_{u \in Fe(v)}$, $(\lambda_{vh})_{h=1}^{24}$, k,

$(\alpha_i)_{i=1}^{k}$, $(M_v^{(i)})_{i=d-k}^{d-1}$, $(Y_v^{(i)})_{i=d-k}^{d-1}$, $\overline{\tau}$

Result: $\delta(v,m)$, $\forall m \in M_v^{(d)}$

Step 1: Generate ITC components for messages:

for $j = 0$ *to* k **do**

    for $m \in M_v^{(d-j)}$ **do**

        $x_i(m) = PLR_{v,\ tu(m)}$

        $x_t(m) = \lambda_{v,\ (tt(m).hour+\overline{\tau})}$

        $x_c(m) = K_f(tu(m), v, m)$

        $\mathbf{x}(m) := (x_i(m), x_t(m), x_c(m))$

    end

end

Step 2: Train daywise classifiers:

for $j = 1$ *to* k **do**

    $X_v^{(j)} := \left\{ \mathbf{x}(m) \,\middle|\, m \in M_v^{(d-j)} \right\}$

    Train classifier h_{vj} using $X_v^{(j)}$ and $Y_v^{(d-j)}$

end

Step 3: Predict for messages $M_v^{(d)}$

for $m \in M_v^{(d)}$ **do**

    $\delta(v,m) = g\left(\mathbf{x}^{(j)}(m) \right)$

end

Algorithm 2: Prediction

to understand the spread of a variety of tweets written in different styles with different contents.

After crawling the profile information of the users–their name, gender, verification status, number of followers, number of followees, etc.–we finally collected a set of latest 500 messages (both tweets and retweets sent). For analyzing the retweeting behavior, we first restored the tweets posted by the followees of central nodes sorted by their posting time from more to less recently posted. We calculate 'retweet fraction' for all users using time of posts of retweets.

8.2   Baseline Methods

To examine the efficiency of our model, we evaluate the following methods on the constructed dataset

1. **SVMRank:** Luo, *et al.* [7] used the family of features explaining microblog content, retweet history, follower status, followers active time, and followers' interests and incorporated it in learning-to-rank framework to predict retweet decisions.

2. **LRC-BQ:** Zhang *et al.* [10] introduced a new notion 'social influence local-ity' to explain retweet decisions. It combines the influence locality func-tion with additional features including personal attributes, instantaneity, and topic similarity.
3. **ITC-F1:** This represents our model, where k is fixed by cross-validation and α_i are modeled using Equation 2
4. **ITC-F2:** Same as **ICT-F1**, except that α_i are fixed using Equation 3
5. **ITC-AR:** Parameters are fixed using the AR model, and all the features used.
6. **IC-AR:** Same as **ITC-AR**, but without the temporal aspect.
7. **CT-AR:** Same as **ITC-AR**, but without the parameter, "Influence".
8. **IT-AR:** Same as **ITC-AR**, but without incorporating content properties.

9   Experimental Results and Discussion

Prediction Results: Our model gives improved performance over all baseline models. Table 1 shows precision, recall and F1-measure of basline methods. The models ITC-F1, ITC-F2 and ITC-AR have performed significantly better than other classifiers. There are high fraction of ordinary users in central nodes. All the methods perform well for active users. However ordinary user prediction is better in our model. As LRC-BQ is not user centric (more prone to incomplete reading behavior of users), it has shown relatively poorer results than that men-tioned in their paper. We also conducted leave-one-feature-out for classification experiments (IC-AR, TC-AR and IT-AR models). It gives best results for IT-AR model for all the classes of users. Influence and Temporal aspect could be seen to be important predictors of retweet decisions. The models IC-AR and TC-AR are found to affect the prediction results on ordinary users. We briefly investigate the importance of features of the model further. Our model ITC-AR has performed significantly better than all other models.

Parameter k of AR Model: Optimal k value for AR model shows some distinguishing characteristics of active and ordinary users. Figure 3(a) shows the optimal k values vs the corresponding fraction of active users and ordinary users. The active users are found to have low k values. It is intuitively true as

Table 1. Comparison of prediction results of various models

	all users			active users			ordinary users		
	Precision	Recall	F1-measure	Precision	Recall	F1-measure	Precision	Recall	F1-measure
LRC-BQ	0.1039	**0.7348**	0.1820	0.4331	**0.8211**	0.5670	0.1241	**0.6655**	0.2091
SVMRank	0.1951	0.0975	0.1322	0.3841	0.1514	0.2171	0.1489	0.2057	0.1727
ITC-F1	0.5133	0.4237	0.4642	0.5823	0.4798	0.5261	0.4216	0.4238	0.4226
ITC-F2	0.5833	0.4607	0.6151	0.6023	0.4928	0.5450	0.4166	0.4238	0.4201
ITC-AR	**0.6719**	0.6601	**0.6659**	0.6825	0.7045	**0.6933**	0.6162	0.5865	**0.6009**
IC-AR	0.5214	0.5649	0.5422	0.6208	0.6794	0.6487	0.3995	0.4275	0.4130
TC-AR	0.4837	0.5414	0.5109	0.5747	0.6675	0.6176	0.3651	0.3902	0.3772
IT-AR	0.5894	0.6171	0.6029	0.6825	0.7045	0.6933	0.4811	0.5122	0.4961

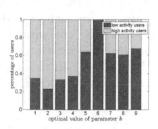
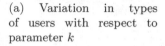

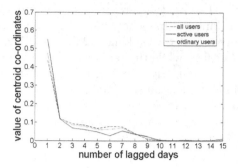

(a) Variation in types of users with respect to parameter k

(b) Plot of α_i vs. number of lagged days

Fig. 3. Inferences on parameters of AR model

a few recent days can be good estimators for the activity of a specific day for the active users with less noisy data. The optimal value of k is found to be high for the ordinary users as the ordinary users have many consecutive days with no activity (OFF state). To capture the interstate pattern of behavior, the value of k is justified to be high.

Nature of Parameters α_i: We analyzed the nature of parameter α_i estimated by AR model for all the users, chosen for the model ITC-AR. We calculate centroid of α_i. We append zeros at the tail of α_i vectors for consistency of length. Let $\alpha_{u1}, \alpha_{u2}, \ldots, \alpha_{uk}$ be alpha parameters of the classifiers of user u. We calculate centroid as

$$\overline{\alpha_i} = \frac{\sum_{u \in U} \alpha_{ui}}{|U|} \qquad for \; i = 1, 2, \ldots, k$$

The centroid is shown in Figure 3(b). It can been that the weight of the classifiers has decreasing values with respect to difference between current and the prior day. It is a decreasing function for the values of α_i. We find Pearson correlation coefficient for all the users and the centroid. It is a measure of rank correlation, i.e., the similarity score for the ordering of elements of two data vectors when ranked by each of the quantities. The mean of pearson correlation coefficients of parameters α_{ui} of every user and the centroid is calculated and the value is 0.73. Almost all the users have the same decreasing nature of α_{ui}, explaining collective nature of users.

Influence, Time and Content Features: We have a series of daywise classifiers trained by Logistic Regression. The absolute normalized values of coefficients of features generated by classifiers indicate their importance as predictors. We rank these values of co-efficients of all the daywise classifiers generated during training. For instance, [2 1 3] is a tuple showing ([I T C]) ranks of influence, time and content as second, first and third respectively (i.e. time is more significant than influence and influence contributes more that content as predictors).

Figure 4 shows the percentage of classifiers vs a specific ranking for co-efficents of ITC model (Influence, time and content respectively) exhibiting it. It can be observed that the ranking [1 2 2] and [2 1 2] are mostly found in the coefficients of classifiers while the ranking [2 2 1] with content feature at the highest rank, is more frequently found in the classifiers by ordinary users. By the nature of coefficients, we can conclude that the ordinary users are topically motivated for retweeting a tweet more often than an active user. Decisions of all the users are found to depend on social aspects (Influence) and temporal aspect significantly too.

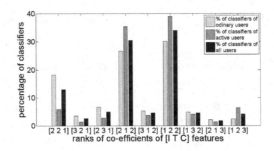

Fig. 4. Importance of features [I T C] for predicting retweet decision

10   Conclusion

Most of the users of the microblogging sites like Twitter do not exhaustively read all the messages. Previous works predict poorly for the users who have this incomplete reading behavior. We realised that a simple classification method will not be able to model such user centric time dependent decisions. Simple feature engineering is also not enough to address this problem. A model should incorporate timely attentiveness of users. We define temporal behavior of users as the 'retweet fraction'. It is the fraction of tweets a user retweeted. We analysed the behavior of users by looking at the daywise retweet fraction values. It follows an autoregressive model. We suggest an incremental model using temporal locality based decisions that are observed in activity pattern of the users. Our model also considers the online nature of problem. We evaluate our model on a real world Twitter data and show that it improves overall baseline prediction models for the users with different patterns of activity. Influence is strongly correlated to retweet decisions of users. The classifiers corresponding to ordinary users show correlation between content similarity and retweet decisions. Ordinary users are more topically motivated than active users. We analysed parameter k. It can be infered that only recent data is useful for predicting retweet decisions. It confirms our analysis of temporally local retweet behavior of users. It will be interesting to check how the model works after incorporating the virality of message content and influence from activated neighbors.

References

1. Yang, J., Counts, S.: Predicting the speed, scale, and range of information diffusion in twitter. In: ICWSM, vol. 10, pp. 355–358 (2010)
2. Suh, B., Hong, L., Pirolli, P., Chi, E.H.: Want to be retweeted? large scale analytics on factors impacting retweet in twitter network. In: 2010 IEEE Second International Conference on Social Computing (Socialcom), pp. 177–184 IEEE (2010)
3. Kupavskii, A., Ostroumova, L., Umnov, A., Usachev, S., Serdyukov, P., Gusev, G., Kustarev, A.: Prediction of retweet cascade size over time. In: Proceedings of the 21st ACM International Conference on Information and Knowledge Management, pp. 2335–2338. ACM (2012)
4. Petrovic, S., Osborne, M., Lavrenko, V.: Rt to win! predicting message propagation in twitter. In: ICWSM (2011)
5. Kong, S., Feng, L., Sun, G., Luo, K.: Predicting lifespans of popular tweets in microblog. In: Proceedings of the 35th International ACM SIGIR Conference on Research and Development in Information Retrieval, pp. 1129–1130. ACM (2012)
6. Hong, L., Dan, O., Davison, B.D.: Predicting popular messages in twitter. In: Proceedings of the 20th International Conference Companion on World Wide Web, pp. 57–58. ACM (2011)
7. Luo, Z., Osborne, M., Tang, J., Wang, T.: Who will retweet me?: finding retweeters in twitter. In: Proceedings of the 36th International ACM SIGIR Conference on Research and Development in Information Retrieval, pp. 869–872. ACM (2013)
8. Uysal, I., Croft, W.B.: User oriented tweet ranking: a filtering approach to microblogs. In: Proceedings of the 20th ACM International Conference on Information and Knowledge Management, pp. 2261–2264. ACM (2011)
9. Chen, K., Chen, T., Zheng, G., Jin, O., Yao, E., Yu, Y.: Collaborative personalized tweet recommendation. In: Proceedings of the 35th International ACM SIGIR Conference on Research and Development in Information Retrieval, pp. 661–670. ACM (2012)
10. Zhang, J., Liu, B., Tang, J., Chen, T., Li, J.: Social influence locality for modeling retweeting behaviors. In: Proceedings of the Twenty-Third International Joint Conference on Artificial Intelligence, pp. 2761–2767. AAAI Press (2013)
11. Bild, D.R., Liu, Y., Dick, R.P., Mao, Z.M., Wallach, D.S.: Aggregate characterization of user behavior in twitter and analysis of the retweet graph 2014. arXiv preprint arXiv:1402.2671
12. Lagnier, C., Denoyer, L., Gaussier, E., Gallinari, P.: Predicting information diffusion in social networks using content and user's profiles. In: Serdyukov, P., Braslavski, P., Kuznetsov, S.O., Kamps, J., Rüger, S., Agichtein, E., Segalovich, I., Yilmaz, E. (eds.) ECIR 2013. LNCS, vol. 7814, pp. 74–85. Springer, Heidelberg (2013)
13. Box, J., Jenkins, G.M., Reinsel, G.C.: Time series analysis, forecasting and control (1994)
14. Luo, Z., Wang, Y., Wu, X.: Predicting retweeting behavior based on autoregressive moving average model. In: Wang, X.S., Cruz, I., Delis, A., Huang, G. (eds.) WISE 2012. LNCS, vol. 7651, pp. 777–782. Springer, Heidelberg (2012)

Choosing the Right Home Location Definition Method for the Given Dataset

Iva Bojic[1,2]($\boxtimes$), Emanuele Massaro[1], Alexander Belyi[2],
Stanislav Sobolevsky[1], and Carlo Ratti[1]

[1] Senseable City Lab, Massachusetts Institute of Technology, Cambridge, MA, USA
{ivabojic,emassaro,stanly,ratti}@mit.edu
[2] SMART Centre, Singapore, Singapore
alex.bely@smart.mit.edu

Abstract. Ever since first mobile phones equipped with Global Position System (GPS) came to the market, knowing the exact user location has become a holy grail of almost every service that lives in the digital world. Starting with the idea of location based services, nowadays it is not only important to know where users are in real time, but also to be able predict where they will be in future. Moreover, it is not enough to know user location in form of latitude longitude coordinates provided by GPS devices, but also to give a place its meaning (i.e., semantically label it), in particular detecting the most probable home location for the given user. The aim of this paper is to provide novel insights on differences among the ways how different types of human digital trails represent the actual mobility patterns and therefore the differences between the approaches interpreting those trails for inferring said patterns. Namely, with the emergence of different digital sources that provide information about user mobility, it is of vital importance to fully understand that not all of them capture exactly the same picture. With that being said, in this paper we start from an example showing how human mobility patterns described by means of radius of gyration are different for Flickr social network and dataset of bank card transactions. Rather than capturing human movements closer to their homes, Flickr more often reveals people travel mode. Consequently, home location inferring methods used in both cases cannot be the same. We consider several methods for home location definition known from the literature and demonstrate that although for bank card transactions they provide highly consistent results, home location definition detection methods applied to Flickr dataset happen to be way more sensitive to the method selected, stressing the paramount importance of adjusting the method to the specific dataset being used.

1 Introduction

A high portion of our daily lives is also happening in digital world. People can express themselves using microblogging platforms such as Twitter; stay in a contact with their family and friends using Online Social Networks (OSNs) such as Facebook; check in at different locations using Location Based Social Networks (LBSN) such as Foursquare or share their photos from vacations using Flickr.

© Springer International Publishing Switzerland 2015
T.-Y. Liu et al. (Eds.): SocInfo 2015, LNCS 9471, pp. 194–208, 2015.
DOI: 10.1007/978-3-319-27433-1_14

All of those online platforms offer people to create their own digital content that is very often public and consequently to a certain extent available not only for scholars in their scientific studies, but also for city policy makers on urban scales or for marketing purposes world-wise. With mobile phones equipped with a highly precise Global Position System (GPS) modules becoming omnipresent and the increase in sharing our location within different mobile apps, world has become richer for an enormous source of data on human mobility. Digital data not only does play a crucial role in research on human behavior and human mobility [6,12,17,21], but also opens new unprecedented opportunities for supporting planning decisions, such as regional delineation [30,35,38] or land use classification [14,28,34,36,37], as well as for instance paving a way towards a smarter urban transportation optimization [31].

Although the amount of publicly available data is rising, a lot of information is still missing and has to be inferred from the data that is shared. In that sense, a lot of related work is dedicated towards building different *location to profile* frameworks using spatial, temporal and location knowledge for example to infer demographic attributes of users sharing their location [42] or labeling important places in somebody's life [20]. In this paper we focus on predicting user home locations for different online platforms, as well as for everyday traces that people leave in digital world outside OSNs (e.g., bank card transactions, call records). Namely, the exact locations where people live or work are very often ambiguous and are missing along with other pieces of human mobility puzzle. As mentioned in [23] only 16% of Twitter users reported their city level location. Moreover, it was shown that 34% people reported invalid information or they even put sarcastic comments [16]. Inferring home location is just one special case of the process in which semantic place labels such as home, work or school are given to geographic locations where people spend their time. Information on home location is not only important for investigating mobility patterns, but also for delivering localized news, recommending friends or serving targeted ads [22]. It can be also helpful when designing more personalized urban environments (e.g., pollution management, transportation systems) or modeling outbreaks and disease propagation [40].

The contribution of our work is twofold. First, we show how the basic mobility patterns quantified by means of the radius of gyration are different in cases of digital trails of bank card transactions and Flickr. Second, we show how different nature of the datasets affects the applicability of various methods for inferring user home locations by comparing the outcomes of five simple home definition methods applied to the two aforementioned datasets. The rest of the paper is structured as follows. Section 2 summarizes related work on methods for determining home locations, while Section 3 introduces two datasets used in this work: Flickr dataset and dataset of bank card transactions. In Section 4 we present results of radius of gyration applied to the two aforementioned datasets and then in Section 5 we compare results of five different home definition methods for those two datasets. Finally, Section 6 concludes the paper and gives guidelines for future work.

2   Related Work

Related work can be mostly divided into two lines of work: studies that only focus on finding suitable algorithms for predicting where people live or studies that focus on topics for which knowing home locations is the prerequisite. In the latter group of studies scholars mostly use simple methods for determining where people live such as maximal number of geolocalized tweets/photographs/check-ins, which might work well for some of the datasets, but not necessarily for all of them. In that sense, the context of the dataset is often not really considered. With this work we want to raise the concern that before using simplified home definitions, one must be aware of the said context, as otherwise using an inappropriate method can cause uncontrollable errors in inferring home locations.

In this paper we consider three different types of OSNs: microblogging platforms such as Twitter, LBSNs such as Foursquare and photo-sharing sites such as Flickr. In addition to them, we also distinguish digital traces that people leave and that do not have a direct social component such as Detail Call Records (DCRs) or bank card transactions. As mentioned before, the most commonly used method to infer where people live is to assume that it is the location from which they sent the maximal number of tweets in case of studies on Twitter [15,33], where they had the highest cell phone communication activity in case of DCR datasets [25] or where they did the maximal number of check-ins in case of LBSNs [24]. In addition to this max^* method, in some cases together with the max number of '*', time also plays an important role when inferring home locations for Flickr [5,26], Gowalla and Brightkite [10] or DCRs [3,21].

2.1   Estimating User Locations: Content Approach

Even before trying to extract information about user location using content approach, one must face with a challenge of detecting the right language for the given text [13]. Once when the right language is detected, the next step is to extract information about user location from the text which is usually done in two ways – searching for geographic hints or building probabilistic language models for a specific location. In the former case words in the text are compared against a specialized external knowledge base, while in the latter case messages posted from the same location are clustered based on the word usage in them.

The problem when matching locations shared in the user generated text with information stored in an external dataset (e.g., gazetteer) is the ambiguity for examples for cities that do not have country/state associated with them and their names are not unique or when non-location related words are matched with cities/towns. Possible solutions for dealing with this are to: 1) use location priors, 2) search for disambigitors or 3) apply spatial minimality method [32]. The first method proposes to infer user location following the simple rule that places with larger populations or places that are more frequently mentioned in the text are more likely to be candidates, while the second and third one propose to use a list of disambigitors for every place or to calculate the minimal bounding rectangle containing all of them and then again choosing the most likely one.

When using probabilistic language models, two approaches can be distinguished – building a language model for a city estimated from tweet messages [11,19] or calculating the city distribution on the use of each word [7,9]. The first approach assumes that users living in the same city use similar language (i.e., language usage variations over cities), while the second one concentrates on calculating spatial word usage. It was shown that selected set of words (i.e., local words) can be used as a stronger predictor of a particular location [9]. The selection process of local words can be done either using a supervised classification method like in [9] or unsupervised one like in [7]. Both approaches achieved the accuracy of around 50% predicting user home locations within 100 miles.

Although text analysis is the most commonly chosen method for estimating user home locations using content approach because text is present in almost all human digital traces, scholars have also used other content sources (e.g., photographs, audio, video) to infer where people live. In that sense in [40] and [41] authors used visual features of photographs to distinguish between "home" and "non-home" photographs. As claimed by authors, unlike photo tags and descriptions, which do not have to be available, visual content is always available for every photograph. It was reported that the proposed method for home prediction achieved an accuracy of 71% with a 70.7-meter error distance.

2.2   Estimating User Locations: Social and Historical Tie Approach

Even when users think they are cautious enough not to reveal not only their locations while using OSNs, but also where they live, their less cautious friends can implicitly *fill out* these gaps for them. Namely, users with known locations can be treated as noisy location sensors of their more privacy-aware friends, as it was observed that likelihood of friendship with a person is decreasing with distance [4]. Moreover, the total number of friends tends to decrease as distance increases. As a result, it was shown that over half of users in Twitter network have one friend that can be used to predict their location within 4 km. This is not surprising given that our social options are not endless, but in fact constrained in sense that it takes time, energy and money to maintain them. However, the interesting finding was that our online activity is still influenced by locality from the physical world [18].

One of the first papers in the field was from Backstrom in 2010 who predicted the location of Facebook US-based users using location information of their friends [4]. His results showed that almost 70% users with 16 and more friends can be placed within 25 miles of their actual home. In later years, by defining a user home location in Twitter network as the place where most of his/her activities happen and assuming that 1) a user is likely to follow and be followed by users who live close to him/her and 2) in his/her tweets he/she may mention some "venues" which may indicate his/her location, in [23] scholars presented their method for which they showed improvement of 13% when compared to state-of-the-art methods. Additionally, Chen et al. showed that these results can even be improved when adding information how strongly users are connected and when trusting more friends with whom users interact more frequently [8].

3   Dataset

In this study we use two different datasets – a complete set of bank card transactions recorded by Banco Bilbao Vizcaya Argentaria (BBVA) during 2011 in
the whole Spain[1] and Flickr dataset for the whole world created by merging two
publicly available Flickr datasets [1,2,39]. In BBVA dataset, transactions were
performed by two groups of bank card users. The first one consists of the bank
direct customers, residents of Spain, who hold a debit or credit card issued by
BBVA. In the considered time period, the total number of active customers was
more than 4 million, altogether they executed more than 175 million transactions in over 1.2 million points of sale, spending over 10 billion euros. The second
group of card users includes over 34 million foreign customers of all other banks
abroad coming from 175 countries, who made purchases at one of the approximately 300 thousand BBVA card terminals. In total, they executed another
166 million transactions, spending over 5 billion euro. Flickr dataset used in our
study has more than 1.25 million users who took more than 130 million geo-
tagged photographs/videos in 247 countries around the world within a ten-year
time window period (i.e., from 2005 and until 2014).

In BBVA dataset for each transaction we know: date and time when it happen associated with the anonymous customer and business IDs and amount
of money that was spent. IDs of customers and businesses are connected with
certain demographic characteristics (e.g., age group and gender) and location
(i.e., latitude and longitude coordinates) where people live or where businesses
are located. Moreover, business IDs are associated with business categories they
belonged to. In total there are 76 categories such as restaurants, gas stations
or supermarkets. In Flickr dataset each photograph/video has its own ID associated with unique user ID, time stamp, and geo coordinates. For some pho-
tographs/videos we have also additional information such as: title, description,
user tags, machine tags, photograph/video page download URLs, license name
and URLs. However, unlike in BBVA dataset, Flickr dataset does not contain
any additional information about users that took photographs/videos.

Data preprocessing for both datasets is implemented as a two-step process:
first we prune the data and then we filter users/customers that are not active
enough, i.e., for whom we do not have enough information. In the case of BBVA
dataset pruning is done in such a way that we omit all customers for whom we
do not have records of their exact home location, and all photographs/videos in
Flickr dataset for which reverse geo-codding did not return any results or was
inconclusive (i.e., we could not determine country where they were taken). Fig. 1
shows resident and foreign customer activity in BBVA dataset and activity of
all users that took at least one photograph/video in Spain during the observed
time window of ten years in Flickr dataset. This was the data that was used to
calculate radius of gyration which results was shown in Section 4.

---

[1] Although the raw dataset is protected by a non-disclosure agreement and is not
publicly available, certain aggregated data may be shared upon a request and for
the purpose of findings validation.

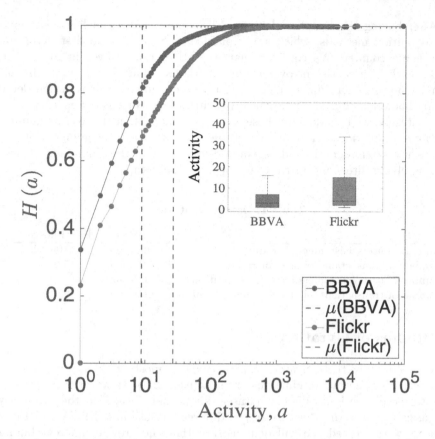

Fig. 1. (*xaxis log*) We report the empirical cumulative density function ($H(a)$) of the users' activity for both the datasets and the statistical boxplots in the top right. The "central box" representing the central 50% of the data. Its lower and upper boundary lines are at the 25%-75% quantile of the data. The central line indicates the median of the data. The average activity in the BBVA dataset (i.e., average number of transactions) is $\hat{a}^{BBVA} \sim 8.75$ while the Flickr users made in average $\hat{a}^{Flickr} \sim 27$ photographs/videos while medians, represented by horizontal dotted lines in the boxplots, are respectively 3 and 4.

Unlike in the case of BBVA dataset, for Flickr dataset we do not have ground truth that would tell us where people live. In order to compare resident activity from BBVA dataset with resident activity from Flickr dataset, we had to put a firm threshold on Flickr user activity to make sure that we include only Spanish users. In that sense, when determining home location, we first choose users that took at least one photograph/video in Spain (105 918 users in total) and then take their activity all around the world which brings us from 5 758 938 photographs/videos that those users took only in Spain to 21 134 113 photographs/videos they took worldwide.

After having the complete data for Flickr dataset, we run five home location definition methods, which are explained in Section 5, to first determine their home country. We consider that a particular user lives in Spain only if the result was consistent over all five methods, which leaves us in the end with 19 336 users for whom Spain is their home country. Finally, when determining home locations on 52 Spanish provinces level for customers/users from both datasets, we include only those who made more than the average number of photographs/videos/transactions. Table 1 shows the total numbers of customers/users/photographs/videos/transactions and how those numbers changed when applying filters and methods for pruning the data.

Table 1. Datasets statistics.

	BBVA	Flickr
Number of users/customers for radius	39.3 mln	105.4 k
Number of transactions/photos for radius	341.4 mln	5.2 mln
Number of users/customers for home definition	1.2 mln	2.5 k
Number of transactions/photos for home definition	141.7 mln	2.4 mln

4   Radius of Gyration

It is undoubtedly that on average people spend most of their time at home as it was also shown by numerous studies (e.g., [29]). However, the question that we pose here is if different datasets show that pattern in the same way. Our assumption is that because of the different nature how BBVA and Flickr datasets were created, we will demonstrate that they reveal different human mobility patterns and consequently that we cannot apply to them the same methods for predicting home locations. In this section we thus present results of radius of gyration applied to BBVA and Flickr datasets where the linear size occupied by each user's trajectory up to time t is defined as [12]:

$$r_g^a(t) = \sqrt{\frac{1}{n_c^a(t)} \sum_{i=1}^{n_c^a} \left(r_i^a - r_{cm}^a\right)^2} \tag{1}$$

where r_i^a represents the $i = 1, 2, ..., n_c^a(t)$ position visited by user a, while $r_{cm}^a = 1/n_a^c(t) \sum_{i=1}^{n_a^c} r_i^a$ is the center of mass of the trajectory.

In general, radius of gyration refers to the distribution of the components of an object around an axis. In particular, we use radius of gyration in this paper as it has already been proven that it can be used as a good proxy for human mobility. Unlike for example in the case of average travel distance, the radius of gyration is smaller for a user who travels in a comparatively confined space even though he/she covers a large distance, but is larger when someone travels with small steps but in a fixed direction or in a large circle. Consequently, it captures exactly the same pattern that we are interested to compare for our two datasets – time dependent dissipation of user movements.

The distribution of the radius of gyration, which is shown in Fig. 2, can reveal interesting human mobility patterns captured in Flickr and BBVA datasets. The most interesting result is that the radius of gyration is higher in Flickr than in BBVA dataset, with the averages 83 km and 54 km respectively. Moreover, from Flickr dataset it can be observed that 30% of users travel more than 100 km compared to only 20% in BBVA scenario. Although both datasets include activity of residents and tourists, differences between their behavior are more emphasized in Flickr dataset supporting our initial assumption. Namely, the travel pattern observed from Flickr can be explained that it is more often used when people travel as it has already been shown that tourists generally have a higher and sparser travel activity during the visit to a new country [15].

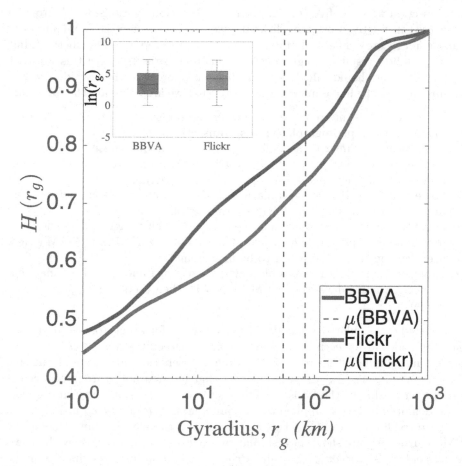

Fig. 2. We report the empirical cumulative density function ($H(r_g)$) for both the datasets (the boxplots of the natural logarithm of the radius of gyration on the top left). The average radius of gyration (i.e., dotted lines in the figure) for the BBVA and the Flickr dataset are respectively $\hat{r}_g^{BBVA} \sim 54\ km$ and $\hat{r}_g^{Flickr} \sim 83\ km$, while medians are respectively 1.67 km and 2.25 km.

In Fig. 3, each point corresponds to the coordinate of the center of mass of r_{cm} for each customer/user from both dataset, while the color highlights its (natural logarithmic) value. From these two maps we can observe the fact that people who have their center of mass in big cities have a smaller radius of gyration than people who live in suburban areas meaning that in general people who lives in urban areas travel less than people who live in rural areas or small villages [27]. This result makes a perfect sense because most of the day life services (e.g., from work to shopping, from free time to education) are concentrated in the cities and people who live outside tend to travel there for most of their activities.

5   Home Detection Methods

In this section we apply five very simple home detection methods, which are very often used in related work, to Flickr dataset and dataset of BBVA bank card transaction records. The goal of our paper is not to propose a new method, but to compare how results change in case of different datasets. With this we want to show that one should calculate in for dataset differences when deciding which method to use. The home inferring methods that we are using are the following:

1. home location is inferred as a place where a user/customer took/made maximal number of photographs/videos/transactions,
2. home location is inferred as a place where user/customer spend the maximal number of active days, where an active day is a day when user/customer took/made at least one photograph/video/transaction,
3. home location is inferred as a place with the maximal timespan between the first and the last photograph/video/transaction,
4. home location is inferred as a place where a user/customer took/made maximal number of photographs/videos/transactions from 7 PM to 7 AM (when users/customers are supposed to be near home)
5. home location is inferred as a place where user/customer spend the maximal number of active days from 7 PM to 7 AM (when users are supposed to be home).

Once when calculated home definition methods for all customers/users, for each method and for each dataset we generate a vector $Prov_{method\_x}$ containing province numbers that represent customers/users estimated home locations. For example, $Prov_{method\_3}[i] = 34$ denotes that inferred home province for customer/user i using method 3 is 34. The results of comparison between different home definition methods applied to the same dataset are reported in Fig. 4. To compare results of two different methods we used Simple Matching Coefficient (SMC). This measure shows how similar two vectors are. In our case we had two vectors of length n: $Prov_{method\_x}$ and $Prov_{method\_y}$ containing province numbers assigned to n customers/users and we calculate SMC as:

$$SMC(method\_x, method\_y) = \frac{\sum_{i=1}^{n} \delta(Prov_{method\_x}, Prov_{method\_y})}{n},$$

where $\delta(x, y)$ equals to 1 if $x = y$ and 0 otherwise.

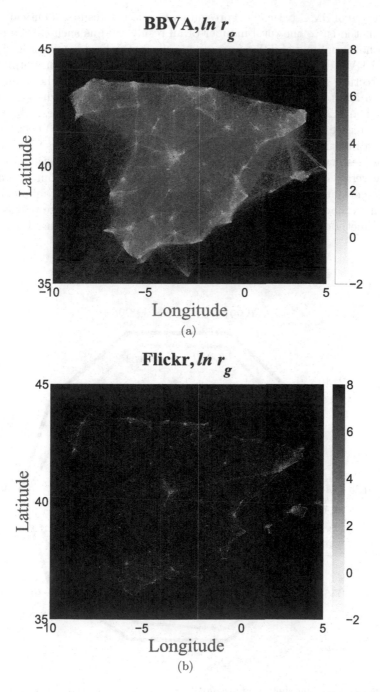

Fig. 3. Radius of gyration $(ln\,(r_g))$ for the users in the (a) BBVA and (b) Flickr datasets. Each point corresponds to the center of mass of the users, while the color represents the value of the radius of gyration.

The value of SMC basically represents a fraction of results, identical for two selected methods, to the total number of all results and as such can be used to compare how applying of different methods affects the final results. Fig. 4 shows that for BBVA dataset differences between comparison of all pairs of five used home detection methods are less than 9% (worse results are when comparing methods 3 and 5) and can go as high as less than 1% in case when comparing methods 1 and 2. In Flickr dataset final results of comparison can differ more than 20% in case of home detection methods 3 and 4 and they are never less than 7% (e.g., for methods 2 and 5). Moreover, an interesting finding is that based on results of these two datasets we cannot conclude which two methods are the most similar/diverse. Nevertheless, results shown in Fig. 4 support our assumption that different datasets are not equally susceptible when applying different home definition methods, providing a proof that it is of vital importance to choose the "correct" method with the respect of the specific dataset being used.

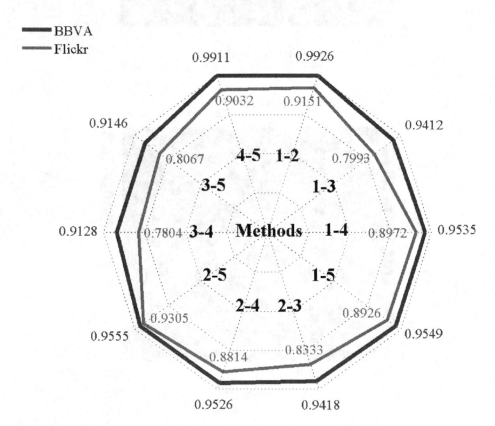

Fig. 4. Radar plot showing comparison of results (in terms of pairwise SMC) of 5 different home detection methods for BBVA and Flickr. Radar axes denote different pairs of home definition methods.

6   Discussion and Conclusions

With the emergence of digital datasets (e.g., bank card transaction records or detail phone records) containing interesting information about human race, scholars can give answers to questions we have never thought could be answered before. However, the available data represents just a certain sample of human activity, and often a very heterogeneous one, which might lead to certain mistakes that can be introduced into the process of discovering new things when not fully understanding used data sources. In this paper we focused on showing how the usage of oversimplified home definition methods can change the final results when semantically labeling most probable places where people live. Although process of inferring people home locations can be a focus of standalone studies, it is in fact very often a (first) part of more complex studies on for example human mobility or personalized services and as such is important for a larger body of related work.

Most of the simple home definition methods rely on the assumption that people spend the maximal amount of their time at home and that it can be inferred in the same way no matter what kind of dataset is used. In order to see if this assumption holds, we chose on purpose two very different digital datasets of Online Social Networks (OSNs) and Banco Bilbao Vizcaya Argentaria (BBVA) bank card transaction records. Namely, if we can show this is true in case of two datasets that are so different, then it will also hold for datasets that are more similar (e.g., two OSNs). To check if there is any difference of how different datasets capture human mobility, we first compared distributions of the radius of gyration and then we applied five different simple home definition methods to the aforementioned datasets.

The distribution of the radius of gyration showed that indeed there are observable differences how and where people use their bank cards and Flickr. Namely, with almost 50% of larger radius of gyration and 10% of more people who traveled more than 100 km, Flickr dataset provides another view on human mobility than BBVA dataset. Moreover, BBVA dataset seems to be more robust to different home definition methods as the results differ only from 1% to 9 % unlike in case of Flickr that are in range of 10%–20%. With this we showed that although for some datasets (such as bank card data) particular choice of a home definition method does not really seem that important, for the other datasets, it might actually affect the results quite a bit. Therefore, the choice of home definition method should be done carefully with respect to the characteristics of the particular dataset being considered.

However, results presented in this paper are only the initial results that pointed towards differences between human mobility patterns captured in two different datasets and should be extended to other datasets. In future work we will first compare the results of different home definition methods against available ground truth for different types of datasets, then classify datasets into different categories and finally give specific recommendations of which method to use for datasets belonging to a particular category.

Acknowledgment. The authors would like to thank Banco Bilbao Vizcaya Argentaria (BBVA) for providing the dataset for this research. Special thanks to Juan Murillo Arias, Marco Bressan, Elena Alfaro Martinez, María Hernández Rubio and Assaf Biderman for organizational support of the project and stimulating discussions. We further thank BBVA, MIT SMART Program, Center for Complex Engineering Systems (CCES) at KACST and MIT, Accenture, Air Liquide, The Coca Cola Company, Emirates Integrated Telecommunications Company, The ENELfoundation, Ericsson, Expo 2015, Ferrovial, Liberty Mutual, The Regional Municipality of Wood Buffalo, Volkswagen Electronics Research Lab, UBER, and all the members of the MIT Senseable City Lab Consortium for supporting the research. Finally, the authors also acknowledge support of the research project "Managing Trust and Coordinating Interactions in Smart Networks of People, Machines and Organizations", funded by the Croatian Science Foundation.

References

1. Flickr dataset. http://sfgeo.org/data/tourist-local
2. Yahoo! Webscope dataset YFCC-100M. http://labs.yahoo.com/Academic-Relations
3. Alexander, L., Jiang, S., Murga, M., González, M.C.: Origin-destination trips by purpose and time of day inferred from mobile phone data. Transportation Research Part C: Emerging Technologies **58**(Part B), 240–250 (2015)
4. Backstrom, L., Sun, E., Marlow, C.: Find me if you can: improving geographical prediction with social and spatial proximity. In: Proceedings of the 19th International Conference on World Wide Web, pp. 61–70 (2010)
5. Bojic, I., Nizetic-Kosovic, I., Belyi, A., Sobolevsky, S., Podobnik, V., Ratti, C.: Sublinear scaling of country attractiveness observed from Flickr dataset, pp. 1–4 (2015). arXiv preprint
6. Calabrese, F., Di Lorenzo, G., Liu, L., Ratti, C.: Estimating origin-destination flows using mobile phone location data. IEEE Pervasive Computing **4**(10), 36–44 (2011)
7. Chang, H.W., Lee, D., Eltaher, M., Lee, J.: @ Phillies tweeting from philly? predicting twitter user locations with spatial word usage. In: Proceedings of the International Conference on Advances in Social Networks Analysis and Mining, pp. 111–118 (2012)
8. Chen, J., Liu, Y., Zou, M.: From tie strength to function: home location estimation in social network. In: Proceedings of the IEEE Computing, Communications and IT Applications Conference, pp. 67–71 (2014)
9. Cheng, Z., Caverlee, J., Lee, K.: You are where you tweet: a content-based approach to geo-locating twitter users. In: Proceedings of the 19th ACM International Conference on Information and Knowledge Management, pp. 759–768 (2010)
10. Cho, E., Myers, S.A., Leskovec, J.: Friendship and mobility: user movement in location-based social networks. In: Proceedings of the 17th ACM SIGKDD International Conference on Knowledge Discovery and Data Mining, pp. 1082–1090 (2011)

11. Eisenstein, J., O'Connor, B., Smith, N.A., Xing, E.P.: A latent variable model for geographic lexical variation. In: Proceedings of the Conference on Empirical Methods in Natural Language Processing, pp. 1277–1287 (2010)
12. González, M., Hidalgo, C., Barabási, A.L.: Understanding individual human mobility patterns. Nature **453**(7196), 779–782 (2008)
13. Graham, M., Hale, S.A., Gaffney, D.: Where in the world are you? Geolocation and language identification in Twitter. The Professional Geographer **66**(4), 568–578 (2014)
14. Grauwin, S., Sobolevsky, S., Moritz, S., Gódor, I., Ratti, C.: Towards a comparative science of cities: using mobile traffic records in New York, London and Hong Kong. In: Computational Approaches for Urban Environments, pp. 363–387. Springer International Publishing (2015)
15. Hawelka, B., Sitko, I., Beinat, E., Sobolevsky, S., Kazakopoulos, P., Ratti, C.: Geo-located Twitter as proxy for global mobility patterns. Cartography and Geographic Information Science **41**(3), 260–271 (2014)
16. Hecht, B., Hong, L., Suh, B., Chi, E.H.: Tweets from Justin Bieber's heart: the dynamics of the location field in user profiles. In: Proceedings of the SIGCHI Conference on Human Factors in Computing Systems, pp. 237–246 (2011)
17. Hoteit, S., Secci, S., Sobolevsky, S., Ratti, C., Pujolle, G.: Estimating human trajectories and hotspots through mobile phone data. Computer Networks **64**, 296–307 (2014)
18. Jurgens, D.: That's what friends are for: inferring location in online social media platforms based on social relationships. In: Proceedings of the AAAI International Conference on Web and Social Media, pp. 273–282 (2013)
19. Kinsella, S., Murdock, V., O'Hare, N.: I'm eating a sandwich in glasgow: modeling locations with tweets. In: Proceedings of the 3rd International Workshop on Search and Mining User-Generated Contents, pp. 61–68 (2011)
20. Krumm, J., Rouhana, D.: Placer: semantic place labels from diary data. In: Proceedings of the ACM International Joint Conference on Pervasive and Ubiquitous Computing, pp. 163–172 (2013)
21. Kung, K., Greco, K., Sobolevsky, S., Ratti, C.: Exploring universal patterns in human home/work commuting from mobile phone data. PLoS One **9**(6), 1–15 (2014)
22. Li, R., Wang, S., Chang, K.C.C.: Multiple location profiling for users and relationships from social network and content. Proceedings of the VLDB Endowment **5**(11), 1603–1614 (2012)
23. Li, R., Wang, S., Deng, H., Wang, R., Chang, K.C.C.: Towards social user profiling: unified and discriminative influence model for inferring home locations. In: Proceedings of the 18th ACM SIGKDD International Conference on Knowledge Discovery and Data Mining, pp. 1023–1031 (2012)
24. Noulas, A., Scellato, S., Lathia, N., Mascolo, C.: A random walk around the city: new venue recommendation in location-based social networks. In: Proceedings of the International Conference on Privacy, Security, Risk and Trust and International Confernece on Social Computing, pp. 144–153 (2012)
25. Onnela, J.P., Arbesman, S., González, M.C., Barabási, A.L., Christakis, N.A.: Geographic constraints on social network groups. PLoS one **6**(4), 1–7 (2011)
26. Paldino, S., Bojic, I., Sobolevsky, S., Ratti, C., González, M.C.: Urban magnetism through the lens of geo-tagged photography. EPJ Data Science **4**(1), 1–17 (2015)
27. Pateman, T.: Rural and urban areas: Comparing lives using rural/urban classifications. Regional Trends **43**(1), 11–86 (2011)

28. Pei, T., Sobolevsky, S., Ratti, C., Shaw, S.L., Li, T., Zhou, C.: A new insight into land use classification based on aggregated mobile phone data. International Journal of Geographical Information Science **28**(9), 1988–2007 (2014)

29. Pontes, T., Vasconcelos, M., Almeida, J., Kumaraguru, P., Almeida, V.: We know where you live: privacy characterization of foursquare behavior. In: Proceedings of the ACM Conference on Ubiquitous Computing, pp. 898–905 (2012)

30. Ratti, C., Sobolevsky, S., Calabrese, F., Andris, C., Reades, J., Martino, M., Claxton, R., Strogatz, S.H.: Redrawing the map of Great Britain from a network of human interactions. PLoS One **5**(12), 1–6 (2010)

31. Santi, P., Resta, G., Szell, M., Sobolevsky, S., Strogatz, S.H., Ratti, C.: Quantifying the benefits of vehicle pooling with shareability networks. Proceedings of the National Academy of Sciences **111**(37), 13290–13294 (2014)

32. Serdyukov, P., Murdock, V., Van Zwol, R.: Placing flickr photos on a map. In: Proceedings of the 32nd International ACM SIGIR Conference on Research and Development in Information Retrieval, pp. 484–491 (2009)

33. Sobolevsky, S., Bojic, I., Belyi, A., Sitko, I., Hawelka, B., Arias, J.M., Ratti, C.: Scaling of city attractiveness for foreign visitors through big data of human economical and social media activity, pp. 1–8 (2015). arXiv preprint arXiv:1504.06003

34. Sobolevsky, S., Massaro, E., Bojic, I., Arias, J.M., Ratti, C.: Predicting regional economic indices using big data of individual bank card transactions. In: Proceedings of the 6th ASE International Conference on Data Science, pp. 1–12 (2015)

35. Sobolevsky, S., Sitko, I., Tachet des Combes, R., Hawelka, B., Arias, J.M., Ratti, C.: Money on the move: big data of bank card transactions as the new proxy for human mobility patterns and regional delineation. the case of residents and foreign visitors in spain. In: Proceedings of the IEEE International Congress on Big Data, pp. 136–143 (2014)

36. Sobolevsky, S., Sitko, I., Combes, R.T.D., Hawelka, B., Arias, J.M., Ratti, C.: Cities through the prism of people's spending behavior, pp. 1–21 (2015). arXiv preprint arXiv:1505.03854

37. Sobolevsky, S., Sitko, I., Grauwin, S., Combes, R.T.D., Hawelka, B., Arias, J.M., Ratti, C.: Mining urban performance: Scale-independent classification of cities based on individual economic transactions, pp. 1–10 (2014). arXiv preprint arXiv:1405.4301

38. Sobolevsky, S., Szell, M., Campari, R., Couronné, T., Smoreda, Z., Ratti, C.: Delineating geographical regions with networks of human interactions in an extensive set of countries. PloS One **8**(12), 1–10 (2013)

39. Thomee, B., Shamma, D.A., Friedland, G., Elizalde, B., Ni, K., Poland, D., Borth, D., Li, L.J.: The new data and new challenges in multimedia research, pp. 1–7 (2015). arXiv preprint arXiv:1503.01817

40. Zheng, D., Hu, T., You, Q., Kautz, H., Luo, J.: Inferring home location from user's photo collections based on visual content and mobility patterns. In: Proceedings of the 3rd ACM Multimedia Workshop on Geotagging and Its Applications in Multimedia, pp. 21–26 (2014)

41. Zheng, D., Hu, T., You, Q., Kautz, H., Luo, J.: Towards lifestyle understanding: predicting home and vacation locations from user's online photo collections. In: Proceedings of the 9th International AAAI Conference on Web and Social Media, pp. 553–560 (2015)

42. Zhong, Y., Yuan, N.J., Zhong, W., Zhang, F., Xie, X.: You are where you go: inferring demographic attributes from location check-ins. In: Proceedings of the 8th ACM International Conference on Web Search and Data Mining, pp. 295–304 (2015)

Proposing Ties in a Dense Hypergraph of Academics

Aaron Gerow[1][(✉)], Bowen Lou[1,2], Eamon Duede[1], and James Evans[1,3]

[1] Computation Institute, University of Chicago, Chicago, USA
gerow@uchicago.edu
[2] Wharton School of the University of Pennsylvania, Philadelphia, USA
[3] Department of Sociology, University of Chicago, Chicago, USA

Abstract. Nearly all personal relationships exhibit a multiplexity where people relate to one another in many different ways. Using a set of faculty CVs from multiple research institutions, we mined a hypergraph of researchers connected by co-occurring named entities (people, places and organizations). This results in an edge-sparse, link-dense structure with weighted connections that accurately encodes faculty department structure. We introduce a novel model that generates dyadic proposals of how well two nodes should be connected based on both the mass and *distributional similarity* of links through shared neighbors. Similar link prediction tasks have been primarily explored in unipartite settings, but for hypergraphs where hyper-edges out-number nodes 25-to-1, accounting for link similarity is crucial. Our model is tested by using its proposals to recover link strengths from four systematically lesioned versions of the graph. The model is also compared to other link prediction methods in a static setting. Our results show the model is able to recover a majority of link mass in various settings and that it out-performs other link prediction methods. Overall, the results support the descriptive fidelity of our text-mined, named entity hypergraph of multi-faceted relationships and underscore the importance of link similarity in analyzing link-dense multiplexitous relationships.

1 Introduction

High impact research commonly spans fields of science, leading universities to increasingly focus on ways to catalyze cross-disciplinary collaborations. Some institutions have sought to overcome challenges of disciplinary compartmentalization by implementing research networking and researcher profiling systems[1]. However, there is little evidence to suggest that such systems generate new, effective collaborations that span traditional boundaries [32]. Inter-disciplinary institutes may be positioned to play the match-maker, but these are also difficult to systematically qualify [15]. An important challenge, then, is to represent relationships among academics from which novel, productive links can be reliably modeled. The challenge is compounded by the fact that actual collaborations are a sparse structure: most people do not collaborate with most others. This paper

---

[1] Profiles.catalyst.harvard.edu is one example.

© Springer International Publishing Switzerland 2015
T.-Y. Liu et al. (Eds.): SocInfo 2015, LNCS 9471, pp. 209–226, 2015.
DOI: 10.1007/978-3-319-27433-1_15

proposes a method of mining a hypergraph with a large number of hyper-edges — 25 times the number of nodes — and a model which uses an edge-based, distributional measure of link similarity to propose relationships among academics.

Network mining and modeling bears on a range of problems in computer systems, power distribution, cell-biology, cognition, organizational structure and even terrorist networks [3,12,33]. Mining social structure, particularly from text, is an important task for automated recommendation systems [18]. Perhaps the simplest form of social recommendation is proposing ties between people in a social network. This task, known generally as link prediction, can lend insights to how networks grow and change over time, and is the subject of a great deal of research [28–30,42]. However, link prediction has been almost exclusively explored in the context of single-mode, unipartite networks where edges only connect two nodes [5], which over-simplifies the multiplexity of social dynamics. In reality, nearly all personal relationships consist of multiple, sometimes diverse connections that vary in strength, breadth and meaning. Here, we show that such multiplexity can be represented using *named entities* (NEs) in faculty CVs. These entities can be interpreted as the relationships in which people participate [13], and can consist of organizations (publishing in the same journal, sharing professional affiliations or committee membership), location (cohabiting an office, building or city) and personal relationships (co-authors, supervisors or mentors). Connecting academics by the co-occurring NEs observed in their CVs forms a complex hypergraph that is edge-sparse (most edges connect only few nodes) and link-dense (most dyads are connected by at least one edge)[2].

The social recommendation task in an edge-sparse, link-dense hypergraph is similar to traditional link prediction given its focus on dyadic links. The goal is to propose ties (in our case, weighted links) between nodes based on the structure of the network. Because most nodes are already connected in a link-dense graph, recommendations can be interpreted as how strong a link *should be* based on a dyad's local structure. As we will see, this density is problematic for link prediction methods that rely on variation in the neighborhood around a dyad. To overcome this density, the proposals should account for both the *strength and similarity* of links through a dyad's neighbors. Fortunately, we can use the density to our advantage by treating edges as points in a space defined by nodes' edge-incidence vectors, allowing similarity to be measured as inverse distance in high-dimensional space. This strategy is particularly advantageous as it allows similarity to be calculated without using external information.

There is likely some interplay between the mass and similarity of links on either side of a neighbor. We hypothesize that the mass of a link through a transit node is only "available" to the extent links are similar. That is, similar links can use most of the edge mass, while dissimilar links use only a fraction. In figure 1b, this is analogous to saying that the mass of link **AB** (edges 1, 2 and 3) plus **BC** (edges 1, 4 and 5) is proportional to the similarity of **AB** and **BC**. We hold that this intuition is not only socially plausible, but that it can be used to more accurately recommend connections.

---

[2] We use *link* to refer to the set of hyper-edges connecting a pair of nodes.

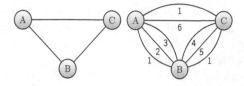

Fig. 1. A triad of nodes A, B and C as a simple network (a; left) and as a hypergraph (b; right). In (b), note that not only can nodes be connected by more than one edge (ie. a multi-mode graph), but edges can connect more than two nodes as with edge 1. We write **AB** to denote the link from A to B, here consisting of edges $\{1, 2, 3\}$.

After reviewing related work and providing some background in the next section, section 3 introduces our model, including definitions of link mass and distributional link similarity on weighted hypergraphs. Section 4.1 presents four experiments where the model is used to recover link mass from four versions of the data, lesioned in different ways. Section 4.2 compares our model to other link prediction methods after which we review the results of the evaluations. The final section discusses the contribution of the representation and associated model, pointing to future work and applications.

2   Background and Related Work

Network analysis is a widely studied and increasingly important area of research in biology, economics, natural language processing, sociology and computer science [12,16,17,37]. In particular, social network analysis seeks to model how people (nodes) develop relationships to one-another (edges). Such social organization tends to form groups where similar kinds of ties are shared among the members [10,11]. To represent social interactions in a simple network, the meaning of edges must be globally defined, perhaps denoting friendship, kinship or shared affiliation. A hypergraph, on the other hand, can represent relations that connect any number of people, who in turn may have any number of connections [4,31,37]. Figure 1 depicts a simple network and a hypergraph with six hyperedges, where each dyadic link consists of a set of hyper-edges. What these edges are, that is to say their semantics, may vary from one graph to another, but as we will see, people, places and organizations offer a reasonable starting-point for defining academic relationships.

Scott Feld [13] suggested that social circles, such as affiliations, beliefs and practices, make up focal points around which people engage and connect. These foci naturally comprise a network structure that interrelates people as nodes connected by the foci in which they participate. The current study operationalizes Feld's *focus theory* as a model on hypergraphs where people are nodes and foci are edges. Our approach effectively extends focus theory to a high-dimensional space where the modes of interaction (edge-types) can greatly out-number the actors themselves. It is important to note that foci need not be restricted to social circles or affiliations — they may also be people themselves. Using people, places and organizations as edges will allow our analysis to retain the intuitions from social network analysis and enable us to represent complex relationships [22]. The challenge will be to propose new, or at least stronger ties in a structure that is already rich with connections.

Recently, some work has explored statistical learning methods on networks to predict academic relationships [21], which combines advances in machine learning with social network analysis in a unipartite setting. Similar to the work here, given the use of NEs as features in the network, is a method of extracting names from websites to mine communities of people [27]. Some work on layered social networks implicitly adopts the hypergraph representation [25, 26, 44], though most social network analysis still relies on single-mode, unipartite graphs. What's more, most research using hypergraphs is framed as n-partite graph analysis, typically with a small number of layers [4, 24]. Here, we explore hypergraphs where the layers far out-number the nodes themselves. Although this certainly compounds the representational complexity of a network, it is considerably more plausible as it allows actors to relate in the myriad ways people do in the real world. Hypergraphs have been used to predict multi-actor collaborations (ie. teams) in academic social networks [36], by predicting the formation of links via new and old hyper-edges. Because academic teams tend to evince relationships in different and systematic ways, the hypergraph is a particularly robust data-structure for modeling team formation [40], recommending new collaborations [43] and as a model of how scientists actually seek out new projects [37]. In this paper, instead of assuming a set of edge-types, we show that a distributional measure of link similarity can help leverage the abundance of edge-types found in a text-mined hypergraph of NEs.

Proposals in our model are effectively an estimate of how connected two nodes *should* be, and can be thought of as link predictions in a static setting. Traditional link prediction is the task of assigning a likelihood of observing new a connection at some future time using information about the current network (see [28] for a review). Getoor et al. [19] introduce a framework for representing and learning probabilistic link structure in arbitrarily complex relational data. Extending the statistical learning approach, Al Hasan et al. [2] develop a supervised learning method that uses structural and "aggregate" features to train a classifier to predict links. Although there are some hybrids, link prediction methods tend either to allow the use of global information about the network (diameter, path-distances, etc.) or limit themselves to local information (degree, shared neighbors, etc.). The method in this paper is more similar to the latter, as it restricts the search space for dyadic proposals to the links through a dyad's shared neighbors.

We hypothesize that links are characterized by two related qualities: their edge-mass and their similarity to other links. Mass accounts for the strength of a link, regardless of its composition. The weights of individual hyper-edges can then be used to discriminate, for example, between participation in a large conference where two people are unlikely to meet, and more intimate relationships, like a workshop or committee.

3   Method

Our goal is to propose ties in a social network that is already quite dense with links. Two nodes are considered linked if they are connected by one or more hyper-edges. A link refers to a dyadic connection whereas a hyper-edge can connect any number of nodes (see figure 1). Proposals, then, can be calculated for any pair of nodes, connected or not, referred to as a dyad.

Proposals are calculated using transitivities, with the assumption that shared neighbors provide a strong indication of how actual people tend to introduce other people to one-another [6,35]. Although restricting the search space to shared neighbors can be problematic, especially in sparse graphs where higher-order structure is important [39], it is both plausible to assume academics limit their search behavior in this manner and it is known to successfully character-ize social networks more generally [38]. Also, this formulation allows users of the model to recover straight-forward justifications for individual proposals — something discussed in the final section. Specifically, a proposal for a pair of nodes, n_1 and n_2 in a hypergraph, G, is calculated as follows:

$$proposal(n_1, n_2; G) = \sum_{t \in \Gamma(n_1) \cap \Gamma(n_2)} S(\mathbf{n_1 t}, \mathbf{n_2 t}) \qquad (1)$$

where $\Gamma(x)$ denotes the neighbors of node x and $\mathbf{n_x t}$ is the set (hence in bold) of edges connecting nodes n_x and t. S is a function relating two sets of weighted hyper-edges to be defined below.

3.1   Link Mass

To account for the strength of participation in a given edge, for example mem-bership in the Association for Computing Machinery (ACM), an edge's node-incidence is weighted by the number times it was mentioned in someone's CV. This helps account for the difference between someone who mentioned the ACM 100 times and someone who mentioned it only once. There is no limit to the frequency of an edge's weight per-node, nor is there a limit to the total num-ber of entities by which a node may be connected. However, to accommodate the inverse relationship between the number of participants in an edge and its importance, we normalize the weights of each edge by their sum. Formally, we take the edge-wise L1-norm of G, which makes edge-weights inversely propor-tional to the number of nodes they connect; an edge that only connects two nodes, receives a weight of .5, whereas edges that connect many nodes receive a weight closer to zero. Although this normalization down-weights predominant edges, because entities are initially coded as occurrence frequencies, there is still room for variation in a how much nodes participate in even the most frequent edges. Figure 2 shows an example of the initial, frequency-weighted matrix and its normalized counterpart.

One simple way to think about the quality of a link is by its mass. Socially, this can be thought of as the sum of all the ways, important and trivial, two people relate. We assume that edges in a graph, G, have weights associated individually with the nodes they connect. That is, each node may "participate" in an edge to a varying degree (ranging from 0 to 1). The weight for a dyadic link, then, is the sum of these edge-incidence weights. The weight for a link between two nodes *through a transit node* is the minimum of the two links' mass, as this represents the maximal relation two people have through either link. This minimum is the weight used to compute dyadic link mass from one node to another, though an intermediate node, t:

$$Mass(n_1, t, n_2) = min(\sum_{i \in n_1 t} w_i, \sum_{i \in n_2 t} w_i) \qquad (2)$$

where w is the edge's weight as described above. Link mass only accounts for the combined strength of edges through a transit node. Referring to figure 1, link mass from A to C through B would be the sum of the edge-weights in **AB** and **BC**. While this may be a plausible way to model how people recommend relationships to one another (by the magnitude of their relationship to each person), it fails to account for the edges themselves.

3.2   Link Similarity

Link similarity is an important component when examining the social closure of transitivities: two people are more likely to be introduced if they have *similar* links through a shared acquaintance than if those links are dissimilar. To address link similarity, we conceive of edges as points in the normalized space defined by edges' node-incidence. A single edge is represented as the coordinate given by its weighted node-incidence vector in G, which in the data for our experiments is highly dimensional. The similarity of two links, $n_1 t$ and $n_2 t$ (ie. the connection from n_1 to n_2 through t), is defined as one minus the Euclidean distance between each edge-sets' centroid, $C_{n_1 t}$ and $C_{n_2 t}$:

$$C_{n_1 t}^{(d)} = \frac{1}{|n_1 t|} \sum_{i \in n_1 t} w_i \in [0, 1]^d$$

$$C_{n_2 t}^{(d)} = \frac{1}{|n_2 t|} \sum_{i \in n_2 t} w_i \in [0, 1]^d \qquad (3)$$

$$Sim^{(d)}(n_1, t, n_2) = 1 - \sqrt{(C_{n_1 t}^{(d)} - C_{n_2 t}^{(d)})^2}$$

where d is the dimensionality of the normalized node-space. Using nodes to define a normalized space in which to compare edges allows a completely internal, data-driven conception of similarity. This feature of our model not only frees it from potential problems with extrinsically defined similarity metrics, such as under-specification, incompleteness or incorrectness, it also scales well to graphs with arbitrarily large number of edge-types. While there is no reason an externally defined similarity measure could not be used in its place, the distributional measure allows our proposals to be derived within-model.

With these definitions of link mass and link similarity, we can test our hypothesis that the strength of links through a transit are in fact dependent on their similarity. Further, because we assumed the node-incidence weights used to define each edge are normalized from 0 to 1, we can use them to scale the link mass by multiplying the values of Sim by $Mass$. The intuition is that greater participation in similar, intimate venues increases the likelihood two people connect. Substituting this combination of Sim and $Mass$ for S in the first equation, we can define the strength of a proposal between two nodes, n_1 and n_2:

$$proposal(n_1, n_2; G) = \sum_{t \in \Gamma(n_1) \cap \Gamma(n_2)} Mass(n_1, t, n_2) * Sim(n_1, t, n_2) \quad (4)$$

After generating proposals, p, the values are normalized: $p_i \leftarrow \frac{p_i - p_{min}}{p_{max} - p_{min}}$.

4   Results

To evaluate the method outlined above, we retrieved a set of 2,511 CVs of faculty members at several large, research-intensive universities[3]. Each CV was coded by department[4]. The Stanford Named Entity Recognizer (NER) was used to extract NEs from the texts of each CV [14,41]. The Stanford NER extracts people, locations and organizations with a reported F_1-score of 0.93 (precision = 0.93, recall = 0.92). These entities define the hyper-edges in the network. Note this means an academic may be a node *and* a hyper-edge. This process yielded 802,131 unique hyper-edges (27,982 locations, 142,350 persons, 230,739 organizations). All NEs that connected at least two people in the network were retained. This has the side-effect of filtering out spurious entities that were only observed once. Though most edges do not connect most nodes, only 14 nodes were not part of the largest connected component, to which we restrict our analysis. In fact, the hypergraph was almost fully connected: most nodes

	Observed					Normalized				
	n_1	n_2	n_3	n_4	n_5	n_1	n_2	n_3	n_4	n_5
W3C	1	2	0	5	2	0.1	0.2	0.0	0.5	0.2
ACM	1	0	0	3	0	0.25	0.0	0.0	0.75	0.0
IEEE	0	0	2	0	0	0.0	0.0	1.0	0.0	0.0
SIGKDD	0	0	1	1	0	0.0	0.0	0.5	0.5	0.0

Fig. 2. Example $N \times E$ matrix defining G. Nodes / CVs are represented in each column with four example NEs / edges as rows. Shown are the initial matrix (left), where the numbers represent an entity's frequency in each CV, and the normalized matrix (right).

---

[3] Available at klab.ci.uchicago.edu/data/CV_data.tar.gz
[4] 122 faculty members had appointments in more than one department.

were connected to all other nodes by at least one edge, which is not surprising given that each university constitutes an ORGANIZATION edge connecting a large number of academics. In total, there were 3,116,256 dyads of which 66% were directly connected by an average of 2.2 hyper-edges. This hypergraph comprises the $N \times E$ matrix referred to as G, an example of which is depicted in figure 2. After link-mining, the matrix was weighted and normalized by the method described above. A preliminary evaluation of this representation was performed where the node-incidence vectors $\{E \in G\}$ were used as features to classify academics in their departments. We found that, when restricted to the same number of features, token-based feature-sets are significantly outperformed by our representation (see the appendix).

4.1   Recovering Link Mass

The goal of the proposal model described in section 2 is to generate recommendations of stronger ties. Because actual relationships are formed over the course of time, proposing ties in networks is often framed as predicting whether or not a link will form in the future [28]. In the absence of temporal data, we evaluated our model in a static setting. We describe its performance in four settings where the observed graph, G, was systematically lesioned, after which proposals were calculated on the resulting graph, G'. In each experiment, the goal is for proposals generated on G' to replicate or "recover" the link mass observed in G. Because we make no assumptions about the distribution of link-masses in G, we use Spearman's rank-correlation to evaluate whether the *ordering* of dyadic proposals on G' is similar to that observed in G. By computing this correlation at rank from the strongest to weakest links in G, we can also test whether our proposals tend to be more or less accurate as a function of the initial link strength. These experiments exhaust a reasonable space of evaluation, showing that our

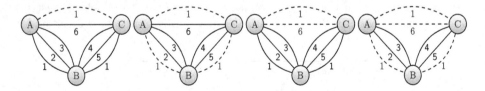

Fig. 3. Four versions of G' corresponding to four evaluation schemes, derived from the example triad in figure 1b. Dashed lines represent deleted hyper-edges that were initially present in G. From left to right: (a): Local edge deletion, where one component of a hyper-edge in a dyad is deleted. (b): Global edge deletion, where a sample of hyper-edges are deleted from the entire graph (here hyper-edge 1 is deleted). (c): Local link deletion, where all edges connecting a dyad are deleted. (d): Global link deletion, where the edges comprising the chosen dyadic link are deleted from the entire graph (here, because link $\mathbf{AC} = \{1, 6\}$, hyper-edges 1 and 6 are deleted globally). In all evaluations, dyadic proposals are generated from nodes analogous to $\mathbf{AC} \in G'$ over all shared neighbors (those in a similar position to B).

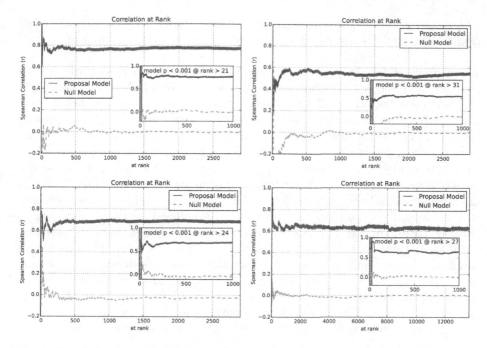

Fig. 4. Correlation at rank for the four link-mass recovery experiments in section 4.2. Results for each lesioning scheme (figure 4) clock-wise from top left: local hyper-edge deletion, global hyper-edge deletion, local link-deletion and global, like-selected hyper-edge deletion. In each, rank refers to the inclusion of the n strongest dyadic links in G (equivalent to **AC** in figures 1 and 3) of which in-lays show the top 1,000. The shaded region in all plots denotes $+/-$ 1 SD of the mean over 10 folds with Fisher's z-transformation. In all experiments the model results (solid blue line) are significantly better than a null model (dashed grey line).

model is robust to local and global lesioning of both links and hyper-edges (figure 4). In all cases, we compare the model to a random model.

Local Hyper-Edge Deletion. The first experiment of the recommendation model tests how well the proposal scores can compensate for locally deleted hyper-edges. To simulate a world where two people in our network failed to relate in a particular way, a random hyper-edge is deleted from a dyad, for which a proposal is then generated using the rest of the graph. In figure 4a, this is analogous to removing hyper-edge 1, locally from the **AC** link. The proposal for $A \rightarrow C$ should compensate for the loss of the mass of hyper-edge 1. The proposals correlated positively with a mean Spearman $r = 0.77$ ($p < 0.0001$; mean of 10, 10% samples; using Fisher's z-transformation) for all dyads. If we use only transitive link mass to generate proposals (without link similarity) the correlation is positive but less strong at $r = 0.69$ ($p < 0.0001$): the use of link similarity improved the model by about 12%. Additionally, there was no discernible trend in the correlation at rank (figure 5a) from the strongest to weakest links in the original graph, G, implying the model performs comparably well for strong and weak links alike.

The model also out-performs a null model that draws a random dyadic mass from G, showing the results are not an artifact of the mass distribution.

Global Hyper-Edge Deletion. In the second experiment, the model was used to generate proposals on a globally, edge-lesioned version of G. In this task, 20% of the hyper-edges in the observed network were removed, and proposals were generated on the remainder. Even after this lesioning, most dyads were linked by at least one hyper-edge. Figure 4b shows a triad in which hyper-edge 1 has been removed, which affects each dyad because it connects all three nodes. Each run of the model on this evaluation generates a proposal for every dyad, but the edge-removal process selects a random 20% of hyper-edges, therefore, a 10-fold cross-validation scheme was employed. The model was judged to be successful to the extent that the proposals drawn on the lesioned graph, G' correlate to the original link masses in G. On average, the proposals were found to correlate with original link masses at a Spearman $r = 0.55$ ($p < 0.0001$, 10 folds). With proposals that use link mass alone, the correlation is still positive, but less strong at $r = 0.49$ ($p < 0.0001$). This indicates the link similarity is approximately as important on this task as it was in the first evaluation. Figure 5b shows the correlation at rank. The lack of trend in this statistic implies the model did as well for strong links as for weak. The model also out-performs the null model.

Local Link Deletion. This evaluation tests the ability of the model to recover the mass of entire links. Here, a dyad is completely disconnected, for which a proposal is then generated. In Figure 4c, this is equivalent to removing hyper-edges 1 and 2. The strength of the generated proposals, then, should correlate to the original dyadic link masses. On the local link deletion task, the proposals correlated to the original dyads with a Spearman $r = 0.62$ ($p < 0.0001$; 10, 10% folds). Using mass alone produced a slightly weaker correlation of $r = 0.61$ ($p < 0.0001$) which shows that on this task, link similarity is less important than on the previous two evaluations, providing less than 1% performance gain. Figure 5c shows the correlation at rank. Overall, the model performs significantly better than the null model, reiterating that the findings are not an artifact of the distribution of link masses in G.

Global, Link-Selected Hyper-Edge Deletion. The final experiment tests the ability to compensate for hyper-edges selected from links and deleted globally. For randomly selected dyads in G, the hyper-edges comprising the link are removed from the entire graph. In figure 4d, nodes A and C are connected by hyper-edges 1 and 6, which are then deleted throughout the network to yield the lesioned G'. Proposals were generated on G' and their correlation to the the original dyadic masses is reported. This experiment is effectively a combination of experiments 2 and 3: the model must provide proposals that compensate for an entirely missing link (experiment 3) using a graph with fewer edges than initially observed (experiment 2). The proposals in this configuration correlated positively with the link masses in G: Spearman $r = 0.69$ ($p < 0.0001$; 10, 10% folds). This was stronger than when using link mass alone, which yielded $r = 0.57$ ($p < 0.0001$).

Figure 5d shows the correlation at rank for this experiment, in which there is no discernible trend: the model is consistent across strong and weak links in G.

4.2 Proposals as Predictions

The preceding experiments show that our model can recover the dyadic link mass in G after it has been systematically lesioned in different ways. This section tests our model as a method of link prediction. Link prediction is typically thought of as the task of predicting a link at some future time given historical data. Because our mined hypergraph lacks temporal data, we employ a static variant of link prediction. As in other formulations, a link refers to a single dyadic connection. Although some work has explored predicting hyper-edges that connect more than two nodes [21,40], in our model, though it uses hyper-edges to define link similarity, it generates dyadic proposals more analogous to links in a unipartite graph. To compare our proposals to other link prediction methods, we simplify G into a unipartite graph, $\hat{G}$, by treating links (sets of edges in G) as unique, un-labeled edges. Additionally, because our proposals are not simply yes or no binary suggestions as with typical link-prediction tasks, evaluations like precision, recall and ROC / AUC are less well-suited to assessing the proposal scores. As such, we again use Spearman rank-order correlation to compare predictions that preserve the rank of link-strengths observed in G.

We compare our model to three other link prediction methods: the Jaccard coefficient, preferential attachment and the Adamic-Adar method [1,28]. The Jaccard coefficient is a structural measure defined as the size of the intersection of a dyads' common neighbors over that of their union. Preferential attachment metricates the intuition that highly connected nodes are more likely to form links than less connected nodes. Adamic-Adar predictions account for the size of shared neighbors' neighbors, defined as the inverse log of the number of every shared neighbor's neighbors. Each method is defined as follows:

$$\text{Jaccard Coefficient:} \quad \frac{|\Gamma(n_1) \cap \Gamma(n_2)|}{|\Gamma(n_1) \cup \Gamma(n_2)|}$$

$$\text{Preferential Attachment:} \quad |\Gamma(n_1)| \cdot |\Gamma(n_2)|$$

$$\text{Adamic-Adar:} \quad \sum_{x \in \Gamma(n_1) \cap \Gamma(n_2)} \frac{1}{\log |\Gamma(x)|}$$

We compared our model to the three link prediction methods described above using variants of the link-centric experiments in the preceding section (figure 3c and d). In the first link prediction task, we remove a random 10% of the links in G, to yield G', on which which our model is run. For the other link prediction methods, G' is converted to a unipartite graph, $\widetilde{G}'$ from which predictions are generated. Because $\widetilde{G}'$ includes even the weakest links in G', it is quite dense. We varied the threshold required for a link to be considered and report performance on resulting graph. That is, the weakest dyadic ties (the same links in G' as in $\widetilde{G}'$) were successively removed over 40 cuts until the graph was empty. Figure 5a shows the results of each method under this local, link-deletion scheme.

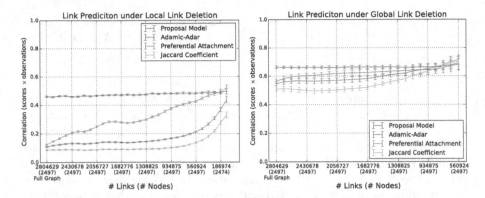

Fig. 5. (a; left): Spearman rank-correlation of predictions vs. observations on a randomly removed selection of 10% of the links. The lesioned graph, G' for the proposal model and $\widetilde{G}'$ for the other methods, was used to generate predictions. As weak links are successively removed from consideration (horizontal axis), the graph becomes increasingly sparse. After 36 cuts, the graph fractured into separate components. **(b; right)**: Analogous correlation on a random selection of 10% available links having been removed with their comprising hyper-edges (figure 4d). Here, the graph fractured after 30 cuts. All error-bars are $+/-$ 1 SE of the mean across 10 runs.

A high correlation implies that a method's predictions preserved the ranking of links in the initial graph, which is a stronger indication of performance than a binary distinction for extant links.

Figure 5a shows that when all links, weak and strong, are considered, the graph's density is detrimental to the performance of the unipartite methods. The performance of these methods, which do not account for link similarity, underscores its importance in dense settings where structural information is more uniform. The results also show that our model is considerably more stable with regards to the effect of weak links. The Jaccard coefficient and the Adamic-Adar method perform weakly until the number of links is reduced to the strongest 15%, after which they begin to rise. Preferential attachment exhibits similar behavior to the other models at either end of the link density spectrum, but performs steadily better as weak links are pruned. Our model is not only more consistent, its performance is only matched (by preferential attachment) when the links are reduced to their strongest 8%.

In a second link prediction experiment, the task is designed similarly to the final experiment in section 4.1. The setup is the same as the link prediction experiment above, except that when a link is selected for removal, all the hyper-edges comprising it are removed from the graph (see figure 3d). After this lesioning yields G', it is again simplified to $\widetilde{G}'$ for use with the other link prediction methods. Figure 5b shows the rank-correlations of the four methods in this task over the increasingly sparse graph. The results show that the added sparsity, due to the removal of additional hyper-edges during the lesioning process, gives all methods a boost. The proposal model still out-performs the other methods until the graph consists of only the strongest links.

4.3   Discussion of Results

Altogether, our model performs well across a range of evaluations. However, we make no claim that the algorithm is generalizable to hypergraphs of arbitrary structure, but for the purpose of recommending ties in an already densely connected social network, it is quite successful. Some of the model's strength is due to the data-structure itself, which is not only a novel operationalization of Feld's focus theory of social interaction [13], but also accurately realizes departmental structure (see the appendix). The four experiments in section 4.2 show that in many conceivable situations, corresponding to the respective lesioning schemes, the proposals can account for situations where connections either were not properly mined / observed, or have actually not been made. On the link prediction tasks in section 4.3, our model out-performs three other methods that use local structure. This setting is slightly unfair to the traditional link prediction methods because they are unable to account for link similarity. Because the unipartite graph $\widetilde{G}$ is a simple weighted graph that is exceptionally dense, structural information like a node's degree and neighborhood are relatively unhelpful. It might be that precisely the information that is lost in simplifying G to $\widetilde{G}$ is the very information making our model more accurate. It remains to be seen whether our model performs well in a dynamic setting where historical data are used to predict future links. In a static setting, however, our results not only support our model and representation, they also point to an important social dynamic: that in multi-faceted relationships, both the magnitude *and* similarity of ties are important in developing connections.

Though the results here support the veracity of proposals from our model, they could be evaluated more qualitatively by presenting experts with recommendations with the greatest difference between proposed scores and observed mass. Though this evaluation would be expensive, it would potentially uncover dimensions of scholarly relationships not accounted for in our representation. Another area for future exploration is dimensionality reduction. Unlike a number of recent approaches to tasks like collaborative filtering and community detection, our model does not employ dimensionality reduction on the observed structure (eg. [20]). Given the cell-wise sparsity of G (99.8% zero-values), we explored using sub-spaces reduced with PCA and tSNE. On the experiments in section 4.2, we found a steady degradation of performance as target the dimensionality approached 0, which implies there is little to no additional information embedded in lower-dimensional projections. This reduction amounts to a blind compression of the edges based on their points in node-space — a space we have no reason to believe is adequately represented by its most prominent components. Alternatively, one could imagine performing such compression in a semantically principled manner, for example, by collapsing similar publication venues or geographical locations, though we leave this to future work.

5   General Discussion

People are embedded in a deeply social, highly connected, dense social structure. As representations grow to account for the multiplexity of social ties, they will naturally begin to exhibit this density — something that is surely not unique to academics. Social media, for instance, has greatly added to the available ways for people to connect, ostensibly confounding traditional social network analysis. The hypergraph structure explored in this paper is able to capture the emergent multiplicities apparent in relationships, specifically among academics. By using the mass of links through two people's shared neighbors, scaled by the similarity of the same links, scores from our model were able to correlate strongly to initial observations. This confirms our hypothesis that link mass and similarity mutually influence how likely transit nodes are to connect neighbors.

One critique of the model is that it overlooks higher-order structure, which is often exploited using spectral methods [8,34] or exponential random graph models [23,35]. However, we found that the average dyad shares 74% of all other nodes as neighbors:

$$\frac{1}{|\mathbf{G}|} \sum_{n_1,n_2 \in \mathbf{G}; n_1 \neq n_2} \frac{|\Gamma(n_1) \cap \Gamma(n_2)|}{|\mathbf{G}| - 2} = 0.74 \tag{5}$$

This means that accounting for higher-order structure is largely unnecessary. Also, by limiting our model to first-order transitivities [38], one can recover a ranking of the shared neighbors that contributed the most to a proposal. This kind of ranking would be particularly important for a qualitative analysis that sought to justify *why* proposals were made.

The model explored in this paper is primarily a social one, where academics are placed in the space of locations, organizations and people observed in CVs. It would be possible to expand the set of edge-types that populate the hypergraph considered by our method. There is also, of course, no inherent reason to restrict analysis to CVs, especially as more robust edge-types are explored that may be found elsewhere. Other sources of relational information about academics might include research statements, grant proposals and research publications, which might encode topical and methodological connections among researchers. One could imagine recovering such information from text about research topics and their respective methods of inquiry using a mixture of human annotation, machine learning and crowd curating. Though these further sources of relational information remain unexplored, we hold that the representation and model presented here provide a strong foundation — one that is data-driven and that can account for the complex, multi-dimensional nature human of relationships.

Acknowledgments. Thanks to Alex Dunlap, Joshua Beck, Ariel Gans and Michael Hochman for help gathering, cleaning and analyzing data, as well as to Bill Shi, John Goldmsith and Birali Rusheda for advice on the model. Thanks to the SWIFT team (swift-lang.org) for help parallelizing various aspects of the model and to the Open Computing Consortium for computing resources. This work was supported by the

Neubauer Collegium at the University of Chicago and by a grant from the Templeton Foundation to the Metaknowledge Research Network.

Appendix: Classifying Academics

As a preliminary task, the veracity G was tested by using it to classify academics into their known departments. The expectation is that the closer two academics are in G, the more likely it should be they share a department. In this experiment, all NEs containing forms of the word *department* were discarded. Proximity was defined as the cosine similarity between two nodes' weighted edge-incidence vectors (rows in figure 2). All pairs of academics were ordered by decreasing distance and the percent matched was calculated at every rank (precision at rank; figure 6). Though some academics hold appointments in more than one department, which makes a random guess slightly easier than $\frac{1}{departments=90}$, a null model that guesses the most frequent department (Economics, $N = 194$) provides a baseline. Note that this classification model is not learned or fit to the data, rather, it simply shows that academics with similar relationships, defined by the weighted edge-incidence vectors, tend to be in the same department. While there is some difference in performance of each edge-type, they all do significantly better than the null model.

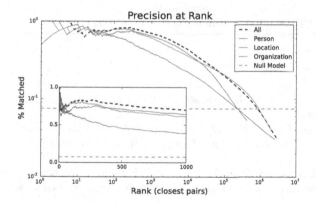

Fig. 6. The precision at rank for pairs of academics in decreasing order of similarity. Precision is defined as the pair being in the same department. In-set is a linear plot for the top 1,000 closest pairs.

To assess how useful the extracted NEs are as features in a statistical learning setting, they were used to train a multi-label classifier over departments. Table 1 shows the performance of the three feature-sets on a linear-kernel SVM and a random forest with 10 estimators [7,9]. To make the comparison fair, the other models were restricted to using the same number of features (the number of edges in the graph, $|E \in G|$). Each model was trained in one-vs-all scheme according to each departmental label held by a faculty member (max = 5). Validation consisted of 20 folds on 20% held-out data. In all configurations, the node-incidence vectors, $\{E \in G\}$, outperformed the unigram models. Due to the multi-label scheme, the test-performance was averaged over all labels (for a single fold), which could falsely inflate the performance on held-out data. However, held-out performance was not our goal: the features in G were not mined for the purpose of training a classifier, nor was the weighting-scheme designed for such a task. Instead, we note its performance against relatively simple unigram models to underscore that G does indeed encode departmental structure, indicating it reliably represents academic relationships.

Table 1. Multi-label classifier performance on three feature-sets: the weighted edge incidence vectors in G, count- and TF*IDF-weighted unigrams. Each set was used to train a linear-kernel support vector machine and a 10-estimator random forest. Reported are the test-scores averaged over 20 folds with randomized 80/20 train-test splits.

Feature-set	Model	Precision	Recall	F_1 Score
$\{E \in G\}$	Linear-SVM	**0.53**	**0.95**	**0.66**
Count Unigrams	Linear-SVM	0.15	0.27	0.19
TF*IDF Unigrams	Linear-SVM	0.45	0.66	0.51
$\{E \in G\}$	Random Forest	**0.62**	**0.92**	**0.74**
Count Unigrams	Random Forest	0.60	0.89	0.69
TF*IDF Unigrams	Random Forest	0.61	0.92	0.71

References

1. Adamic, L.A., Adar, E.: Friends and neighbors on the web. Social Networks **25**(3), 211–230 (2003)
2. Al Hasan, M., Chaoji, V., Salem, S., Zaki, M.: Link prediction using supervised learning. In: Workshop on Link Discovery: Issues, Approaches and Applications (2005)
3. Baronchelli, A., Ferrer-i Cancho, R., Pastor-Satorras, R., Chater, N., Christiansen, M.H.: Networks in cognitive science. Trends in Cognitive Sciences **17**(7), 348–360 (2013)
4. Boccaletti, S., Bianconi, G., Criado, R., Del Genio, C., Gómez-Gardeñes, J., Romance, M., Sendiña-Nadal, I., Wang, Z., Zanin, M.: The structure and dynamics of multilayer networks. Physics Reports (2014)
5. Bollobás, B.: Modern graph theory, vol. 184. Springer (1998)
6. Breiger, R.L.: The duality of persons and groups. Social Forces **53**(2), 181–190 (1974)
7. Breiman, L.: Random forests. Machine Learning **45**(1), 5–32 (2001)
8. Chung, F.R.: Spectral graph theory, vol. 92. American Mathematical Soc. (1997)
9. Cortes, C., Vapnik, V.: Support-vector networks. Machine Learning **20**(3), 273–297 (1995)
10. Duan, D., Li, Y., Li, R., Lu, Z.: Incremental k-clique clustering in dynamic social networks. Artificial Intelligence Review **38**(2), 129–147 (2012)

11. Dunbar, R.I., Spoors, M.: Social networks, support cliques, and kinship. Human Nature **6**(3), 273–290 (1995)
12. Easley, D., Kleinberg, J.: Networks, crowds, and markets: Reasoning about a highly connected world. Cambridge University Press (2010)
13. Feld, S.L.: The focused organization of social ties. American Journal of Sociology, 1015–1035 (1981)
14. Finkel, J.R., Grenager, T., Manning, C.: Incorporating non-local information into information extraction systems by gibbs sampling. In: Proceedings of the 43rd Annual Meeting on Association for Computational Linguistics, pp. 363–370. Association for Computational Linguistics (2005)
15. Fischman, J.: Arizona's big bet: The research rethink. Nature **514**(7522), 292 (2014)
16. Gerow, A.: Extracting clusters of specialist terms from unstructured text. In: Proceedings of the 2014 Conference on Conference on Empirical Methods in Natural Language Processing (EMNLP 2014), Doha, Qatar, pp. 1426–1434 (2014)
17. Gerow, A., Evans, J.: The modular community structure of linguistic predication networks. In: Proceedings of TextGraphs-9, Doha, Qatar, pp. 48–54 (2014)
18. Getoor, L., Diehl, C.P.: Link mining: a survey. ACM SIGKDD Explorations Newsletter **7**(2), 3–12 (2005)
19. Getoor, L., Friedman, N., Koller, D., Taskar, B.: Learning probabilistic models of link structure. The Journal of Machine Learning Research **3**, 679–707 (2003)
20. Goldberg, K., Roeder, T., Gupta, D., Perkins, C.: Eigentaste: A constant time collaborative filtering algorithm. Information Retrieval **4**(2), 133–151 (2001)
21. Guns, R., Rousseau, R.: Recommending research collaborations using link prediction and random forest classifiers. Scientometrics **101**(2), 1461–1473 (2014)
22. Heintz, B., Chandra, A.: Beyond graphs: toward scalable hypergraph analysis systems. ACM SIGMETRICS Performance Evaluation Review **41**(4), 94–97 (2014)
23. Holland, P.W., Leinhardt, S.: An exponential family of probability distributions for directed graphs. Journal of the American Statistical Association **76**(373), 33–50 (1981)
24. Lang, J., Lapata, M.: Similarity-driven semantic role induction via graph partitioning. Computational Linguistics **40**(3), 633–669 (2014)
25. Li, D., Xu, Z., Li, S., Sun, X.: Link prediction in social networks based on hypergraph. In: Proceedings of the 22nd International Conference on World Wide Web Companion, pp. 41–42 (2013)
26. Li, L., Li, T.: News recommendation via hypergraph learning: encapsulation of user behavior and news content. In: Proceedings of the Sixth ACM International Conference on Web Search and Data Mining, pp. 305–314. ACM (2013)
27. Li, X., Liu, B., Yu, P.S.: Discovering overlapping communities of named entities. In: Fürnkranz, J., Scheffer, T., Spiliopoulou, M. (eds.) PKDD 2006. LNCS (LNAI), vol. 4213, pp. 593–600. Springer, Heidelberg (2006)
28. Liben-Nowell, D., Kleinberg, J.: The link-prediction problem for social networks. Journal of the American Society for Information Science and Technology **58**(7), 1019–1031 (2007)
29. Lichtenwalter, R.N., Lussier, J.T., Chawla, N.V.: New perspectives and methods in link prediction. In: Proceedings of the 16th ACM SIGKDD International Conference on Knowledge Discovery and Data Mining, pp. 243–252. ACM (2010)
30. Ma, H., Yang, H., Lyu, M.R., King, I.: Sorec: social recommendation using probabilistic matrix factorization. In: Proceedings of the 17th ACM Conference on Information and Knowledge Management, pp. 931–940. ACM (2008)

31. Mika, P.: Ontologies are us: A unified model of social networks and semantics. Web Semantics: Science, Services and Agents on the World Wide Web **5**(1), 5–15 (2007)
32. Mitchum, R., Brand, A., Transande, C.: White paper: Information, interaction, influence: Research information technologies and their role in advancing science (2014)
33. Newman, M.E.: The structure and function of complex networks. SIAM Review **45**(2), 167–256 (2003)
34. Ng, A.Y., Jordan, M.I., Weiss, Y., et al.: On spectral clustering: Analysis and an algorithm. Advances in Neural Information Processing Systems **2**, 849–856 (2002)
35. Robins, G., Pattison, P., Kalish, Y., Lusher, D.: An introduction to exponential random graph (p*) models for social networks. Social Networks **29**(2), 173–191 (2007)
36. Sharma, A., Srivastava, J., Chandra, A.: Predicting multi-actor collaborations using hypergraphs. arXiv preprint arXiv:1401.6404 (2014)
37. Shi, F., Foster, J.G., Evans, J.: Weaving the fabric of science: Dynamic network models of sciences unfolding structure. Social Networks (forthcoming, 2015)
38. Sintos, S., Tsaparas, P.: Using strong triadic closure to characterize ties in social networks. In: Proceedings of the 20th ACM SIGKDD International Conference on Knowledge Discovery and Data Mining, pp. 1466–1475. ACM (2014)
39. Snijders, T.A.: Markov chain monte carlo estimation of exponential random graph models. Journal of Social Structure **3**(2), 1–40 (2002)
40. Taramasco, C., Cointet, J.-P., Roth, C.: Academic team formation as evolving hypergraphs. Scientometrics **85**(3), 721–740 (2010)
41. Tjong Kim Sang, E.F., De Meulder, F.: Introduction to the conll-2003 shared task: language-independent named entity recognition. In: Proceedings of the Seventh Conference on Natural Language Learning at HLT-NAACL 2003, vol. 4, pp. 142–147. Association for Computational Linguistics (2003)
42. Walter, F.E., Battiston, S., Schweitzer, F.: A model of a trust-based recommendation system on a social network. Autonomous Agents and Multi-Agent Systems **16**(1), 57–74 (2008)
43. Xia, F., Chen, Z., Wang, W., Li, J., Yang, L.T.: Mvcwalker: Random walk based most valuable collaborators recommendation exploiting academic factors. IEEE Transcactions on Emerging Topics in Computing **2**(3), 364–375 (2014)
44. Zhang, Z.-K., Liu, C.: A hypergraph model of social tagging networks. Journal of Statistical Mechanics: Theory and Experiment **2010**(10), P10005 (2010)

Photowalking the City: Comparing Hypotheses About Urban Photo Trails on Flickr

Martin Becker[1]([✉]), Philipp Singer[2], Florian Lemmerich[2], Andreas Hotho[1],
Denis Helic[3], and Markus Strohmaier[2,4]

[1] University of Wuerzburg, Wuerzburg, Germany
{becker,hotho}@informatik.uni-wuerzburg.de
[2] GESIS - Leibniz Institute for the Social Sciences, Mannheim, Germany
{philipp.singer,florian.lemmerich,markus.strohmaier}@gesis.org
[3] Graz University of Technology, Graz, Austria
dhelic@tugraz.at
[4] University of Koblenz-Landau, Mainz, Germany
strohmaier@uni-koblenz.de

Abstract. Understanding human movement trajectories represents an important problem that has implications for a range of societal challenges such as city planning and evolution, public transport or crime. In this paper, we focus on geo-temporal *photo* trails from four different cities (Berlin, London, Los Angeles, New York) derived from Flickr that are produced by humans when taking sequences of photos in urban areas. We apply a Bayesian approach called HypTrails to assess different explanations of how the trails are produced. Our results suggest that there are common processes underlying the photo trails observed across the studied cities. Furthermore, information extracted from social media, in the form of concepts and usage statistics from Wikipedia, allows for constructing explanations for human movement trajectories.

1 Introduction

Understanding the way people navigate urban areas represents an important problem that has implications for a range of societal challenges such as city planning and evolution, public transportation or crime. Recent research in the computational social sciences has studied human movement trajectories in cities through a variety of data sources including mobile phone data [15,35], GPS tracking [41], WiFi tracking [29], location-based social media platforms [8], online photo sharing sites [11,13,14] and others. Such studies have provided a number of insights into human movement trajectories. For example, past work has indicated that human mobility exhibits regularities [15,35] and spatio-temporal patterns [8]. Research has also shown that we can successfully leverage these patterns for certain tasks such as constructing high quality travel itineraries [11]. Yet, little is known about how the corresponding trails materialize, i.e., what factors play a role when people move through urban spaces. Thus, in this paper, we extend the stream of research on human movement by assessing different

© Springer International Publishing Switzerland 2015
T.-Y. Liu et al. (Eds.): SocInfo 2015, LNCS 9471, pp. 227–244, 2015.
DOI: 10.1007/978-3-319-27433-1_16

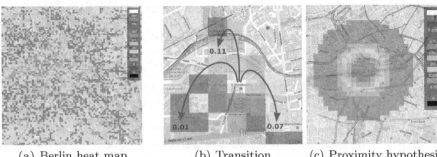

(a) Berlin heat map   (b) Transition   (c) Proximity hypothesis
                             probabilities

Fig. 1. *Main concepts of this work.* In (a), we visualize a cell-based grid layout of Berlin with photo frequencies visualized in a heatmap format, as derived from our data at interest. (b) depicts an example of transition probabilities between cells. In particular, it depicts the cell where the "German Bundestag" is located, and visualizes the transition probabilities to subsequent cells, i.e. cells where people take photos after they photographed the Bundestag. For instance, with a probability of 0.07, people take a picture at the "Brandenburg Gate" after they have taken one at the "Bundestag". As we are interested in gaining insights into the processes producing these trails, we formulate hypotheses based on belief in transitions (parameters of the Markov chain). In (c), we depict an exemplary proximity hypothesis which represents a belief that people successively take photos in proximate areas of a city; higher proximity refers to higher belief.

explanations (hypotheses) of *how* urban photo trails are produced. A better understanding of this process is relevant for a series of practical problems, such as local recommendations of picturesque locations, studying touristic movement patterns or movement of people in urban environments in general.

Problem and Objective. To that end, we aim to *compare a set of different hypotheses about how urban photo trails are produced* given actual sequential photo data from four cities (Berlin, London, Los Angeles and New York) as derived from Web data (Flickr). In particular, we want to assess the plausibility of different potential explanations (hypotheses) for the trails of photos that we can observe and compare them across cities. We define an *urban photo trail* as a sequence of spatial positions in a city over a period of time as, e.g., obtained from the geo-temporal metadata of photos. Then, *hypotheses* can be expressed as different beliefs about transitions between spatial positions. For example, we might want to compare a *proximity hypothesis*—which represents a belief that people frequently take subsequent photos in geographically close regions of a city— with another *points of interest (POI) hypothesis* that represents the belief that humans take subsequent photos of POIs.

Approach. To tackle these challenges, we resort to a Bayesian approach called *HypTrails* [32] that allows for relative comparison of hypotheses given sequential categorical data. The approach is based on Markov chain modeling and Bayesian inference; hypotheses can be seen as Markov transitions and our belief in them. For working with HypTrails, we map each photo of a trail to a cell of a geo-spatial grid over the city. For constructing hypotheses, we utilize information extracted from the social semantic web, specifically, from Wikipedia, DBpedia and YAGO and corresponding usage statistics. For illustration, we visualize and explain the main concepts of this work in Figure 1.

Contributions and Findings. The main contribution of this work is a systematic evaluation of hypotheses for explaining how urban photo trails are produced in four different cities. We find interesting commonalities, in particular we find that the partial ordering of evidence for different hypotheses is quite stable across the cities we have investigated. Furthermore, information extracted from social media, in the form of concepts and usage statistics from Wikipedia, allows for finding advanced explanations for human movement trajectories. Most prominently, our results suggest that humans seem to prefer to consecutively take photos at proximate POIs that are also popular on Wikipedia. In addition, we can also observe differences between cities: For example, proximity is less relevant for Los Angeles, which is a plausible finding given the unique topology of the city among the studied datasets.

The findings of our work can enable photo sharing websites to offer localized recommendations of picturesque photo spots according to actual tourist trajectories, city planners to explore human movement patterns of its inhabitants in general or tourist organizations to facilitate and optimize tours.

Structure. We start by giving an overview of related work in Section 2. Section 3 describes the utilized HypTrails method as well as the Flickr data at interest. In Section 4, we present our hypotheses of interest. Section 5 summarizes our experiments and presents the corresponding results. We discuss our work in Section 6 and conclude it in Section 7.

2   Related Work

This section covers related work from three areas of research: (i) studies on geo-spatial trails, (ii) studies on Flickr, and (iii) studies on human trails on the Web in general.

Geo-Spatial Trails. In the past, researchers have studied geo-spatial trails originating from diverse sources along several dimensions. For example, Song et al. [35] as well as Gonzales et al. [15] showed that geo-spatial trails (derived from mobile phone data) indicate high predictability with a lack of variability between humans. Cho et al. [8] supplemented mobile phone data with check-in data from social networks and found that human mobility is not only driven by periodicity, but also by social ties. Generally, social check-in data on the Web has supplemented our understanding of human mobility as e.g., shown in [9,25,26]. Further works have studied human mobility as derived from e.g., Twitter [12,16], taxi data [27] or bike data [18].

Flickr. Geo-spatial trails have also been studied on Flickr; e.g., De Choudhury et al. [11] aimed at leveraging photo trails for automatically constructing travel itineraries through cities by utilizing popularity of POIs. Similarly, Tai et al. [36] used past landmarks photographed by users for recommending sequences of new landmarks as derived from sequential information by other users on Flick. Girardin and colleagues have conducted several studies on Flickr photo trails. In [13], they studied digital footprints and in [14] they focused on tourist dynamics based on concentrations and spatio-temporal flows revealing popular points of interests, density points and common trails tourists follow. Apart from trails, Flickr has also been studied in other contexts specifically regarding tagging [10,23,31] and social network properties [5,24].

Human Trails on the Web. Our research community has studied sequential interactions of humans with the Web in various settings. While some work has mainly focused on modeling (e.g., [2,3,7,28,30,33]), others have been more interested in investigating regularities, patterns and strategies (e.g., [17,38,39]) that emerge when humans sequentially engage with the Web. This kind of work has been interested in answering similar objectives as this paper, i.e., understanding sequential steps by humans.

Many researches have been studying human navigational trails on the Web [4,17,33,39]. This line of research has inspired other works to improve the Web, based on these found patterns by e.g., better website design [6]. Similarly, a better understanding of human mobility might lead to improvements in urban city design. Other works have focused on e.g., studying search trails [40], diffusion trails [1] or ontology editing trails [37].

3 Methods and Material

For comparing hypotheses about how humans choose their next location to take a photo, we resort to an approach called *HypTrails* [32] and study data derived from *Flickr*, a social photo sharing platform. This section describes both the method and the data set.

3.1 Methodology

HypTrails [32] is an approach utilizing first-order Markov chain models [33] and Bayesian inference [20] for expressing and comparing hypotheses about human trails. Since a Markov chain model is a stochastic system that models transition probabilities between a set of discrete states, it is not directly applicable to the continuous space of geo-spatial data. To address this issue, we discretize the geo-spatial regions by defining a grid over the target area where the grid cells then can be interpreted as states in the Markov chain. This step allows us to use HypTrails in the context of geo-spatial data.

With HypTrails, hypotheses can be expressed as beliefs in Markov transitions, i.e., assumptions on common and uncommon transitions at individual states; Section 4 presents several such hypotheses about human photo trails. To obtain insights into the relative plausibility of a set of hypotheses given data, HypTrails resorts to Bayesian inference and specifically to the marginal likelihood—also called *evidence*—denoting the probability of the data D given a hypothesis H. HypTrails utilizes the sensitivity of the marginal likelihood on the conjugate Dirichlet prior for comparing hypotheses. The main idea is thus, to incorporate hypotheses as Dirichlet priors into the inference process. The hyperparameters of Dirichlet distributions can be interpreted as pseudo counts; higher pseudo counts refer to higher beliefs in a specific transition.

Thus, for each hypothesis, we need to provide a hypothesis matrix Q that captures our generic beliefs in transitions (see Section 4) based on our geospatial states. HypTrails then internally elicits proper Dirichlet priors from these expressed hypotheses matrices by setting the pseudo counts of the priors accordingly. An additional parameter K steers the total number of pseudo counts assigned; the higher we set it, the more we believe in a given hypothesis. Basically, this means, that with higher K, the stronger we believe in the single parameter configuration specified in Q. With lower values of K, the Dirichlet prior also assigns more probability mass to other, similar parameter configurations, thus giving the hypothesis some "tolerance".

Based on the elicited priors for different hypotheses, HypTrails determines the marginal likelihood with respect to the empirical (observed) data. Then, we can judge the relative plausibility of two hypotheses by comparing their evidences for the same K; higher evidences refer to higher plausibility. The fraction of the evidence of two hypotheses (priors), called Bayes factor [19], is then used for determining the strength of evidence for one hypothesis over the other. A Bayes factors can be directly interpreted as the Bayesian equivalent to a frequentist's significance value. In this article, all Bayes factors for reported results are decisive. Therefore, we refrain from explicitly reporting them and, for simplicity, we can state that one hypothesis H_1 is more plausible compared to another hypothesis H_2, if the evidence of H_1 is higher than the one of H_2 for the same value of K. Thus, the partial ordering based on the plausibility of respective hypotheses $\mathbf{H} = \{H_1, H_2, ..., H_n\}$ can be determined by ranking their evidences from largest to smallest for single values of K. In this work, we report evidences on a log scale. We present corresponding results in Section 5.

Table 1. Bounding boxes and center coordinates used for data collection and hypothesis creation.

	min lon.	min lat.	max lon.	max lat.	center lon.	center lat.
Berlin	13.088400	52.338120	13.76134	52.675499	13.383333	52.516667
London	-0.5103	51.2868	0.3340	51.6923	-0.1280	51.5077
Los Angeles	-118.6682	33.7037	-118.1552	34.3368	-118.2450	34.0535
New York	-74.2589	40.4774	-73.7004	40.9176	74.0071	40.7146

3.2   Datasets

In this work, we study human phototrails through cities by analyzing data from the social photo-sharing platform Flickr[1]. In this section, we first describe the data collection and its transformation into the required representation of trails and state transitions. Then, we highlight some basic characteristics of our datasets.

Data Collection. Our datasets[2] contain metadata—i.e., user, temporal and geo (latitude and longitude) data—about images uploaded to the Flickr platform. In particular, we focus on pictures taken in the cities of Berlin, London, Los Angeles and New York between January 2010 and December 2014. For each city, we define a bounding box, see Table 1. We acquired corresponding data by crawling Flickr's public API. Since our analysis requires an exact position, we remove pictures with less than street-level accuracy (level 16 on the Flickr scale[3]).

For our analyses, we interpret the sequence of all photos of a single user as a *phototrail* ordered by the time each photo was taken, regardless of the time difference between the photos, see also Section 6. Each photo in a trail is mapped to a cell of a grid that we place over the respective city according to its geo-reference. The grid cells are then used as the discrete states required by the HypTrails approach. Since we want to capture how people move between places in a city, such as sights or train stations, we choose a cell size of 200m x 200m for our experiments. From our experience, this cell size is small enough to distinguish places close to each other and it is large enough to aggregate movement at a single place as well as to reduce the sensitivity due to GPS inaccuracies. Figure 1 shows cells of such a grid on Berlin to give an idea about the chosen granularity.

Finally, since we are only interested in the sequence of different places people photograph, we remove all self-transitions (i.e., transitions from one grid cell to itself) from the photo trails in order to account for people taking several photos at one place. Basic statistics of the processed datasets are summarized in Table 2.

Points of Interest. We work with hypotheses (see Section 4) that utilize information about points of interest (POIs) in a city. We query these POIs from

Table 2. Basic dataset statistics.

city	years	photos	cells	trails	covered cells	avg. trail length	POIs	avg. view counts
Berlin	2010-11	60,978	43,052	4,364	6,343	13.97	1,085	1,240
London	2010-14	794,535	66,444	35,101	23,694	22.64	7,228	1,272
Los Angeles	2010-14	300,373	84,014	15,357	25,834	19.56	1,462	3,654
New York	2010-14	714,549	58,065	31,246	15,232	22.87	6,002	1,511

---

[1] https://flickr.com

[2] Dataset access can be requested via e-mail.

[3] see https://www.flickr.com/services/api/flickr.places.findByLatLon.html

the social semantic web, in our case DBpedia [21] and YAGO [22]. For each city, the POIs are filtered by bounding box. Also, area concepts such as "Germany" or "Berlin", which do not correspond to actual locations (*rdf:type* equal to *yago:District108552138*), are removed.

Additionally, we quantify the importance of a POI in some hypotheses. As an approximate measure of importance we take pageview counts of the Wikipedia articles describing the POIs. For that purpose, we extracted view counts from data available at the Wikimedia download page[4]—in this work, we use the view counts for January 2012. Table 2 shows the number of POIs per city and their average view count.

4 Hypotheses

The HypTrails approach allows to compare hypotheses about human trails on discrete states—see Section 3 for details. Hypotheses are expressed by constructing matrices that reflect beliefs about transitions between such states. This section describes how several intuitions about phototrails can be expressed as hypothesis matrices Q.

For our geo-spatial setting, we define states by discretizing the continuous geo-spatial area using a grid. Then, we elicit the belief in the transition between each ordered pair of grid cells. That is, given a grid cell s_i of a user's last photo, we specify how likely her next photo will be taken in every other cell s_j. In the HypTrails approach, higher values correspond to higher beliefs in the corresponding transitions. In this paper, we express our beliefs as local transition probabilities $P(s = s_j|s_i)$. For example, if a hypothesis assumes that a user, who took her last photo in cell s_1, will take the next photo in cell s_2 with probability 0.5, then we set $P(s = s_2|s_1) = 0.5$. We assign these transition probabilities between states as the values of the hypothesis matrix Q: $Q(i,j) = P(s = s_j|s_i)$. Please note that the hypothesis matrix is a stochastic matrix since each row i of Q sums to 1, i.e. $\sum_j Q(i,j) = 1$.

For simplicity, we do not directly express transition beliefs as probabilities, but rather in the form of a belief function $\bar{P}(i,j)$. This function can then be transformed into a probability distribution by multiplying it with a normalization factor $\frac{1}{Z}$ obtained by summing over all values of $\bar{P}$ with regard to the source cell s_i:

$$P(s = s_j|s_i) = \frac{1}{Z}\bar{P}(i,j), \quad Z = \sum_{j=1}^{n} \bar{P}(i,j)$$

In this paper, we apply Gaussian distributions for weighting transition probabilities or factors. In this context, the elements of a hypothesis matrix Q often take very small values. For practical computational reasons, we set the value for a belief in a transition $Q(i,j) = \bar{P}(i,j)$ to 0 if the transition probability falls below a threshold. For our experiments we use 0.01 as a threshold. Furthermore, we set the beliefs in self-transitions to zero ($\bar{P}(i,i) = 0$) for all hypotheses as we

---

[4] http://dumps.wikimedia.org/other/pagecounts-raw/

are more interested in modeling actual movement. This is in accordance to the removal of self-transitions in the observed data for our experiments. Next, we describe the hypotheses that we analyze in this paper.

Uniform Hypothesis. This hypothesis believes that each transition is equally likely assuming that users take pictures uniformly at random anywhere in the city regardless of the previous location:

$$\bar{P}_{uniform}(i, j) = 1$$

We use the uniform hypothesis as a baseline hypothesis: an informative hypothesis should at least be more plausible than the uniform hypothesis in order to express valid notions about the processes underlying human photo trails.

Center Hypothesis. Typically, the city center is the most lively part of a city and this hypothesis assumes that a user always takes her next picture near the city center regardless of the location of her last picture. To formalize this hypothesis, we use the geographic center C of the city (as listed in Table 1) and lay a two-dimensional Gaussian distribution centered at this point over the corresponding grid. Given the geographic (haversine [34]) distance $dist(C, j)$ between the city center C and the central point of the grid cell s_j, we calculate the entries of the hypotheses matrices from the following distribution:

$$\bar{P}_{center}(i, j) = e^{-\frac{1}{2\sigma^2}dist(C,j)^2}$$

We parametrize the center hypothesis with the standard deviation σ (e.g., in kilometers). A small value of σ indicates the belief that most pictures are taken very close to the city centre. If σ approaches infinity the hypothesis approximates the uniform hypothesis.

Points of Interest (POI) Hypothesis. Previous work on photo trails has shown that it is possible to automatically construct travel itineraries through a city by analyzing Flickr users behavior [11]. This suggests that humans favor points of interests—including not only tourist attractions, but also important public transportation spots or the locations of government institutions—when taking photos throughout major urban tourist areas. Thus, the POI hypothesis captures the intuition that people take a majority of pictures near such POIs. Similar to the center hypothesis, we express the attraction force of each POI with a two-dimensional Gaussian distribution. Formally, for each cell s_j and each POI $q \in Q$, we get an attraction value $G(q, j)$ that corresponds to the likelihood that q generates a picture in cell s_j according to their distance:

$$G(q, j) = e^{-\frac{1}{2\sigma^2}dist(q,j)^2}$$

Here, as before, $dist(q, j)$ describes the haversine distance between POI q and cell s_j. Then, for each cell, we aggregate the distance weighted attraction values of all POIs.

$$\bar{P}_{poi}(i, j) = \sum_{q \in Q} G(q, j)$$

In doing so, cells that contain multiple POIs have a stronger attraction to their neighboring cells. Again, we have to choose an appropriate standard deviation σ; a small σ assumes that photos are taken directly at the point of interest, whereas a larger σ assumes that pictures are taken in the surroundings of a POI. Larger values of σ may represent the fact that people do not take pictures directly at a POI, e.g., to cover an architectural attraction fully in one picture, or they find something interesting to photograph nearby. In this work, we utilize points of interest extracted from DBpedia, see Section 3.2.

Weighted POI Hypothesis. Each city contains a large amount of potential POIs, however, not all of these are equally important. For example, the "Brandenburg Gate" is more likely to influence human trails in Berlin than the less known "Charlottenburg Gate". We capture this intuition in a weighted POI hypothesis by approximating the importance of a POI q by the view count $views(q)$ of the Wikipedia article corresponding to this POI. If the view count of an article is very high (as e.g., for the "Brandenburg Gate"), we expect the respective POI to have a stronger influence on the sequence of image locations. We quantify this hypothesis by weighting each term of the POI hypothesis:

$$\bar{P}_{weighted\_poi}(i,j) = \sum_{q \in Q} (views(q) \cdot G(q,j)).$$

Since we expect view counts to follow a power law, we also apply a sub-linear weighting scheme to avoid overemphasizing the importance of large points of interest:

$$\bar{P}_{log\_weighted\_poi}(i,j) = \sum_{q \in Q} (\log(views(q)) \cdot G(q,j)).$$

Proximity Hypothesis. This hypothesis—motivated by findings of previous work [8,15,32]—expresses the belief that the next image of a user will be taken nearby the last image. To formalize this hypothesis, we consider the haversine distances $dist(i,j)$ between the center points of two cells s_i, s_j. Then, we can again specify the respective transition probabilities by applying a two-dimensional Gaussian distribution:

$$\bar{P}_{prox}(i,j) = e^{-\frac{1}{2\sigma^2}dist(i,j)^2}$$

As before, a standard deviation σ must be specified; a small value of σ suggests a photo is more likely to be taken very close to a user's previous photo. An example for this hypothesis is depicted in Figure 1(c) where we visualize our beliefs in transitions from one state to other states (i.e., one row of Q).

Mixture of Hypotheses. Finally, we are interested in studying the effects of a mixture of two hypotheses. Technically, we mix two hypotheses by element-wise multiplication of the corresponding hypothesis matrices. In this paper, we study two mixtures combining the intuition that people are likely to take pictures at POIs or close to the city center on one hand, but at the same time stay close to their current location for their next photo on the other hand. We can capture this by combining the POI or center hypotheses with the proximity hypothesis. Please note that other kinds of combinations are also conceivable.

Proximate weighted POI hypothesis. First, we are combining the POI hypothesis with the proximity hypothesis, i.e., we assume that people will move to a POI to take their next photo but, instead of moving to a random POI, they choose one close by.

$$\bar{P}_{prox\_(log\_)weighted\_poi}(i,j) = \bar{P}_{prox}(i,j) \cdot \bar{P}_{(log\_)weighted\_poi}(i,j.)$$

Proximate center hypothesis. Similarly, the following formulation expresses the belief that the next picture is likely taken closer to the city center, but limits the area to move to to a location close to the current one.

$$\bar{P}_{prox\_center}(i,j) = \bar{P}_{prox}(i,j) \cdot \bar{P}_{center}(i,j.)$$

5   Experiments

In Section 4, we introduced a set of hypotheses that express beliefs on where people take their next picture while moving through a city. In this section, we compare these hypotheses with each other based on empirical trails derived from four different cities—Berlin (Germany), London (United Kingdom), Los Angeles (USA), and New York (USA) (see Section 3.2)—by employing the HypTrails approach as outlined in Section 3.1.

First, we focus on Berlin as a representative example in Section 5.1. We report in-depth experimental results for different parameter settings of each hypothesis. Afterwards, we report results for the other three cities in Section 5.2 focusing on the individually best parameter settings and highlight prominent differences between them.

5.1   Berlin

In this section, we thoroughly study each hypothesis and their different parameterizations in the same order as they have been introduced in Section 4, focusing on Berlin.

Center Hypotheses. For Berlin, the most photos are clearly centered around the cultural center as shown in Figure 1(a). Thus, we expect the center hypothesis—i.e., the belief that people move towards the city center and stay there for taking photos (see Section 4)—to be a better explanation of human photowalking behavior than our baseline (uniform) hypothesis. We consider the center of Berlin (see Table 1) and four different standard deviations σ: 1km, 3km, 5km and 10km. The results are depicted in Figure 2(a).

As expected, the results show that for all considered values of $K > 0$ and all parameterizations of the hypothesis, the center hypothesis is more plausible than the uniform hypothesis (higher evidences). The best center hypothesis is

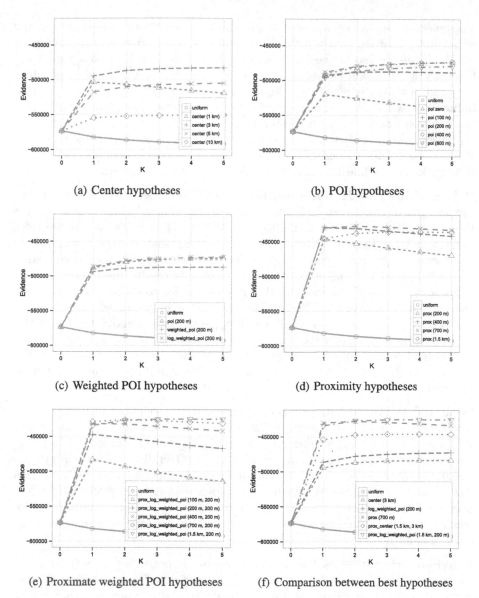

(a) Center hypotheses

(b) POI hypotheses

(c) Weighted POI hypotheses

(d) Proximity hypotheses

(e) Proximate weighted POI hypotheses

(f) Comparison between best hypotheses

Fig. 2. *Berlin.* This figure visualizes the results for our hypotheses on human photo trails on Berlin. First, for each type of hypotheses at interest, we compare various parameter configurations (a-e). Then, in (f) we compare the best hypotheses from each set. Overall (f), a combination of proximity and weighted POIs provides the best hypothesis. This suggests that people prefer to subsequently take photos at important, yet proximate POIs in a city (c.f., Section 5.1).

based on $\sigma = 3$km and the worst on $\sigma = 10$km. Standard deviations of 1km and 3km are mediocre and cross for increasing K. The initially high evidence values of 1km mean that this hypothesis covers an important aspect of the data.

The quickly dropping values, however, are an indicator that it also fails to model important transitions outside the 1km radius. This is because with increasing K, HypTrails decreases the tolerance of a hypothesis, cf. Section 3.1 and [32]. Contrary, for $\sigma = 5$km low values of K show lower evidence, but it does not drop as quickly, eventually resulting in higher evidence values than for $\sigma = 1$km. This indicates that the 5km standard deviation covers most transitions, but fails to model the strong focus on the center aspect.

Overall, we find that the center hypothesis is a reasonable explanation for photowalking trails in Berlin. In detail, of the investigated standard deviations, 3km works best, while 1km is too specific and 5km is too broad.

Points of Interest Hypotheses. For the POI hypothesis (see Section 4), we consider five different standard deviations: 0m (only considering the grid cell the POI is located in), 100m, 200m, 400m, and 800m. The results (see Figure (2b)) suggest that the POI hypothesis provides good explanations about how people photowalk a city as all parameterizations indicate higher evidence compared to the baseline (uniform) hypothesis. In detail, the results show that the POI hypothesis without Gaussian spread performs inferior to those POI hypotheses allowing their influence to spread. The two rather close-ranged spreads 200m and 400m perform the best implying that people indeed move towards POIs. The worse performance of too narrow and too wide ranges is an indicator that people tend to visit places and take photos of the place at a close range, but not necessarily directly at the POI. For example, a minimum range might be required to capture a large building in one picture.

Weighted Points of Interest Hypotheses. In the weighted POI hypotheses, more popular POIs have a stronger influence on the transitions. For tractability, we focus on the best spreading parameter σ for the unweighted POI hypothesis from the previous paragraph, i.e. $\sigma = 200$m. Overall (see Figure (2c)), the hypothesis that people prefer to take pictures at places with many important POIs (here, derived from Wikipedia) provides a reasonable explanation for how people photowalk a city. By using online usage statistics from Wikipedia (view counts), we can strengthen the evidence of the hypothesis by a small, but significant amount if we use logarithmic scaling.

Proximity Hypotheses. For the the proximity hypothesis (see Section 4), we use four different standard deviations σ: 200m, 400m, 700m and 1.5km. Overall, the results shown in Figure (2d) demonstrate that the hypothesis that people prefer to consecutively take pictures in their proximity captures an important aspect of the production of human photo trails; $\sigma = 700$m produces the highest evidence for all considered values of $K > 0$. For standard deviations of 200 and 400m, a similar situation occurs as for the center hypotheses with a standard deviation of 1km: They seem to concentrate their belief on a too narrow proximity leading to decreasing evidence values for higher values of K. Contrary, the proximity hypothesis with $\sigma = 1.5$km is too broad, somewhat neglecting the centralized character of the proximity aspect.

Mixtures of Hypotheses. To evaluate the mixture of POI and proximity hypothesis, we focus on the logarithmically weighted POI hypothesis with $\sigma =$ 400m since it was one of the best performing hypotheses so far. This is combined with different standard deviations for proximity, i.e., 100m, 200m, 400m, 700m and 1.5km. The results shown in Figure (2e) indeed demonstrate that adding the proximity aspect to the POI hypothesis strongly improves the evidence of the corresponding belief how people consecutively take pictures in a city. The best results can be achieved with larger standard deviations σ, i.e., $\sigma = 700$m and $\sigma = 1.5$km.

We also investigated different parametrization for the mixture of the proximity and the center hypothesis (not visualized in this paper due to limited space). The best parameter setting was a standard deviation of 3km for the center and a standard deviation of 1.5km for the proximity hypothesis (see Figure (2f)).

Comparison. For a direct comparison of the different hypotheses we are taking the most plausible ones (best parameters) of each set as elaborated beforehand. The results are shown in Figure (2f). We can see that the center and the *weighted* POI hypothesis perform quite similar which may be due to the larger number of (important) POIs in the city center. At the same time, the proximity hypothesis performs very well and combining it with the other hypotheses improves them strongly. Indeed, the combination of the proximity hypothesis and the *weighted* POI hypothesis provides the best explanation of how people move around Berlin while taking photos. This result suggests that information extracted from the social semantic web, in the form of concepts and usage statistics from Wikipedia, allows for finding advanced explanations for human movement trajectories.

5.2   Los Angeles, London and New York

To further augment the results from Section 5.1, we analyze three more cities, namely, Los Angeles (USA), London (United Kingdom) and New York City (USA). We show similarities and highlight some differences between the cities. For a concise presentation, we focus on the best parameter settings for each hypothesis only. The best parameters were determined separately for each city. Results are depicted in Figure 3. For most parts, all hypotheses perform very similar and the best parametrizations are consistent. This indicates that the hypotheses about photo trails in Berlin can be generalized to other cities quite well, implying that some basic patterns exist that even hold across countries.

However, there are two exceptions which are worth mentioning. First, in Los Angeles (see Figure (3c)), the most plausible center hypothesis has a standard deviation of 10km instead of 3km. This indicates that LA either has a very large center or none at all—arguably, LA is a spread out city which may cause this divergence. Additionally, in LA higher standard deviations for the POI hypothesis, i.e., 400m instead of 200m, are favored compared to the other cities. Also, even the best performing hypotheses are strongly decreasing with increasing K. This further supports the idea that LA is structurally different from the other cities. Second, the linearly weighted POI hypothesis in London is superior to the

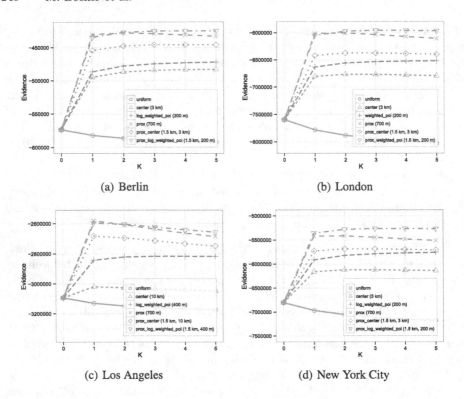

Fig. 3. *Comparison of cities.* This figure visualizes the results for our studies on human photo trails derived from Berlin(a), London (b), Los Angeles (c), and New York City (d). We only present a comparison of the best hypotheses for each type of hypotheses for each city. We can identify similar explanations across cities, but there are some differences for LA and London (c.f., Section 5.2).

logarithmically weighted one. This may be due to different view count distributions and has to be further investigated in the future.

6   Discussion

In this work, we have conducted extensive experiments to gain a better understanding of the underlying processes that are employed when people take photos while moving through cities. We dedicate this section to discuss characteristics specific to our approach and corresponding results and to highlight some potential limitations.

Data Characteristics. Next, we shortly discuss four relevant aspects regarding our data: (i) splitting photo trails due to time constraints, (ii) observation sparsity, (iii) Flickr movement characteristics and (iv) state granularity.

(i) We have considered the sequence of *all* photos of a user as a single phototrail regardless of the time span in between two photos. However, if such a time span is too long (e.g., a week or even a few hours), the corresponding two photos are most likely unrelated. Thus, in additional experiments, we have removed transitions with time intervals exceeding 6 or 24 hours respectively The results are very similar to the ones reported in Section 5 which is why we refrain from explicitly reporting them.

(ii) Since we are using grids with 200m by 200m cells over relatively large areas, the number of observations for corresponding transitions is limited. However, as HypTrails automatically focuses on observed states, the sparsity of the data does not randomly bias our results. Thus, no further testing of possibly derivable predictors is necessary since all available information is drawn directly from the data via Bayesian inference.

(iii) Due to our focus on studying Flickr, we are only able to make judgments about behavioral aspects that emerge when people move through a city and take photos as captured by Flickr. Studying other forms of mobility data might reveal different results. However, we might assume that certain behavioral aspects are similar, regardless which type of data we look at as suggested in [8]. This may be verified, for example by contrasting different cities or consider different kinds of movement data—e.g., social check-in data, mobile phone data or business reviewing data.

(iv) While the focus on an intra-city level in this work has allowed us to gain insights into urban behavior, we might observe different movement patterns if we extended our scope of interest. To give an example, by looking at cities, we constrain our studies to a small geographic area which might favor proximity based hypotheses. If we extended the scope, by e.g., looking at a country or continent level, the results might largely differ. However, then, other kinds of hypotheses might be more plausible to study.

Choice of Hypotheses. The observations in this work are limited by our choice of which hypotheses to study and how to express them; they have mostly been motivated by related work. Many other kinds of hypotheses are conceivable and can be investigated with HypTrails and our data. We suggest some potential candidates: (i) A hypothesis expressing the belief that a river is a natural barrier in a city. (ii) Also, district boundaries may be some kind of barrier. Additionally, (iii) demographic aspects (such as crime rates) might influence movement patterns in a city.

Tourists and Other User Groups. Previous work has suggested that the photographing behavior on Flickr differs between tourists and residents in a city [11,14]. The authors of [11] argue that residents are not under the direct pressure of visiting as many POIs within a certain time span as tourists are. Thus, we might also see differences in their behavioral aspects producing the human photo trails studies in this article. The same may apply for a number of other user groups or sub-groups, such as visitors from different countries or users from different generations. While we have focused on an aggregated view in this paper, the distinction between such user groups might be an additional highly interesting layer which we leave open for future work.

7   Conclusion

In this paper, we have investigated and compared hypotheses about urban photo trails across different cities by analyzing sequences of geotagged photos uploaded to the Flickr platform using the Bayesian HypTrails approach. For the informed specification of hypotheses, we have utilized additional data sources such as DBpedia, YAGO and view counts of Wikipedia articles which has allowed us to find advanced explanations for human movement trajectories. Our results suggest that cities share interesting commonalities and differences. For example, while proximity is an overall good explanation across all cities, for the city of Los Angeles we observe movement patterns on a different scale. Most prominently, our results suggest that humans seem to prefer to consecutively take photos at proximate POIs that are popular on Wikipedia.

In future work, we plan on extending our work by looking at additional cities. Furthermore, it would be interesting to expand the current city-level analysis to a larger scale, e.g., trails across different cities or countries. Finally, improved tool support for the interactive exploration of location sequences and hypotheses would be helpful.

Acknowledgments. This work was partially funded by the DFG in the research project "PoSTs II". Special thanks goes to Georg Dietrich from Chair 6 of Computer Science at the University of Würzburg for crawling the Flickr data.

References

1. An, J., Quercia, D., Crowcroft, J.: Partisan sharing: facebook evidence and societal consequences. In: Conference on Online Social Networks, pp. 13–24. ACM (2014)
2. Borges, J., Levene, M.: Evaluating variable-length markov chain models for analysis of user web navigation sessions. IEEE Transactions on Knowledge and Data Engineering **19**(4), 441–452 (2007). http://dx.doi.org/10.1109/TKDE.2007.1012
3. Brin, S., Page, L.: The anatomy of a large-scale hypertextual web search engine. In: International Conference on World Wide Web, pp. 107–117. Elsevier Science Publishers B. V. (1998)
4. Catledge, L.D., Pitkow, J.E.: Characterizing browsing strategies in the world-wide web. Computer Networks and ISDN Systems **27**(6), 1065–1073 (1995)
5. Cha, M., Mislove, A., Gummadi, K.P.: A measurement-driven analysis of information propagation in the flickr social network. In: International Conference on World Wide Web, pp. 721–730. ACM (2009)
6. Chi, E.H., Pirolli, P., Pitkow, J.: The scent of a site: a system for analyzing and predicting information scent, usage, and usability of a web site. In: Proceedings of the SIGCHI Conference on Human Factors in Computing Systems, pp. 161–168. ACM (2000)
7. Chi, E.H., Pirolli, P.L.T., Chen, K., Pitkow, J.: Using information scent to model user information needs and actions and the web. In: Conference on Human Factors in Computing Systems, pp. 490–497. ACM (2001). http://doi.acm.org/10.1145/365024.365325

8. Cho, E., Myers, S.A., Leskovec, J.: Friendship and mobility: user movement in location-based social networks. In: International Conference on Knowledge Discovery and Data Mining, pp. 1082–1090. ACM (2011)
9. Cramer, H., Rost, M., Holmquist, L.E.: Performing a check-in: emerging practices, norms and'conflicts' in location-sharing using foursquare. In: Proceedings of the 13th International Conference on Human Computer Interaction with Mobile Devices and Services, pp. 57–66. ACM (2011)
10. Crandall, D.J., Backstrom, L., Huttenlocher, D., Kleinberg, J.: Mapping the world's photos. In: International Conference on World Wide Web, pp. 761–770. ACM (2009)
11. De Choudhury, M., Feldman, M., Amer-Yahia, S., Golbandi, N., Lempel, R., Yu, C.: Automatic construction of travel itineraries using social breadcrumbs. In: Conference on Hypertext and Hypermedia, pp. 35–44. ACM (2010)
12. Gallegos, L., Lerman, K., Huang, A., Garcia, D.: Geography of emotion: Where in a city are people happier? arXiv preprint arXiv:1507.07632 (2015)
13. Girardin, F., Calabrese, F., Fiore, F.D., Ratti, C., Blat, J.: Digital footprinting: Uncovering tourists with user-generated content. Pervasive Computing 7(4), 36–43 (2008)
14. Girardin, F., Fiore, F.D., Ratti, C., Blat, J.: Leveraging explicitly disclosed location information to understand tourist dynamics: a case study. Journal of Location Based Services 2(1), 41–56 (2008)
15. Gonzalez, M.C., Hidalgo, C.A., Barabasi, A.L.: Understanding individual human mobility patterns. Nature 453(7196), 779–782 (2008)
16. Hawelka, B., Sitko, I., Beinat, E., Sobolevsky, S., Kazakopoulos, P., Ratti, C.: Geo-located twitter as proxy for global mobility patterns. Cartography and Geographic Information Science 41(3), 260–271 (2014)
17. Huberman, B.A., Pirolli, P.L.T., Pitkow, J.E., Lukose, R.M.: Strong regularities in world wide web surfing. Science 280(5360), 95–97 (1998). http://www.sciencemag.org/content/280/5360/95.abstract
18. Kaltenbrunner, A., Meza, R., Grivolla, J., Codina, J., Banchs, R.: Bicycle cycles and mobility patterns-exploring and characterizing data from a community bicycle program. arXiv preprint arXiv:0810.4187 (2008)
19. Kass, R.E., Raftery, A.E.: Bayes factors. Journal of the American Statistical Association 90(430), 773–795 (1995)
20. Kruschke, J.: Doing Bayesian data analysis: A tutorial introduction with R. Academic Press (2010)
21. Lehmann, J., Isele, R., Jakob, M., Jentzsch, A., Kontokostas, D., Mendes, P.N., Hellmann, S., Morsey, M., van Kleef, P., Auer, S., et al.: Dbpedia-a large-scale, multilingual knowledge base extracted from wikipedia. Semantic Web (2014)
22. Mahdisoltani, F., Biega, J., Suchanek, F.: Yago3: a knowledge base from multilingual wikipedias. In: 7th Biennial Conference on Innovative Data Systems Research, CIDR 2015 (2014)
23. Marlow, C., Naaman, M., Boyd, D., Davis, M.: Ht06, tagging paper, taxonomy, flickr, academic article, to read. In: Conference on Hypertext and Hypermedia, pp. 31–40. ACM (2006)
24. Mislove, A., Koppula, H.S., Gummadi, K.P., Druschel, P., Bhattacharjee, B.: Growth of the flickr social network. In: Workshop on Online Social Networks, pp. 25–30. ACM (2008)
25. Noulas, A., Scellato, S., Mascolo, C., Pontil, M.: An empirical study of geographic user activity patterns in foursquare. ICwSM 11, 70–573 (2011)

26. Noulas, A., Shaw, B., Lambiotte, R., Mascolo, C.: Topological properties and temporal dynamics of place networks in urban environments. arXiv:1502.07979 [cs.SI] (2015)

27. Peng, C., Jin, X., Wong, K.C., Shi, M., Liò, P.: Collective human mobility pattern from taxi trips in urban area. PloS one **7**(4), e34487 (2012)

28. Pirolli, P.L.T., Pitkow, J.E.: Distributions of surfers' paths through the world wide web: Empirical characterizations. World Wide Web **2**(1–2), 29–45 (1999)

29. Rekimoto, J., Miyaki, T., Ishizawa, T.: LifeTag: WiFi-based continuous location logging for life pattern analysis. In: Hightower, J., Schiele, B., Strang, T. (eds.) LoCA 2007. LNCS, vol. 4718, pp. 35–49. Springer, Heidelberg (2007)

30. Sen, R., Hansen, M.: Predicting a web user's next access based on log data. Journal of Computational Graphics and Statistics **12**(1), 143–155 (2003). http://citeseer.ist.psu.edu/sen03predicting.html

31. Sigurbjörnsson, B., Van Zwol, R.: Flickr tag recommendation based on collective knowledge. In: International Conference on World Wide Web, pp. 327–336. ACM (2008)

32. Singer, P., Helic, D., Hotho, A., Strohmaier, M.: Hyptrails: a bayesian approach for comparing hypotheses about human trails on the web. In: International Conference on World Wide Web (2015)

33. Singer, P., Helic, D., Taraghi, B., Strohmaier, M.: Detecting memory and structure in human navigation patterns using markov chain models of varying order. PloS one **9**(7), e102070 (2014)

34. Sinnott, R.W.: Virtues of the haversine. Sky and Telescope **68**(2), 158 (1984)

35. Song, C., Qu, Z., Blumm, N., Barabási, A.L.: Limits of predictability in human mobility. Science **327**(5968), 1018–1021 (2010). http://www.sciencemag.org/cgi/content/abstract/327/5968/1018

36. Tai, C.H., Yang, D.N., Lin, L.T., Chen, M.S.: Recommending personalized scenic itinerarywith geo-tagged photos. In: International Conference on Multimedia and Expo, pp. 1209–1212. IEEE (2008)

37. Walk, S., Singer, P., Noboa, L.E., Tudorache, T., Musen, M.A., Strohmaier, M.: Understanding how users edit ontologies: comparing hypotheses about four real-world projects. In: International Semantic Web Conference (2015)

38. Walk, S., Singer, P., Strohmaier, M.: Sequential action patterns in collaborative ontology-engineering projects: a case-study in the biomedical domain. In: International Conference on Conference on Information & Knowledge Management. ACM (2014)

39. West, R., Leskovec, J.: Human wayfinding in information networks. In: International Conference on World Wide Web, pp. 619–628. ACM (2012). http://doi.acm.org/10.1145/2187836.2187920

40. White, R.W., Huang, J.: Assessing the scenic route: measuring the value of search trails in web logs. In: Conference on Research and Development in Information Retrieval, pp. 587–594. ACM (2010)

41. Wiehe, S.E., Carroll, A.E., Liu, G.C., Haberkorn, K.L., Hoch, S.C., Wilson, J.S., Fortenberry, J.D.: Using gps-enabled cell phones to track the travel patterns of adolescents. International Journal of Health Geographics **7**(1), 22 (2008)

Labor Saving and Labor Making of Value in Online Congratulatory Messages

Jennifer G. Kim[1](✉), Stephany Park[2], Karrie Karahalios[1], and Michael Twidale[3]

[1] Computer Science, University of Illinois at Urbana, Champaign, USA
{jgkim2,kkarahal}@illinois.edu
[2] State Farm, Bloomington, USA
park.stephany@gmail.com
[3] Graduate School of Library and Information Science Department,
University of Illinois at Urbana, Champaign, USA
twidale@illinois.edu

Abstract. Social reminder interfaces on social networking sites (SNSs), such as the Facebook birthday reminder, make sending a congratulatory message easier than ever. However, the lower cost in time and effort can also devalue a simple message, and one-click congratulations may be criticized as impersonal. Nevertheless, they are still widely used. In this paper, we investigate how people find value in short congratulatory birthday messages on Facebook despite the criticism. We conducted interviews with 17 participants with the aid of a reflective prompting tool developed to aggregate participants' previous birthday posts. Participants found the most value in personalized birthday posts, posts for reconnecting with dormant ties, and in the presentation of public affirmation posts for their imagined audience. We use signaling theory to interpret our findings and explain how a social reminder interface that lowers the cost in time for sending congratulatory messages can be both beneficial and problematic.

Keywords: Social reminder interfaces · One-click congratulatory messages

1 Introduction

A growing number of social reminder interfaces — the birthday reminder on Facebook and Google Plus, the wedding anniversary reminder on Facebook, the work anniversary reminder on LinkedIn — on social networking sites (SNSs) greatly reduce the effort required to acknowledge a person's life events such as birthdays, graduations, and wedding anniversaries. With a single click, these reminder services allow users to send a congratulatory message to people in their social network. However, this growing number of abstracted one-click congratulatory systems are often criticized as "masquerading as a social network" [36], "programmed, canned," and "sent without thought or personal feeling" [27]. Perhaps predictably, more and more short, recurrent messages are migrating to one-click interactions (e.g., the iPhone "I can't talk right now..." response while dismissing a call). Burke likens short, recurring written

© Springer International Publishing Switzerland 2015
T.-Y. Liu et al. (Eds.): SocInfo 2015, LNCS 9471, pp. 245–260, 2015.
DOI: 10.1007/978-3-319-27433-1_17

birthday greetings to one-click actions [2]. Despite these and other concerns, one-click congratulatory messages are still widely used in SNSs [15, 37].

In this paper, we address the reasons behind these criticisms, and the value in one-click congratulatory messages despite the criticisms. We define value as what people particularly appreciate while sending and receiving congratulatory messages on SNSs. By understanding the problems and values that people experience around the congratulatory message interfaces, we present design considerations for appropriate social reminders.

To do this, we consider sending congratulatory messages as *signals* [10]. In social signaling theory, a *signal* is a perceivable action to indicate otherwise unperceivable quality about the signaler [9]. In the case of congratulatory messages, sending a congratulatory message is a visible indicator that signals the sender's affection for the receiver [7, 24]. Thus, resources such as time, money, and cognitive effort expended in producing and receiving congratulatory messages can provide important information about the signaler's and receiver's level of motivation [9]. Such resources required to produce and receive a signal are called *costs* [9, 10]. The Facebook birthday reminder, however, is designed to decrease these costs for the sender by reducing the time and effort previously required to archive birthday dates in a calendar, visit a store to buy a card, write in it by hand, and mail it. Costs for the receiver may vary depending on the number of messages; in general, the ease of seeing consolidated birthday greetings in one interface often results in a low-cost viewing experience. Since the cost as a measure of the sender's level of motivation is decreased, congratulatory messages on SNSs are criticized as being impersonal compared to other media.

To investigate what people particularly value from the sending and receiving of Facebook birthday posts, we conducted interviews with 17 participants and collected their received birthday posts data using a reflective prompt tool called "My Birthday Reflection". We contribute quantitative and qualitative findings about the values in Facebook birthday posts and outline future design consideration for a social reminder.

2   Literature Review

Birthdays[1] are notable occasions recognized through normative rituals. These rituals serve to celebrate the specialness of the individual, who is annually honored on the day of his/her birth [7, 17, 24]. In countries where they are celebrated, people expect family and friends to remember their birthday and to acknowledge the occasion by spending time with them and giving gifts. A common gift is a birthday card where a sincere message expresses affection [7, 24].

Sending birthday messages has become easier and faster as new media has decreased the effort of sending birthday messages. Before the first commercial cards appeared in the 1800s, families laboriously crafted charms and block-printed cards on special days for a very special few [3]. Around 1910, the card industry began mass-producing cards containing pre-written greetings [30]. Sending birthday messages

---

[1] Birthdays are cultural rituals. There are some countries that do not celebrate birthdays.

required less time and possibly less writing. Cognitive and time costs remained; people invested time archiving birth dates and contact information (e.g., maintaining up-to-date contact books including birthdays) and then directly contacting the recipient of the greeting via postal mail, email, phone calls, or personal visits. In 1994, the first electronic card, The Electric Postcard [11], further reduced the effort of buying, hand-writing, and sending a card. Some recipients reacted negatively believing this new medium lacked authenticity [38]. With the coming of the Internet Age and widespread use of SNSs, in 2006, Facebook introduced a birthday reminder interface. Over the history of the Facebook birthday posts (2006 - Present), redesigns of the interface have continued to reduce the cost in time of sending birthday greetings. In 2006, users could see a list of birthdays from their friend network in the lower right portion of the Facebook homepage. This birthday reminder feature was responsible for over *half* of the direct interactions between users who infrequently interacted on Facebook [37]. In September of 2011, the Facebook birthday feature was prominently placed in the upper right corner of the home page (see Fig. 1). Clicking this window revealed a list of friends with birthdays and text fields, allowing the user to directly send a congratulatory message from his or her Facebook home page — thereby eliminating the time and effort required to visit a friend's Timeline[2] and perhaps encouraging a quick note.

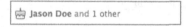

Fig. 1. The Facebook Birthday Reminder (2011).

Changes in signal costs and subsequent effects have been studied by biologists, game-theorists, economists, and designers via signaling theory [10, 23, 32]. We believe it explains the continued use of generic birthday greetings today. Sending a birthday message typically signals how important the receiver is to the sender [7, 17, 24]. Thus, strengths of relationships between senders and receivers play an important role when deciding how much time and effort the sender expends [17]. The receiver also expects a different cost from the sender depending on the relationship. People give an expensive birthday gift to close friends but a cheap birthday card to acquaintances. Similarly, Facebook birthday posts may act as a signal of attention to the receiver [13]. Burke alluded that different lengths and content in posts may have different effects in increasing the strengths of relationships [2]. However, these studies have not shown why people send and receive birthday posts of varying content with people of differing relationship strengths.

The audiences of birthday messages are crucial in signaling. When sending birthday messages via a private channel such as a letter, birthday card, electronic card, text message, or phone call, the sender typically considers the receiver as the only audience. On Facebook, a birthday message sent via the birthday reminder interface is posted on the receiver's public Timeline. Depending on the receivers' privacy

---

[2] The Facebook Timeline contains stories sent by and to the user. Based on the user's privacy setting, the user's friends or other Facebook users can read and write a post on the Timeline.

settings, the audience could extend from the receiver's friends to people outside of his or her social network. Thus, the public birthday posts not only act as a signal to the receivers, but also to the imagined audience [22], including others that can access the contents on the receiver's Timeline.

We explored birthday post content and how the participants in our study interpreted the posts as signals based on their costs and the relationship strength with the sender. We aimed to uncover the following research question: *What do people value in sending and receiving Facebook birthday posts?*

3   Study Design: Exploring Value Understanding

We recruited 17 participants (10 female; aged 19 to 49, mean age 24) from seven departments in a large Midwestern university. Each participant received $10 upon completion. The participants consisted of one faculty member, one staff member, nine graduate students, and six undergraduate students. All participants regularly used Facebook in their everyday lives (14 participants logged in to Facebook several times a day and three participants logged in to Facebook two or three times a week) and had experience sending and receiving Facebook birthday posts for at least four years. Combined, the participants received a total of 5,439 Facebook birthday posts and on average 65 posts per person per year.

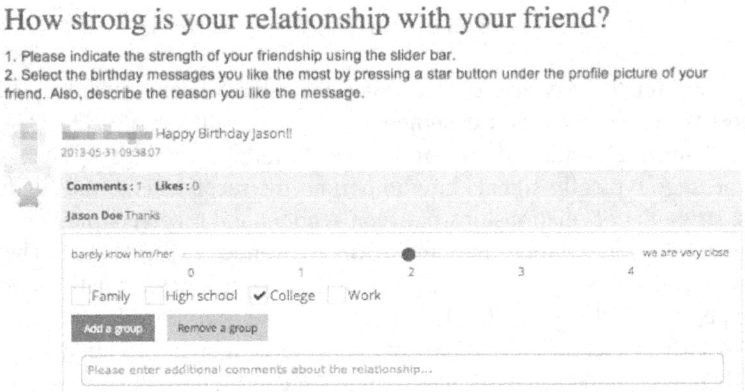

Fig. 2. My Birthday Reflection: Pressing the start icon indicated a post was a favorite.

My Birthday Reflection. The reflective prompting tool (Fig. 2) was designed to enable participants to recollect some of the birthday posts they had received on Facebook and share their response. The tool gathered up to one hundred of the most recent birthday posts and displayed information surrounding each post including the sender, "comments", "likes", and the dates and time of the post. It was designed to display the contents as similarly as possible to the Facebook Timeline interface. Participants were asked to rate: "*How strong is your relationship with this person?*" on a scale from 0 (barely know him/her) to 4 (we are very close). Following Gilbert's methodology in probing tie strength on Facebook [14], they were advised to choose a 4 rating if they could ask this person to loan them $100 or more. Participants were also asked to annotate if a post was one of their favorite few and place the friend in a personally labeled community group of their choice. Commonly selected groups were family, high school, college, and work.

The annotated favorite posts and an explanation supporting each selection were collected. A total of 174 birthday posts were selected, with each participant selecting an average of 11.6 posts. At the end, the tool showed participants a summary visualization of the collected birthday posts categorized by the labeled community groups and the rated relationships strengths over the years. Approximately thirty minutes were spent to complete the first part of the study.

Interviews. We next conducted an hour-long three section semi-structured interview, which was recorded and transcribed. First, we showed them the summary visualization and asked them if they could see patterns in it. This visualization highlighted the number of friends who sent the birthday posts corresponding to the strengths of relationships. Then we asked why those friends would choose Facebook to send the birthday messages and how their interpretation of the birthday posts differed based on the strengths of the relationships. In the second section, we asked participants to further describe their motivation for selecting the favorite posts and what in particular raised the value of these posts compared to other posts. We asked how they perceived the value, including the advantages and disadvantages of receiving birthday messages on Facebook compared to other media. Finally, we asked if they responded differently to messages from friends of different relationship strengths. In the third section, we asked about their experiences sending birthday posts including how participants decided whether to send birthday posts, why they chose Facebook as a medium, and if and why they employed different writing strategies in their posts.

Data Analysis. Three researchers coded the transcribed interview data using open coding [31]. Each researcher reviewed the data, extracted statements of appreciation from Facebook birthday posts, and grouped them according to relevant themes. Then the researchers met as a group over a period of 2 months to evaluate and refine the themes. We identified three core value themes from this analysis: (a) the content of birthday posts, (b) relationships between senders and receivers, and (c) imagined audience of Facebook birthday posts. Two of the authors individually conducted an initial round of open coding of the explanations supporting the favorite birthday post selections. They then met together and iteratively clustered the initial codes into higher-level category groupings until arriving at two themes: (a) the content of birthday posts and (b) special relationships between senders and receivers. We performed a quantitative analysis on the content of birthday posts and proportions of birthday posts received from different relationship strengths. Our goal was to examine if the quantitative results supported our qualitative findings. In the following section, we discuss the themes resulting from the qualitative and quantitative analysis.

4   Findings

4.1   Personal Birthday Posts

Personalization played the largest role in annotating a post as a favorite (61% of the favorite posts). The personalized messages made participants laugh, reminisce about the past, and feel cared for.

Perceived Value of Receiving Personal Birthday Posts. Personal messages referred to shared experiences between senders and receivers, used nicknames, and included humorous messages.

Reminiscing and Feeling Cared For. Participants valued reminiscing with birthday posts that referenced shared experiences such as remembering the times when the sender and receiver struggled all night in the library, traveled together, or had fun at a party. In the 2012 post below, P5 reminisced about times when he had fun with the sender in a bar. *"Happy Birthday P5!! Hope you have a great one and wish we could join you for a drink at Murphy's! (tie strength = 4).*

Participants described the favorite birthday posts as "sincere," "genuine," "thoughtful," "loving," or "emotional." These posts included messages expressing gratitude for teaching the sender and affective messages including the words "love" and "miss" from their close friends, parents, and children. For example, P17 selected the following 2012 birthday post from his grandmother as one of his favorites and described the post as "personal and genuine", *"Hi My Darling - A GREAT BIG HAPPY BIRTHDAY to my fabulous grandson. Try to find some time to enjoy yourself. You most certainly deserve it. Love. Grandmother." (tie strength = 4).*

Entertaining. Participants also valued humorous birthday posts that had inside jokes, nicknames, and teasing messages. Those posts made participants laugh and respond with another joke, often starting a conversation. Because Facebook is public and messages appear on friends News Feeds, sometimes others also join in. P7 found the following 2012 post from his sister very funny. *"Ahhh wait you're old tomorrow....I don't know how I feel about this!!! 22 is a socially acceptable age to get married!!! Stop this aging immediately!!! (aka love you big brother :))" (tie strength = 4).* P7 marked this as a favorite post and stated, "[It is] very personal. The comments are hilarious from my other sister and family friend. Love newsfeed stuff like this."

Better than Generic Birthday Posts. Another reason for valuing the personal birthday posts was the stark contrast to generic, impersonal birthday posts. Participants received a significant number of generic "Happy birthday" posts (about 40%). In contrast with generic birthday posts, personal birthday posts signaled more effort from the sender. P13 stated, "I normally expect 'Happy birthday,' maybe just one more sentence like, 'How are you doing?' That would be nice, but just straight up 'Happy birthday,' it feels like they had to do it because it [the reminder] was there. They didn't really want to, but Facebook told them to, so they had to put up a message" (P13).

Participants noted that the birthday reminder lowered the cost in time and cognitive effort of sending the posts. For example, P2 described, "I think it [Facebook birthday posts] changes the nature of how intentional and thoughtful we have to be. Even with our closest friends, if you forget to deal with sending them a card a couple of days in advance, it's fine. [...] It feels like Facebook will do the remembering and the effortful part of it for you. I think it makes us less responsible for acknowledging and investing in our friends lives" (P2).

Sending Personal Birthday Posts. Participants wanted to do more for close friends, by making an extra effort to express their closeness. This included writing longer or more personalized messages, using inside jokes, sharing memories, adding photographs, and carefully timing the sending of the message so it arrived early on the recipient's birthday (just after midnight). Six participants reported that they intentionally read others' birthday posts on the participants' News Feed or the receiver's Timeline in order to write something different. Those participants noted that unique messages represent more effort: "I try and leave birthday out because I know that everyone else says happy birthday. I try and say something else like, 'Happy, happy. It's officially your day today.'" (P11). Participants had concrete rules when deciding the amount of personal content to put into the birthday posts depending on the strength of their relationship with the receiver. "If the person is [tie strength] 4 or 5, I would try to think of something more thoughtful for them. If the person is [tie strength] 2 or 3, I would do little bit more than happy birthday. If it's just a person who is [tie strength] 1, I usually do happy birthday and [a] couple exclamation points after it" (P5).

Quantitative Analysis of Personalized Birthday Posts Across Tie Strengths. We examined a relationship between the extent of personalization of the birthday posts and the strength of a relationship. Using Linguistic Inquiry and Word Count (LIWC) [33] text analysis software, we first evaluated the personalization level of birthday posts by measuring the percentage of words in the following LIWC categories: Family, Friends, Home, Sexual, Swears, Work, Leisure, Money, Body, Religion, and Health. We selected the LIWC categories containing personal messages that our participants referred to as shared activities, personal conversations, and funny messages.

Table 1. Example of birthday posts and their TF-IDF values

Birthday posts	TF-IDF value
Happy birthday	0
Happy Birthday. hope you have a wonderful day	11.0
Happy Birthday Nick Lol Ball Out!!!	23.8
Happy birthday to my beautiful. kind. and scarily hip mama!! Nothings the same without you!!! Xoxoxo	79.6

We also measured the uniqueness of each word in the birthday post compared to the entire collection of birthday posts, using Term Frequency-Inverse Document Frequency (TF-IDF) [20]. We then calculated average TF-IDF values of the birthday posts in each tie-strength category. Intuitively, words with low TF-IDF values indicate that the word is used frequently throughout the collection of birthday posts, while higher TF-IDF values indicate the word is rare. For example, the words "happy" and "birthday" have the lowest TF-IDF value of zero. For this reason, we refer to "happy birthday" as generic birthday post, and label posts with a TF-IDF value higher than zero as personal birthday posts. Table 1 shows examples of birthday posts and their TF-IDF values.

Participants received more personal birthday posts from closer friends, supporting our findings from the qualitative analysis. Table 2 shows how the percentage of personal words and average TF-IDF values gradually increase as the relationship strength between the sender and receiver increases.

Table 2. T-test of average % of personal words and TF-IDF values in Facebook birthday posts across different strengths of relationships. A mean reported with an asterisk (*) indicates a significant difference with the tie strength value before. * = p < .05, ** = p < .01, *** = p < .001

Tie strength	weak 0	1	2	3	strong 4	Graph
% of Personal Words	1.06%	2.42%	3.17%	4.36%	5.44%	
Agerage of TF-IDF Values (SD)	7.26 (21)	10.06 (17)***	16.08 (22)	19.49 (24)	31.89 (32)***	

4.2   Sending and Receiving Birthday Posts with Dormant Ties

39% of the favorite birthday posts were selected because participants have a special relationship with the sender. Special relationships included close friendships, *dormant ties*, and *special weak ties*. We define *special weak ties* as people with whom the participants have never been close but they nevertheless respect, admire, or like. *Dormant ties* are people with whom the participant was once close, still desires to be close, but has ceased contacting [21]. Making contact with old friends via Facebook is quite common [12, 19], but it happens more frequently on or near one's birthday. The Facebook birthday reminder functions as a social catalyst [18], and creates an opportunity to establish a connection that would otherwise not exist.

Receiving Birthday Posts from Dormant Ties and Special Weak Ties. Because participants normally did not expect to receive birthday posts from dormant ties, they perceived the messages from them as very meaningful, even if the birthday posts were the generic "Happy birthday". P6 explained, "It meant something to me that Steve took the time to write 'Happy birthday' even though we haven't spoken in at least 6 years" (P6). Moreover, P15 described that she could reminisce just by seeing an old friend's name in a Facebook birthday post. "What I most liked in the survey is that the birthday greeting came from an old family friend. He basically just said happy birthday. I was like, oh, I remember some good times with you, so just remembering him again to kind of boost [my] memory just from the fact of seeing him on Facebook" (P15).

Participants also found value in birthday posts from *special weak ties*. For example, P2 said, "Over the ones I just read, I think the only one that have my eyes starred that friends of me that I'm not particularly close to, as a coworker who just said she missed me. I was, 'Oh that's so sweet. I miss working with you too.' It's not something I would expect from her, right?" (P2).

Sending Posts to Reconnect with Dormant Ties. From the senders' perspective, we discovered that etiquette around Facebook birthday posts make *dormant ties* easier to contact. On Facebook, it is easy to find a lost contact and become a "friend" on Facebook [12]. Further, the Facebook birthday reminder provides an excuse to contact others. Direct contact from people with whom one has not contacted in a while via phone calls or text messages may be awkward. With Facebook, however, *dormant ties* could initiate conversation easily without fuss. While the public visibility of writing a personal birthday post can be intimidating, generic birthday posts are commonplace on Facebook and for some may be a better option. "I can see its potential for more bonding over Facebook in ways it couldn't happen before because if someone from a high school I went to high school with called me, it would be a bit awkward, if I haven't spoken [to] them in 23 years. In Facebook, it is perfectly acceptable. And it's nice because you have [the] option to start the conversation if you want" (P1).

4.3   The Imagined Audience of Facebook Birthday Posts

Sending an electronic or a paper card can be a semi-private ritual where the sender typically intends the card to be viewed only by the recipient. It is possible recipients may display the cards in their homes, but it would be unlikely that all senders would be able to see all the cards sent by others. Facebook birthday posts, however, are public and Facebook friends can see all of birthday posts that are not explicitly made private. As a result, users become aware of third-party observers.

Receivers Feeling Special. On average, participants received 65 Facebook birthday posts per year. Although each birthday post was perceived as costing less effort than a post card or electronic card, they had an aggregate value: participants felt special when they received a large number of birthday posts. For example, P12 said, "I value it [Facebook birthday posts] just because I can see how popular I am maybe" (P12).

Sending Public Birthday Posts as a Gift of Status. As senders, participants were concerned about their friends' public presence. They wanted to make others aware of their friend's birthday so that the friend would receive more birthday posts on Facebook. Essentially, participants wanted others to see their friend as popular: "I think part of it is I want my friends to have something public on their wall. I want other people to say 'Oh look at how many birthday wishes that this person got'" (P6).

The Facebook timeline interface also explicitly presents a total number of birthday posts that person received such as "10 people posted on your timeline for your birthday." This interface design might have encouraged senders to send a public birthday post to make their friend receive more birthday posts.

Sending Posts to Publicly Confirm a Close Relationship. Even if participants had greeted a close friend face-to-face on their birthday, they also acknowledged their friend's birthday as a public performance on Facebook. More than half of participants said that, for close friends, they would send a public Facebook birthday post as well as birthday greetings via private media such as a phone call, SMS, email etc. Friends use multiple media to show their efforts to the recipient; Facebook's labor-saving technology thus becomes a labor-making technology [4]. "I think part of it is the publicized aspect of it because people see you wishing them a happy birthday and then people, they're like, 'Oh, they're so close. They're such close friends'" (P8).

5   Limitations

One of the limitations of this study is that we interviewed 17 participants in a university setting — an elite tech-savvy group of users. While an acceptable starting point, a wider demographic of subjects would highlight greater variance in Facebook birthday posts adoption, sense making, and changes in practice. Our Facebook application collected received posts, therefore, participants may have emphasized received posts over sent posts. We asked participants about their sending practices, but an analogous probe highlighting sent posts would provide symmetrical analysis. Finally, a longitudinal approach to explore practices over several years would lead more insight into how people's practice evolves with the changing birthday reminder interfaces.

6   Discussion

Using the visualization of Facebook birthday posts, participants found it easy to discuss issues of their value. While we were originally concerned that it would be an onerous task, participants appeared to enjoy going through up to a hundred previous Facebook birthday posts. One said, "It was interesting to see people who have written messages a long time ago that you haven't talked to or messages recently I missed. Kind of refer to parts of your life that you forgot about" (P7).

Our findings suggest value in Facebook birthday posts for reminiscing about past memories, entertainment, and feeling cared for which parallels findings in a study probing text messages as gifts [34, 35]. Especially, the value of connecting with dormant ties resonates with a use of electronic post cards as a gift for keeping in touch without having to say anything [11, 29]. However, the Facebook News Feed interface, publicly showing others' birthday posts, added its own value in feeling special, improving their friends' reputation, and exhibiting the closeness of their friendship to imagined audiences.

6.1   Facebook News Feed as a Social Signifier of Facebook Birthday Posts

Norman defines social signifiers as signals created or interpreted by people signifying normative social behaviors and activity [25]. On Facebook, the News Feed displays

Facebook birthday posts to the receivers' other Facebook friends. Thus, birthday posts from other senders can act as a social signifier to send a birthday post to the receiver. Burke also found that Facebook users who see their friends contributing on Facebook are more likely to share more content themselves [1]. Our participants noted that dormant ties can easily send Facebook birthday posts because they often see a large quantity of birthday posts that have already been sent to the receiver. Adding another birthday post to this collection follows this existing social pattern. It is also possible, however, that this volume of birthday posts can create sender obligation: an expectation that sending a Facebook birthday post is the least they can do and not sending the birthday post may be perceived as rude. "It [sending a Facebook birthday post]'s better than nothing. Even just for a split second because other then you forget all the people that didn't" (P17). This negative impact of sending obligatory birthday posts resonates with Dindia's holiday greeting card study findings [5]. She emphasized the hygienic factors inherent in holiday greeting cards; their presence does not positively affect relational maintenance, but their absence may have a negative effect on relational maintenance.

6.2   A Minuscule Sociotechnical System

Even in our small sample of 17 people, many issues emerged around sending, receiving, and replying to Facebook birthday messages. Participants could articulate rather complex social patterns of behavior, even informal rules of etiquette that they and others had evolved and seem to keep on evolving. Of course our society has seen this kind of thing emerge many times before. Social commentators, such as Emily Post, the famous writer on etiquette, noted the rules followed by some members of society, making explicit to aspirational groups how to emulate the frequently unspoken rules followed by certain social elites. Emily Post also noted the disruptive influence of a particular kind of social media - the telephone - on the social rules around writing letters and sending telegrams [28].

Our participants seemed to find discussing their actions and their meanings pleasurable. Their reaction could even be one of amusement that they were following 'rules' (often self-created) that they weren't necessarily conscious of – in a similar way to which native speakers can be asked to reflect on how they consistently follow certain conventions in speech that they may not have been aware of. Such social norms can emerge without people explicitly realizing. They may only become visible when someone violates them. There do not seem to be Emily Post-like rigid rules of social etiquette. As one participant explained her detailed behaviors: "That's just me." However, that does not mean that people act randomly. There are patterns. It's just that they may not be visible or explicit. One way to uncover norms is by breaking them – even in thought experiments.

The Facebook birthday message set of features can be thought of as a sociotechnical system – albeit a really tiny one. This can help both in understanding what we have seen and why, but also as a way to inform broader sociotechnical systems analysis and design. Most sociotechnical systems are far larger and many are far more important. But their size and complexity makes them exceptionally difficult to analyze, let alone to design or improve without a considerable risk of actually making things worse. By studying use and change in a microscopic sociotechnical system and (as we do below, undertaking a few thought experiments about designs that uncover norms by breaching them) we believe we can contribute to the analysis of much larger and more important sociotechnical systems.

In the most literal sense the Facebook birthday message setup is obviously a kind of sociotechnical system (STS) – it involves the use of particular computer technologies embedded in a particular social media application (Facebook as a whole) in order to send socially and culturally appropriate messages (birthday greetings). But it is an STS in a more subtle sense too and problems occur if designers think of the setting as a purely technical or individual-technical optimization problem. We speculate that certain Facebook developers have over the years looked at the birthday message application in precisely that manner. If you think that the 'problem' is that first remembering and then sending birthday messages is rather effortful, and that forgetting a birthday is a rather embarrassing social faux pas, then naturally the design challenge is to create an application and associated interface that makes sending birthday messages as easy as possible. With such a (perfectly reasonable) interaction design worldview, you might expect that your innovative redesigns to reduce the cost of sending messages would save people some time and effort. You might also expect that your redesigns could lead to more messages being sent, probably to a wider circle of acquaintance. That follows a classic rational economic model. What you might not expect from this individual-technical perspective would be what we saw – people deliberately making their lives more difficult by creating more personalized messages in order to show that they care about the recipient.

Another example of this effect is carefully timing the sending of the message so it arrives early on the recipient's birthday (just after midnight). As one participant noted: "If they are really close, I login to Facebook to write them a message at midnight. If they are my best friends, I'll wake up and post on their wall when their birthday happens and then send them a text and go to sleep." Again, it is perfectly possible to redesign in order to reduce that cost – but that assumes that cost reduction is actually desirable when in fact it may not be.   For example, the birthdayFB [40] application lets you save greeting messages in advance, and then automatically posts the messages at a random time on the morning of their friend's birthday. This saves effort, but that saving may destroy the meaning of a friend taking special trouble to directly send the message at a significant time.

6.3   The Virtual Social Secretary

Participants often found Facebook to be a very reliable up-to-date contact book, and Facebook birthday practices allow easy contact with people who are not close, and reconnection with people who used to be close without the awkwardness associated with similar face-to-face encounters. Effectively, Facebook becomes our social secretary, reminding us of a friend's birthday, and offering to obtain, write and deliver a card for us. This sounds very convenient, low cost and highly desirable. But that very lowering of effort can make people feel a little uncomfortable. It may feel less personal if we don't do all the work ourselves. Most of us do not have a (human) social secretary; so we have to figure out what that means in our day-to-day lives and how others may interpret our actions. We may have seen TV shows where a male executive has a secretary who reminds him about the birthdays of both close friends and business acquaintances and buys and sends the cards, or even the gifts, on the boss's behalf, generally minimizing the bother of these social exchanges. This delegation to a (human) agent certainly increases efficiency, but seems uncaring.

As designers, if we see our design objective as improving efficiency by minimizing bother, surely having a social secretary is an unalloyed good. But if that bother has actual value in showing that you care, what does minimizing bother really mean? We suspect this issue is not unique to sending greetings. Within HCI we rarely question the worth of making something easier. Designing to reduce effort can be difficult to achieve, and we may be caught in an invidious tradeoffs. But of course we all agree that easier is better. Here we have at least one case where making something easier degrades its value. Part of the point of the activity was its cost. Fortunately (or ironically), in our study we found participants ingeniously create ways to make the sending of greetings more effortful in order to recover their value.

6.4 Sociotechnical Design Implications and Anti-implications

When designing technologies, Donath warned against removing costs that previously signaled value [9]. Let us consider some Bad Idea [6] design anti-recommendations as a way to help to explore views about this kind of interaction. Imagine if Facebook provided an option to...

- Not only remind you of a birthday but also composed a draft message for you so all you had to do was click. (Indeed Facebook already suggests a generic message of "Happy Birthday.")
- Select from a list of carefully crafted, more 'personalized' birthday messages.
- Buy a personalized message from a professional writer.
- Automatically send happy birthday messages to your friends every year from now on so you don't have to bother.
- Automatically send happy birthday messages to your friends with carefully chosen phrases (different each year) so they look like you bothered to compose them.

In all these cases we suspect that many of our readers' reactions would be negative. This reaction reveals norms being violated – norms we may not have ben aware of before. They are currently unacceptable, but will that change? Are they too similar to the stereotypical 1960s male executive deputizing his secretary to buy an anniversary present for his wife? Activities can be violations of social norms for one group while they are acceptable or even expected in another.

This is not unique to technology design. A good example comes from a study of the sending of physical greetings cards. West [39] notes how various commentators have been rather critical of their use. In particular, they complain about the use of pre-written texts in the card as "lazy and less authentic substitute for the best form of interpersonal communication, which would be a handwritten note or face-to-face talk." We see allusions to this view of authenticity in some of our participants' comments, although in a far less pejorative manner.

West explores the idea of authenticity as a classed concept. For some a pre-printed greeting card sentiment is artificial [16], while for others it is about carefully selecting a professionally written phrasing that exactly expresses what the sender would like to say, but may not have had the words for. West examines this difference using the concepts of high and low cultural capital, as assessed mainly by formal education. Those with high cultural capital prefer short messages or even a blank card, and then personalize it by writing their own message. Those with low cultural capital may be less comfortable composing their own unique message, or don't consider their own words as adding much value, but are happy to invest considerable effort into finding and selecting a message that exactly conveys the sentiment they wish to express. For them a blank card is meaningless and a brief message is overly terse and implies a lack of selective care.

Facebook posts have to be composed, but our participants made similar distinctions between the more formulaic messages and those more carefully crafted and personal. All our participants fit into West's high cultural capital category, but it is important to think about other kinds of Facebook users if this work is to be extended.

These issues create a design challenge for social media technology designers. It is certainly possible to create the application features we described above. But do they meet a need? Do they potentially cheapen or pollute the activity? Donath noted that she received many requests for an automated birthday-postcard-sender feature for her Electronic Postcard system [8]. But while it would have been easy to include, she didn't do it because that wasn't the etiquette she wanted around the system. She felt it was a bad design decision to automate a birthday or holiday greeting.

Hallmark manages opposing needs for greetings cards (blank or terse for those wishing to personalize, and carefully crafted messages for those wishing to select a phrasing that exactly captures what they would like to say) by creating different sub-brands [26]. In social media design, it is possible that creating useful options for some may immediately devalue the experience for others, so what is to be done?

Another norm-challenging thought experiment is to consider inappropriate metaphors. We already have Customer Relationship Management (CRM) software that contains details about contacts including personal details like interests, children, birthdays etc. What if we brand certain Facebook features as Friend Relationship Management (FRM) doing essentially the same thing for our numerous Facebook friends? Does even talking about FRM software seem to violate certain norms and expectations? Will our friends feel less valued if Facebook is acting like a social secretary whispering their personal details into the ear of the Important Person who feigns to remember the last time she shook their hand? Speculations about possible Google Glass applications are already raising such scenarios. Or, will our discomfort change as we get used to the idea of lots of people having virtual social secretaries?

References

1. Burke, M., et al.: Feed Me: Motivating Newcomer Contribution in Social Network Sites. CHI (2009)
2. Burke, M., Kraut, R.: Growing Closer on Facebook: Changes in Tie Strength Through Social Network Site Use. CHI (2014)
3. Chase, E.D.: The romance of greeting cards; an historical account of the origin, evolution, and development of Christmas card, valentine, and other forms of engraved or printed greetings from the earliest days to the present time. Rust Craft, London (1956)
4. Cowan, R.: More Work For Mother: The Ironies Of Household Technology From The Open Hearth To The Microwave (1985)
5. Dindia, K.: The function of holiday greetings in maintaining relationships. J. Soc. Pers. Relat. 21(5), 577–593 (2004)
6. Dix, A., et al.: Why bad ideas are a good idea. In: Proceedings of HCIEd.2006-1 Inventivity: Teaching theory, Design and Innovation in HCI, pp. 23–24 (2006)
7. Dodson, K.J., Belk, R.: The Birthday Card Minefield. Adv. Consum. Res. 23, 14–20 (1996)
8. Donath, J.: Personal Communication (2013)
9. Donath, J.: Signals in Social Supernets. J. Comput. Commun. 13(1), 231–251 (2007)
10. Donath, J.: Signals, Truth and Design. MIT Press
11. Donath, J.S.: Inhabiting the virtual city, PhD thesis. MIT (1997)
12. Ellison, N.B., et al.: The Benefits of Facebook "Friends:" Social Capital and College Students' Use of Online Social Network Sites. Journal of Computer-Mediated Communication, 1143–1168 (2007)
13. Ellison, N., Vitak, J., Gray, R., Lampe, C.: Cultivating Social Resources on Social Network Sites: Facebook Relationship Maintenance Behaviors and Their Role in Social Capital Processes. J. Comput. Commun. 19(4), 855–870 (2014)
14. Gilbert, E., Karahalios, K.: Predicting tie strength with social media. In: Proc. CHI 2009. ACM Press, New York (2009)
15. Heffernan, V.: The Social Economics of a Facebook Birthday (2011). http://opinionator.blogs.nytimes.com/2011/08/14/the-social-economics-of-a-facebook-birthday/?ref=virginiaheffernan
16. Jaffe, A.: Packaged Sentiments: The Social Meanings of Greeting Cards. J. Mater. Cult. 4(2), 115–141 (1999)
17. Joy, A.: Gift Giving in Hong Kong and the Continuum of Social Ties. J. Consum. Res. 28(2), 239–256 (2001)
18. Karahalios, K.: Social Catalysts: Enhancing Communication in Mediated Spaces (2004)
19. Lampe, C. et al.: Changes in use and perception of facebook. In: CSCW 2008, p. 721 (2008)
20. Leskovec, J., et al.: Mining of Massive Datasets. Cambridge University Press
21. Lim, E., et al.: Reviving Dormant Ties in an Online Social Network Experiment. Assoc. Adv. Artif. Intell. (2013)
22. Litt, E.: Knock, Knock. Who's There? The Imagined Audience. J. Broadcast. Electron. Media. 56(3), 330–345 (2012)
23. Maynard-Smith, J., Harper, D.: Animal signals. Oxford University Press (2003)
24. Mooney, L., Brabant, S.: Birthday Cards, Love, and Communication. Sociol. Soc. Res. (1988)
25. Norman, D.A.: THE WAY I SEE IT: Signifiers, not affordances (2008)

26. Papson, S.: From Symbolic Exchange to Bureaucratic Discourse: The Hallmark Greeting Card. Theory, Cult. Soc. **3**(2), 99–111 (1986)
27. Plotz, D.: My Fake Facebook Birthdays (2011). http://www.slate.com/articles/technology/technology/2011/08/my_fake_facebook_birthdays.html
28. Post, E.: Etiquette in Society, in Business, in Politics, and at Home. Funk and Wagnalls, New York (1922)
29. Rosen, J.R.: The Early Days of E-Cards (2011). http://www.theatlantic.com/technology/archive/2011/12/the-early-days-of-e-cards/250492/
30. Ruth, J.A., et al.: Gift Receipt and the Reformulation of Interpersonal Relationships. Journal of Consumer Research (1999)
31. Seidman, I.: Interviewing As Qualitative Research: A Guide for Researchers in Education And the Social Sciences. Teachers College Press (2006)
32. Spence, M.: Job Market Signaling. Q. J. Econ. **87**(3), 355–374 (1973)
33. Tausczik, Y.R., Pennebaker, J.W.: The Psychological Meaning of Words: LIWC and Computerized Text Analysis Methods. J. Lang. Soc. Psychol. **29**(1), 24–54 (2009)
34. Taylor, A.S., Harper, R.: Age-old practices in the "New World": A study of gift-giving between teenage mobile phone users. CHI (2002)
35. Taylor, A.S., Harper, R.: The Gift of the Gab?: A Design Oriented Sociology of Young People' s Use of Mobiles. J. Comput. Support. Coop. Work. **12**, 267–296 (2003)
36. Thurston, B.: A personal post on Twitter, September 28, 2011. https://plus.google.com/109716233093030928975/posts/Dq1HtxSvNQ3
37. Viswanath, B., et al.: On the evolution of user interaction in Facebook. In: Proc. 2nd ACM Work. Online Soc. Networks, WOSN 2009, p. 37 (2009)
38. West, E.: Digital Sentiment: The "'Social Expression'" Industry and New Technologies. J. Am. Comp. Cult., 316–326 (2002)
39. West, E.: Expressing the self through greeting card sentiment: Working theories of authentic communication in a commercial form. Int. J. Cult. Stud. **13**(5), 451–469 (2010)
40. birthdayFB. http://birthdayfb.com/

Banzhaf Index for Influence Maximization

Balaji Vasan Srinivasan[1,2](✉) and Arava Sai Kumar[2]

[1] Adobe Research Big Data Intelligence Labs, Bangalore, India
balsrini@adobe.com
[2] Adobe Systems India Private Limited, Bangalore, India
arakumar@adobe.com

Abstract. Social media has changed the way people communicate with each other and has brought people together. Enterprises are increasingly using it as a medium for marketing activities. However, due to the size of these networks, marketers often look for key customers (influencers) to drive the campaign to the community. In this paper, we take a game theoretic approach to identify key influencers in a network. We begin with defining coalition games to model the social network and then use the concept of Banzhaf index to measure the utility of each user to the coalition. We further extend this concept towards identification of influencers and compare the resulting algorithm against existing works on influence maximization on several datasets. Improvements are observed.

1 Introduction

Social media has changed the way people interact with each other and has greatly shrunk the online world. Surveys estimate over 27% of time spent online to be spent on social media and big enterprises are looking to leverage this power of social platforms for business promotions and improving brand value. Social marketers often reach out to a small set of people (influencers) who have the potential to further influence/reach out to the targeted customers.

This core problem is widely studied as "Influence Maximization". Domingos et al. [5] first studied this problem by assigning an influence score to customers based on Markov random fields for designing viral marketing strategies. This has since been extensively studied however scalability of the algorithms is a major concern. Kempe et al. [10] posed this problem as a discrete optimization problem and show that natural greedy algorithm approximates the solution with $(1 - \frac{1}{e} - \epsilon)$. [11] developed an efficient algorithm by utilizing the sparsity in the underlying objective function and by reducing the number of function evaluations using the submodularity of the influence functions. Chen et al. [3] presented an

B.V. Srinivasan and A.S. Kumar—Both authors had equal contribution to the paper.

B.V. Srinivasan and A.S. Kumar—The authors would like to acknowledge Prof. Y. Narahari, Department of Computer Science and Automation, Indian Institute of Science, Bangalore, India for his valuable feedback and guidance during the course of this work.

© Springer International Publishing Switzerland 2015
T.-Y. Liu et al. (Eds.): SocInfo 2015, LNCS 9471, pp. 261–273, 2015.
DOI: 10.1007/978-3-319-27433-1_18

efficient algorithm to find the top nodes in a social network and improves upon the [10] and [11].

An alternative to graph approximations are centrality measures [6] that ranks users of a network based on their importance. However, standard centrality measures do not construct a consistent ranking of individual nodes using group importance. They fail to recognize the "utility" of the nodes when combined with others. Game theoretic network centrality measures address this issue by allowing consistent ranking of individual nodes while considering the various possible multiple interactions between the nodes. In this context, Grofman et al. [9] first applied game theory to the topic of centrality. Narayanam et al. [13] have defined a cooperative coalitional game and used Shapley value to measure the influence. Shapley value can be interpreted as a centrality measure that represents the average marginal contribution made by each node to all other possible coalitions of other nodes taking into consideration the order in which each coalition is formed. While Shapley value confers a high degree of flexibility to the network centrality, it places emphasis to the order in which the individual entered the coalition. However, once the coalition is formed, the order may not matter much. This is captured by the solution concept "Banzhaf index".

With this as our hypothesis, we propose Banzhaf index to measure the importance of an agent to a coalition. As we show later, Banzhaf index outperforms Shapley value in influence computations validating our hypothesis. Banzhaf index was first defined by [2] to measure the criticality of an agent in a voting game. Banzhaf index differs from Shapley value in the sense that the order of the players in which coalitions are formed does not matter. Hence Banzhaf index represents the average marginal contribution made by each node to all other possible coalitions without considering the order of coalition .

The rest of the paper is organized as follows. In Section 2, we give a brief overview of coalition games and introduce Banzhaf Index in the context of coalition games. We then introduce a way to compute Banzhaf index for a generalized game in Section 3 and then utilize this for influence maximization. We finally evaluate the performance of the Banzhaf index against the Shapley value based approach [13] as well as other standard techniques like CELF [11] and Maximum Influence Arborescence [3] in Section 4. Section 5 concludes the paper.

2   Coalition Games and Banzhaf Index

Game theory provides rich mathematical frameworks to model interactions between individual agents involved in a game. In the context of social networks, each node/user in a social network could be thought of as a 'player' in the 'game' of social interactions. Games could be cooperative or non cooperative. Cooperative games are those where players involved in coalitions try to achieve something more than what they could have achieved individually. Social networks can be modelled as a cooperative game over a network in which the agents are the nodes, coalitions are the group of nodes, edges represent the strength of relationship between the nodes, values of coalitions defined as per the requirements of the game. For instance, in a social network like Facebook, maximizing influence can be characterized as a cooperative game where Facebook users (nodes) form a coalition (group/community) to have increased social interactions and knowledge exchange (which is value of the coalition).

The value each user brings to the coalition is a measure of the player's prominence to the game. Several game theoretic concepts have been introduced that deals with dividing the gains of a coalition to all the players in a fair manner, e.g. Shapley value, nucleolus, Banzhaf index, core, Gately Point. These concepts could be used as a measure of a player's influence in a game. Shapley Value measures the gain of each agent as the average of their contribution over all the possible permutations in which the coalition can be formed. Nucleolus is an allocation of gain that minimizes the dissatisfaction of the players from the allocation they can receive in a game. The Gately point of a game is the imputation which minimizes the maximum propensity to disrupt i.e.it minimizes the incentive to move away from the coalition. Banzhaf index is similar to that of Shapley value except for the fact that the order in which an agent arrives in a group to form a coalition does not matter in the former case. We studied Banzhaf index in the light of influence maximization and found it to be equally tractable as Shapley value.

A coalitional game consists of a set of n players and a function v mapping any subset of the agents(coalition) to its "value" ($v : 2^N \to \mathbb{R}$) to the game. A player is "critical" for a coalition if removing the player from coalition causes loss to the coalition. Banzhaf index is a measure of power of a player based on the number of coalitions in which the player is critical out of all possible coalitions containing the player. Banzhaf Index for a player i in the game G above is given by:

$$B_i(v) = \sum_{C \subset N | i \in C} \frac{(v(C) - v(C \backslash i))}{2^{n-1}} \tag{1}$$

Banzhaf index for a node is thus the average of marginal contributions of the node to coalitions formed with all other nodes. A higher Banzhaf index for a node corresponds to the criticality of the node in more coalitions and therefore, a ranked order list of Banzhaf indices would yield the top influencers.

3   Banzhaf Index Computation

The influence maximization problem can be modeled as a game where each individual is an agent, the amount of influence each individual has over his neighbor as an edge between them, a group of individuals as a coalition and the value the group achieves on a whole as the value of the coalition. Thus, the first step towards Banzhaf index based influence scoring is to define a "game" and its properties for which the index can be computed. In a social network sense, a game can be defined via the *sphere of influence* for a group of nodes C which is the boundary within which their influence is effective.

The group of nodes C indicate the coalition that work together for increased individual benefits. Each node v_i in the network is a player in the game and their neighborhood $N_G(v_i)$ represent the neighborhood of vertex v_i i.e. the set of vertices that are at exactly 1-hop away from v_i. $deg(v_i)$ represents the degree of node v_i i.e. the number of nodes adjacent to v_i. A game should have pre-defined rules that dictate the interactions and formation of coalitions. Once a node v_i is a part of a coalition, it can contribute another node v_j to the same coalition, if removal of v_i from the coalition results in the removal of v_j from the coalition as well. The value of the coalition in the influence maximization problem can thus be characterized as the total number of individuals influenced by the individuals inside the coalition.

If the game is being played by n agents in a network and if we want to select only top k individuals out of n agents for maximum influence, then we have to select only those k nodes that have more ability to influence others on an average than the other nodes. The measure of the ability to influence of an agent in our case is given by the number of individuals outside the group who are influenced by the agent. So it is intuitive to pick the agents as top k nodes who contribute more to influence spread than the others on an average per coalition. Hence we shall formulate influence maximization problem as a game and pick the top k nodes which have higher Banzhaf index than others.

We explored various games along the lines in [1] and studied them for influence scoring with Banzhaf index. Here, we progressively define these games and the Banzhaf index computation for each of the games. The final game is the most generalized version and is used for further influence scoring.

Game 1: The first game considers the sphere of influence of C as themselves and their immediate neighbors. Given a set of nodes in a graph G, the sphere of influence of a set of nodes C in G is given by the number of the nodes in C and the nodes that are immediate neighborhood to nodes in C. The marginal contribution $MC_C(v_i)$ of v_i to some coalition C is additional number of nodes added solely due to the presence of v_i in coalition. Let $N_G(v_j)$ denote the neighborhood of v_j in G. Banzhaf index calculates the average marginal contribution of a player over all possible coalitions of other nodes.

Given the game setup, a node v_j may be contributed to C by v_i only if $v_i = v_j$ or v_j is a neighbor of v_i and not an existing neighbor of C. Hence v_j may be contributed to C by v_i if and only if there is no common element between C and $(N_G(v_j) \cup v_j)$. From Theorem 1, the number of coalitions in which v_j is marginally contributed by v_i is $2^{n-(deg(v_j)+1)}$.

Theorem 1. *The number of coalitions in which v_j is marginally contributed by v_i are $2^{n-(deg(v_j)+1)}$*

Proof. The total number of coalitions possible with k elements is 2^k. Node v_j contributes to C only when the coalition C does not contain any of the node in $N_G(v_j) \cup v_j$. Hence C cannot contain $deg(v_j) + 1$ elements and the number of such possible C is $2^{n-(deg(v_j)+1)}$

The total average marginal contribution by v_i when summed up through each v_j is given by,

$$B_i(v) = \sum_{v_j \in N_G(v_i) \cup v_i} \frac{2^{n-(deg(v_j)+1)}}{2^{n-1}} = \sum_{v_j \in N_G(v_i) \cup v_i} \frac{1}{2^{deg(v_j)}} \qquad (2)$$

The above exact closed form expression has time complexity as $\mathcal{O}(|V|+|E|)$ because computation of $deg(v_j)$ needs iteration over all edges.

The corresponding expression for Shapley value [13] is given by,

$$\sum_{v_j \in N_G(v_i) \cup v_i} \frac{1}{deg(v_j) + 1} \qquad (3)$$

While both the concepts (Banzhaf index and Shapley value) have indices increasing with the degree of the vertex, Banzhaf index heavily penalizes the vertex that has neighbors with large degrees more than that of Shapley value as it can be seen in the inverse exponential penalty versus the inverse linear penalty. Both the measures tend to gain power to the nodes by connecting to less connected ones.

Game 2: The second game, an extension of first game, considers the sphere of influence of C as themselves and the other nodes that are connected to the coalition by at least k direct edges. In other words, any node outside C should have at least k neighbors already influenced in C.

Given the game settings, the conditions in which v_j is marginally contributed to C by v_i is that,

1. $deg(v_j) \geq k$
2. $deg(v_j)$ should be exactly $k - 1$ in $(C \backslash i)$ for $v_j \neq v_i$
3. $deg(v_j)$ should be at most $k - 1$ in $(C \backslash i)$ for $v_j = v_i$.

A node v_j is contributed by v_i to coalition C only if v_j satisfies conditions 1 and 2 or conditions 1 and 3. The number of coalitions in which v_j is marginally contributed by v_i to (C) under conditions 1 and 2 is

$$\binom{deg(v_j) - 1}{k - 1} 2^{n-(deg(v_j)+1)}.$$

The total sum of marginal contributions achieved by v_i through v_j satisfying conditions 1 and 2 is given by,

$$\beta_{i_{1,2}}(v) = \sum_{v_j \in N_G(v_i)} \binom{deg(v_j) - 1}{k - 1} \frac{2^{n-(deg(v_j)+1)}}{2^{(n-1)}}. \tag{4}$$

The total sum of marginal contributions achieved by v_i through v_j satisfying conditions 1 and 3 is given by

$$\beta_{i_{1,3}}(v) = \sum_{r=0}^{k-1} \binom{deg(v_j)}{r} \frac{2^{n-(deg(v_j)+1)}}{2^{(n-1)}}. \tag{5}$$

The total marginal contribution of v_i to C is given by addition of above two marginal contributions,

$$\begin{aligned}
B_i(v) &= \beta_{i_{1,2}}(v) + \beta_{i_{1,3}}(v) \\
&= \sum_{v_j \in N_G(v_i)} \binom{deg(v_j) - 1}{k - 1} \frac{1}{2^{deg(v_j)}} \\
&+ \sum_{r=0}^{k-1} \binom{deg(v_i)}{r} \frac{1}{2^{deg(v_i)}}.
\end{aligned} \tag{6}$$

The value of k above can also be made node-specific. The time complexity of this algorithm is same as that of that Game 1. To a certain extent, this game mimics the Independent Cascade Model.

Game 3: Although the previous game approximates the real world networks better than the first game, it still works with unweighted graph. In real-world social networks, the edges are often weighted to indicate the influence of the source node on the destination node. This necessitates a generalization of the previous game for weighted graph.

In this game, the value of C depends on adjacent nodes that are connected to the coalition with weighted edges whose sum exceeds a given threshold w_{cutoff}. The sphere of influence is defined as those nodes whose sum of all edge weights from coalition to the node is greater than a given threshold W_{cutoff} for that node. Given a positive weighted network $G(V, E, W)$ and a value $W_{cutoff}(v_i)$ for every node $v_i \in G$, we define $W(v_j, C) = \sum_{v_i \in C} W(v_j, v_i)$ for every coalition C, where $W(v_i, v_j)$ is the weight of the edge between nodes v_i and v_j. Given this notation, we can formalize our game by the characteristic function:

$$v(C) = \begin{cases} 0, \text{if } C = \Phi \\ size(\{v_i : v_i \in C(or)W(v_i, C) \geq W_{cutoff}(v_i)\}), \\ \quad \text{otherwise} \end{cases}$$

A polynomial solution has been found for a special case when the weights are integers. For a particular node to calculate the marginal contribution, we have to find the number of coalitions which favor the above condition. This problem is a subset-sum selection problem where we can find the number of subsets in n

elements having a sum k in $\mathcal{O}(nk)$ time. Provided the weights are integers and the sum of weights is bounded the number of coalitions that can be contributed by v_j can be computed in polynomial time accurately.

For the general case, computing Banzhaf index involves determining whether the sum of weights on specific edges, adjacent to a random coalition, exceeds the threshold. This problem is as hard as computing the Banzhaf index in weighted majority games, which is NP-Complete [12]. Therefore, a computable closed form expression cannot be calculated in this case accurately. However, we derive an approximate formula for the Banzhaf index in this game.

We observe that node v_i marginally contributes node $v_j \in N_G(v_i)$ to the value of coalition C if and only if $v_j \notin C$ and $W_{cutoff}(v_j) - W(v_i, v_j) \leq W(v_j, C) < W_{cutoff}(v_j)$.

Let B_{v_i, v_j} denote the Bernoulli random variable representing this event. Let $N_G(v_j) = \{v_i, w_1, w_2, ... w_{deg_G(v_j)-1}\}$. Let the weights of edges between v_j and each of the nodes in $N_G(v_j)$ be $W_j = \{W(v_i, v_j), W_1, W_2, ... W_{deg_G(v_j)-1}\}$ in the same order. Let α_j be the sum of all the weights in W_j and β_j be the sum of the squares of all the weights in W_j. Let k_{ij} be the number of nodes of $N_G(v_j)$ that occur before v_i in C.

Let X_t^{ij} be the sum of a t-subset of $W_j \backslash \{W(v_i, v_j)\}$ drawn uniformly at random from the set of all such possible t-subsets. Let Y_m^{ij} be the event that $\{k_{ij} = m \wedge v_j \notin C\}$. Then:

$$E[B_{v_i, v_j}] = \sum_{m=0}^{deg(v_j)-1} P(Y_m^{ij}) * P\{X_m^{ij} \in [W_{cutoff}(v_j) - W(v_i, v_j), W_{cutoff}(v_j))\},$$

where $P(Y_m^{ij})$ is obtained from Eq. 4:

$$P(Y_m^{ij}) = \binom{deg_G(v_j) - 1}{m} \frac{1}{2^{deg_G(v_j)}}$$

Since X_m^{ij} is a complex function of the $deg_G(v_j) - 1$ numbers in $W_j \backslash \{W(v_i, v_j)\}$, calculating $P\{X_m^{ij} \in [W_{cutoff}(v_j) - W(v_i, v_j), W_{cutoff}(v_j))\}$ is a difficult problem. However, we can obtain analytical expressions for the mean $\mu(X_m^{ij})$ and variance $\sigma^2(X_m^{ij})$. These are given by:

$$\mu(X_m^{ij}) = \frac{m}{deg_G(v_j) - 1}(\alpha_j - W(v_i, v_j))$$

$$\sigma^2(X_m^{ij}) = \frac{m(deg_G(v_j) - 1 - m)}{(deg_G(v_j) - 1)(deg_G(v_j) - 2)}(\beta_j - W(v_i, v_j)^2 - \frac{(\alpha_j - W(v_i, v_j))^2}{deg_G(v_j) - 1})$$

Thus, we propose the approximation:

$$X_m^{ij} \sim \mathcal{N}(\mu(X_m^{ij}), \sigma^2(X_m^{ij})), \tag{7}$$

where $\mathcal{N}(\mu, \sigma^2)$ denotes the Gaussian random variable with mean μ and variance σ^2. A similar randomised approach has been tested by Fatima et al. [7]. With this approximation, we have:

$$Z_m{}^{ij} = P\{X_m^{ij} \in [W_{cutoff}(v_j) - W(v_i, v_j), W_{cutoff}(v_j))\} \tag{8}$$

given by

$$Z_m^{ij} \approx \frac{1}{2}\left(erf\left(\frac{W_{cutoff}(v_j) - \mu(X_m^{ij})}{\sqrt{2}\sigma(X_m^{ij})}\right) - erf\left(\frac{W_{cutoff}(v_j) - W(v_i, v_j) - \mu(X_m^{ij})}{\sqrt{2}\sigma(X_m^{ij})}\right)\right)$$

Therefore:

$$E[B_{v_i, v_j}] = \sum_{m=0}^{deg_G(v_j)-1} \binom{deg_G(v_j) - 1}{m} \frac{1}{2^{deg_G(v_j)}} * Z_m^{ij}$$

The above equations are true only for $v_j \neq v_i$. For $v_j = v_i$ we have from Eq. 5:

$$E[Bv_i, v_j] \approx \sum_{m=0}^{deg_G(v_i)} \binom{deg_G(v_i)}{m} \frac{1}{2^{deg_G(v_i)}} P\{\mathcal{N}(\mu(X_m^{ij}), \sigma^2(X_m^{ij})) < W_{cutoff}(v_i)\},$$

where,

$$\mu(X_m^{ii}) = \frac{m}{deg_G(v_i)}\alpha_i \tag{9}$$

$$\sigma^2(X_m^{ii}) = \frac{m(deg_G(v_i) - m)}{deg_G(v_i)(deg_G(v_i) - 1)}\left(\beta_i - \frac{\alpha_i^2}{deg_G(v_i)}\right) \tag{10}$$

Finally, the Banzhaf index of node v_i is given by $\sum_{v_j \in v_i \cup N_G(v_i)} E[B_{v_i, v_j}]$. Since the sum of degrees of all vertices is less than twice the number of edges, the time complexity of this approximation algorithm is $O\left(|V| + \sum_{v_j \in N_G(v_i)} deg_G(v_j)\right) \leq O(|V| + |V||E|) = O(|V||E|)$ time. The approximation of X_m^{ij} as a continuous random variable is good only when $deg_G(v_j)$ is large.

This formulation has applications in the analysis of information diffusion, adoption of innovations and viral marketing. For instance, many cascade models of such phenomena on weighted graphs have been proposed [8,10,14] that work by assuming that an agent will change state from "inactive" to "active" if and only if the sum of the weights to all active neighbors is at least equal to an agent-specific cutoff. We use the Banzhaf index based on this game for influence scoring in our experiments.

4   Experiments

The objective of our experiments is to compare the influence spread based on various diffusion model for the proposed Banzhaf index based approach. Since we propose Banzhaf index as an alternative to the already proposed Shapley value based approach [13], we first compare the performance of Banzhaf index over Shapley value for influence maximization. We then compare our algorithm against other centrality measures and the greedy algorithms in [11] and [4].

4.1   Experimental Setup

For all our experiments, we used the 2 collaboration networks from [4]. Each node in the network represents an author and the edges between 2 authors indicate their collaboration.

The first network, **NetHEPT** is the collaboration network from "High Energy Physics - Theory" between 1991 and 2003. This network consists of $15,233$ nodes and $58,891$ edges. The second network, **NetPHY** is the collaboration network from the entire list of papers from "Physics" and contains $37,154$ nodes and $231,584$ edges.

The 2 diffusion models used to evaluate the spreads are the *Independent Cascade (IC) Model* and the *Linear Threshold (LT) Model*. We approximate the Shapley value and Banzhaf value via Monte Carlo sampling to evaluate the indices as described in [1]. We ran $100,000$ Monte Carlo iterations and found that the Banzhaf and Shapley indices converge. We follow the experimental setup in [4] for the diffusion probabilities and compute the degree of spread that is achieved after selecting top k nodes in a network. Degree of spread is measured as the number of nodes to which the influence has been reached in a graph.

4.2   Comparison of Shapley Versus Banzhaf

In our first experiment, we evaluate the performance of the Banzhaf index against the Shapley value [13]. We formulate the influence maximization problem with Shapley value to pick the top k nodes similar to our Banzhaf Index formulation following [1]. We pick a set of $k = 50$ nodes that have top k Banzhaf indices from the collaboration network data and compare it with the nodes with top k Shapley value. Fig. 1 show the spread due to the top k nodes based on the Independent Cascade and the Linear Threshold [10] based diffusion on NetHEPT and NetPHY datasets respectively.

It can be seen from Fig. 1 that Banzhaf index performs well over Shapley in both Independent Cascade and Linear Threshold based diffusions. We have observed this difference consistently across several experiments across different game formulations as well. This suggests that for the influence maximization problem, the order in which a particular node comes into the coalition does not matter. Therefore, Shapley value based approach that considers the permutation of the players in a game, has a poorer performance over the Banzhaf index that only considers combination of the players in a considered coalition.

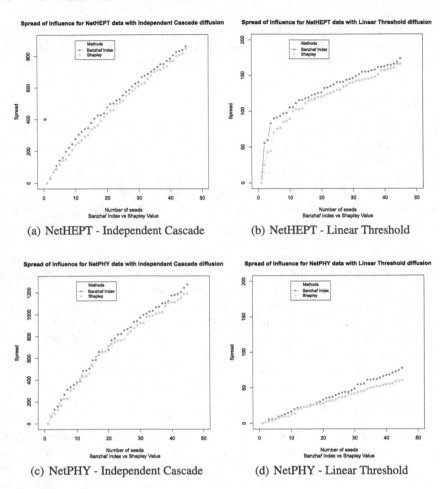

Fig. 1. Degree of spread for Banzhaf index ranked nodes over Shapley value ranked nodes on the NetHEPT & NetPHY collaboration dataset based on Independent Cascade and Linear Threshold diffusion

4.3   Banzhaf vs Other Approaches

Next, we compare the performance of Banzhaf index against other existing influence maximization approaches. We compare the performance against the CELF greedy approach in [11] and 3 greedy approaches in [3]: general, mixed and degree discount along with the Shapley based approach in [1,13]. Fig. 2 show the spread based on the Independent Cascade and Linear Threshold diffusion models for each of the approaches on the NetHEPT and NetPHY datasets.

It can be seen that the influence spreads of Banzhaf and Shapley are the best when Independent Cascade based diffusion is considered. However, when considered for a Linear Threshold based diffusion, the game theoretic centralities do not perform the best. This difference can be attributed to the nature of the game define that is more aligned to cascade diffusion that the linear threshold like diffusion. Despite this difference, the performance of Banzhaf index is comparable to other competing approaches in the NetHEPT dataset and is marginally worse in the NetPHY, where the difference for the corresponding Independent Cascade evaluate is even starker.

Degree centrality is 9.2% and 8.6% lower than best algorithms in both Independent Cascade and Linear Threshold models for NetHept and 39.2% lower in Independent Cascade model for NetPHY. The influence spreads of degree heuristic is very low compared to influence spreads of other approaches in the Indepen-

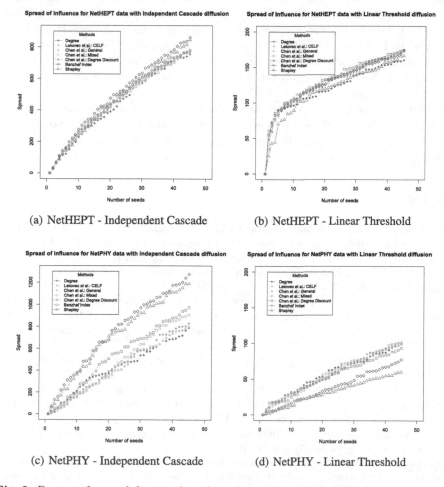

(a) NetHEPT - Independent Cascade   (b) NetHEPT - Linear Threshold

(c) NetPHY - Independent Cascade   (d) NetPHY - Linear Threshold

Fig. 2. Degree of spread for top k nodes selected on the NetPHY & NetHEPT collaboration dataset based on degree centrality, CELF algorithm (Leskovec et al.) [11], Maximum Influence Arborescence (Chen et al.: General, Mixed and Degree Discount [3,4]), Shapley value based approach [1,13] and the proposed Banzhaf index based approach

dent Cascade diffusion for both the collaboration graphs. However, it performs reasonably well in Linear Threshold based diffusion for lower seed sets. Similary, the approaches in CELF algorithm [11] and Maximum Influence Arborescence [3,4] perform comparable to Banzhaf index in Linear Threshold diffusions but are inferior in the Independent Cascade based diffusion.

5   Conclusion

We have explored how Banzhaf index can be used as a centrality measure to rank the nodes in order to evaluate the top k influential nodes for influence maximization problem. We have studied different games to model the influence spread in the social network. We use Monte Carlo approximations in a generic game to approximate the Banzhaf index in our proposed approach. We have compared the performance of Banzhaf index against other algorithms for influence maximization and showed that Banzhaf index performance is better or comparable in most of the cases.

Despite the comparable performance of game theoretic approaches, the computational cost of the solution concepts is high for generic games. Improving the computational costs to make it comparable with state-of-the-art approaches is a subject of further research.

References

1. Aadithya, K.V., Ravindran, B., Michalak, T.P., Jennings, N.R.: Efficient computation of the shapley value for centrality in networks. In: Saberi, A. (ed.) WINE 2010. LNCS, vol. 6484, pp. 1–13. Springer, Heidelberg (2010)
2. Banzhaf, J.F.: Weighted voting doesn't work: a mathematical analysis. Rutgers Law Review 19, 317–343 (1965)
3. Chen, W., Wang, C., Wang, Y.: Scalable influence maximization for prevalent viral marketing in large-scale social networks. In: Proceedings of the 16th ACM SIGKDD international conference on Knowledge discovery and data mining, pp. 1029–1038. ACM (2010)
4. Chen, W., Wang, Y., Yang, S.: Efficient influence maximization in social networks. KDD (2009)
5. Domingos, P., Richardson, M.: Mining the network value of customers. In: Proceedings of the seventh ACM SIGKDD international conference on Knowledge discovery and data mining, pp. 57–66. ACM (2001)
6. Everett, M.G., Borgatti, S.: The centrality of groups and classes. Journal of Mathematical Sociology 23(3), 181–201 (1999)
7. Fatima, S.S., Wooldridge, M., Jennings, N.: A randomized method for the shapley value for the voting game. In: Proceedings of the 11th International Joint Conference on Autonomous Agents and Multi-Agent Systems, pp. 955–92. AAMAS (2007)
8. Granovetter, M.: Threshold models of collective behavior. American Journal of Sociology 83(6), 1420–1443 (1978)
9. Grofman, B., Owen, G.: A game theoretic approach to measuring centrality in social networks. Social Networks 4, 213–224 (1982)

10. Kempe, D., Kleinberg, J., Tardos, É.: Maximizing the spread of influence through a social network. In: Proceedings of the ninth ACM SIGKDD international conference on Knowledge discovery and data mining, pp, 137–146. ACM (2003)
11. Leskovec, J., Krause, A., Guestrin, C., Faloutsos, C., VanBriesen, J., Glance, N.: Cost-effective outbreak detection in networks. In: Proceedings of the 13th ACM SIGKDD international conference on Knowledge discovery and data mining, pp. 420–429. ACM (2007)
12. Matsui, Y., Matsui, T.: Np-completeness for calculating power indices of weighted majority games. Theoretical Computer Science **263**(1–2), 305–310 (2001)
13. Narayanam, R., Narahari, Y.: A shapley value-based approach to discover influential nodes in social networks. IEEE Transactions on Automation Science and Engineering **99**, 1–18 (2010)
14. Young, H.P.: The diffusion of innovations in social networks. In: Proceedings volume in the Santa Fe Institute studies in the sciences of complexity Santa Fe Institute Studies on the Sciences of Complexity, vol. 3, pp. 267–282. Oxford University Press, US (2006)

Modeling Social Media Content with Word Vectors for Recommendation

Ying Ding[(✉)] and Jing Jiang

School of Information Systems, Singapore Management University,
Singapore, Singapore
{ying.ding.2011,jingjiang}@smu.edu.sg

Abstract. In social media, recommender systems are becoming more and more important. Different techniques have been designed for recommendations under various scenarios, but many of them do not use user-generated content, which potentially reflects users' opinions and interests. Although a few studies have tried to combine user-generated content with rating or adoption data, they mostly reply on lexical similarity to calculate textual similarity. However, in social media, a diverse range of words is used. This renders the traditional ways of calculating textual similarity ineffective. In this work, we apply vector representation of words to measure the semantic similarity between text. We design a model that seamlessly integrates word vectors into a joint model of user feedback and text content. Extensive experiments on datasets from various domains prove that our model is effective in both recommendation and topic discovery in social media.

1   Introduction

With the explosive usage of social media, there are many recommendation problems we face on different social media platforms. Recommendation in social media is an important way to improve services and attract more users. For example, on Twitter, followee recommendation can help people find users they are interested in, and thus good followee recommendation can provide users with more sources of interesting information and keep them using the platform. On many other platforms, recommendation is also important, like product recommendation in online review websites, event recommendation in online event websites, etc. To finish a recommendation task in social media well, it is important to understand users' online behaviour and accurately model their preferences and interests.

A traditional solution to these recommendation problems is collaborative filtering. There are generally two kinds of collaborative filtering methods: memory-based methods and model-based methods. Memory-based methods measure the similarity between users or items by directly using the adoption history. Then they perform recommendation based on the calculated similarities. Model-based methods use latent vectors to represent the interests of users and properties of items. They recommend items based on similarities between user latent vectors

© Springer International Publishing Switzerland 2015
T.-Y. Liu et al. (Eds.): SocInfo 2015, LNCS 9471, pp. 274–288, 2015.
DOI: 10.1007/978-3-319-27433-1_19

and item latent vectors. While collaborative filtering can get decent performance, it only uses adoption or rating data but misses other available information.

An important characteristic of social media platforms is that they allow users to contribute content in free-text form. For example, users can publish posts about their daily lives on Twitter, discuss social-political issues in online forums, label web pages with tags in online bookmark websites and write reviews of products in online review websites. These textual data reflects people's opinions, interests and preferences. Presumably it can help improve recommendation. Indeed there has been several recent studies trying to combine textual data with rating or adoption data for recommendation. Among these studies, many use textual information separately from their recommendation model. They first extract useful information from text and then embed such information in their recommendation models [4,30]. Some other work uses a unified, principled model to combine text with rating or adoption data [2,15,19].

A limitation of these recent studies is that their textual similarity is based on lexical similarity only. When two items' descriptions are semantically related but use different words, these models may not consider the two items to be similar. In social media, however, the vocabulary used is very diverse and two pieces of text can be semantically similar even with low lexical overlap, so semantic similarity is especially important when we analyze social media content. For example, in the Meetup dataset we use, which is about online interest groups and organized events, there is a group tagged with "Buddhism" and another group tagged with "vegetarian." If we only consider lexical similarity, these two groups may not be considered related based on the tags. However, we should probably recommend the second group to users who have joined the first group as many Buddhists are also vegetarians. The challenge is how to incorporate the consideration of semantic similarity based on textual descriptions into a traditional collaborative filtering framework in a principled way.

With the recent advances in learning word embeddings from large corpora, we can use vector representation of words to measure the semantic similarity between two pieces of text. Word embeddings are techniques that can project words into vectors carrying their semantic meanings [20,25]. In this paper, we propose a new recommendation model that makes use of word embeddings such that combining content and collaborative filtering becomes more effective. Our model can jointly model ratings, latent factors, topics and word embedding vectors simultaneously. With the help of vector representations of words, the model is able to learn cleaner topics, more accurate latent factors and provide better recommendations. Extensive experiments show that our model outperforms other methods on item recommendation and topic discovery. For example, for the Meetup data, our method can successfully recommend the "vegetarian" group to users who have joined the "Buddhism" group, and based on the ground truth, for such users our method indeed gives better performance than other baseline methods we consider.

In Section 2, we will briefly go through some recent related work. Our model will be introduced in Section 3 and our experiments are described in Section 4. We conclude this study and propose some potential research questions in Section 5.

2   Related Work

With the explosive growth of content in social media, recommender systems are becoming more and more important to users. New techniques adapted to different scenarios have been developed. On Twitter, various types of recommendation has been studied [14], like followee recommendation, tweet recommendation, hashtag recommendation, etc. User network in social media has also been used to improve product recommendation [8,18]. On some platforms, users can form a group for some purpose. Recommendation related to groups in social media has also been studied in some recent work. It includes work on recommendation for groups [1,7], which tries to recommend items to a group of users, group recommendation [33], which recommends groups to users to join, etc.

Traditional recommendation models mainly focus on users' adoption or rating histories [12]. In social media, the user-generated content provides us with useful information that can reflects users' opinions, interests and preferences. It is valuable to model this information to improve recommendation outputs [26]. To this end, people have tried to extract features from text as item representations [13,24]. While these studies use text separately from recommendation models, more work is trying to jointly model text and recommendation in one principled models [2,15,19]. However, these joint models rely on traditional method of modeling text, which uses only lexical similarity to calculate textual similarity.

Word embedding is a recently proposed technique inspired by advances in deep neural networks [20,25]. Based on the learning from large corpora, it can represent words with numerical vectors that carry their semantic meanings. The similarity between vectors of words that are semantically or syntactically similar will be high. It has been used in different applications such as information retrieval [5] and text summarization [11]. Deep learning itself has also been applied to model text in recommendation problems [31]. While this work applies a deep learning model—stacked autoencoder—to learn text representations. It does not make use of word embeddings pre-trained on a large external corpora, which can presumably better represent the semantic meanings of words than representations trained only from texts related to the recommendation problem itself.

3   Method

In this section, we formally formulate our problem and present our proposed model. Based on our model's properties, we denote it by **C**ollaborative filtering with word **E**mbedding-based **T**opic models (CET).

3.1   Problem Formulation and Notation

Suppose we have a collection of N_I items $\mathcal{I} = \{i_1, i_2, \cdots, i_{N_I}\}$ and a collection of N_U users $\mathcal{U} = \{u_1, u_2, \cdots, u_{N_U}\}$. We also observe a collection of ratings[1]

---

[1] For convenience, we assume we have numerical rating data, but the model can be easily generalized for binary adoption data.

$\mathcal{R} = \{r_{ui}\}$ where r_{ui} is the rating of item i by user u. For each item, there is an associated document d_i, which is a sequence of words. This document can be from different sources of user-generated content. For example, in online review websites, we can use the reviews of a product as the document associated with the product. For items that have user-assigned tags, we can use the set of tags as the associated document for an item. The set of all words appearing in our data comprises the vocabulary $\mathcal{V}$ and for each word w of this vocabulary, we assume that we have a pre-trained vector v_w of dimension K, which can be learned by word embedding models [20, 25]. Our task of this work is to recommend items to users according to both their rating histories and the textual data generated by users in social media.

3.2 Collaborative Filtering with Word Embedding-Based Topic Models

Our model is based on matrix factorization, topic modeling and word embedding vectors. On the rating part, we apply matrix factorization as the generative process. On the text part, we design a generative process based on Latent Dirichlet Allocation (LDA) [3]. We also assume that the topic distribution of an item is linked to the item's latent vector used in matrix factorization, which is an idea previously explored in [15, 19, 29]. By doing this, we build a single unified and principled model that combines text and ratings. Similar to [29], we assume that item factors are derived from the corresponding topic distributions instead of setting them to be identical. This renders our model more flexible in modeling latent factors. Different from standard LDA, which treats each word as a single discrete symbol, we use the vector representations of them instead. We still assume that there is a multinomial topic distribution for each document. But for each topic, we assume there is a multivariate Gaussian distribution, which is used to generate word vectors. There are two parameters for each topic t, which are the mean vector μ_t and co-variance matrix Σ_t. To generate a word in a document, we first need to sample a topic according to the document-topic distribution, and then sample a vector from the Gaussian distribution of the sampled topic. The generative process of our model is shown below and we list the used notation in Table 1.

- For each user, sample a bias $b_u \sim \mathcal{N}(0, \sigma_U)$ and a latent vector $p_u \sim \mathcal{N}(0, \Gamma_U)$.
- For each item i, sample a topic distribution $\theta_i \sim \mathrm{Dir}(\alpha)$ for text. Sample a latent vector $q_i \sim \mathcal{N}(\theta_i, \Gamma_I)$ and a bias $b_i \sim \mathcal{N}(0, \sigma_I)$. For each word w in the associated text:
  - Sample a topic $z \sim \mathrm{Multi}(\theta_i)$.
  - Sample a word embedding vector $v_w \sim \mathcal{N}(\mu_z, \Sigma_z)$.
- For rating of item i by user u, sample a numerical value $r_{ui} \sim \mathcal{N}(b_u + b_i + p_u^\mathsf{T} q_i, \sigma_R)$.

With this model, we can find the underlying topics of words based on their semantic meanings. This can help us recommend items to users even if the used text is very diverse, which is common in social media. For example, although "fitness" and "exercise" are two different words, in pre-trained word embeddings, their distance is smaller than a random pair of words, so they are more likely to be generated by the same multivariate Gaussian than from different Gaussian distributions. In our model, items whose descriptions contain "fitness" and items whose descriptions contain "exercise" will have similar topic distributions and so are their latent factors. Then for a user who has adopted items with the word "fitness," our model is more likely to recommend items with the word "exercise" to him. Unfortunately, traditional models may not achieve this as they do not consider the semantic meaning of words.

It is worth pointing out that in our CET model, the modified LDA component, which generates word embedding vectors from a mixture of multivariate Gaussian distributions, is almost the same as in a recent work by Das et. al. [6]. However, we developed our model independently and our focus is to apply the model for the purpose of recommendation. Note also that although here we assume the text is associated with each item, our model is not restricted to this setting. If there is text associated with users, our model can also be directly applied by switching the generative process of items with that of users.

Table 1. Notation of our model.

Variable	Description
r_{ui}	Rating of item i by user u
v_w, v_{wi}	Vector of word w learned by word embeddings and corresponding value at the ith dimension
α	The hyper-parameters for the Dirichlet distribution
$\sigma_U, \sigma_I, \sigma_R$	The standard deviation for univariate Gaussian distributions
Γ_U, Γ_I	The covariance matrices for multivariate Gaussian distribution
b_u, b_i	The rating bias of user u and rating bias of item i
p_u, q_i	The latent factor of user u and latent factor of item i
θ_i	The topic distribution of item i
μ_t, Σ_t	The mean and covariance matrix for the multivariate Gaussian distribution of topic t
$\text{Dir}(\alpha)$	A Dirichlet distribution with hyper-parameter α
$\mathcal{N}(\mu, \Sigma)$	A Gaussian distribution with mean μ and covariance matrix Σ
$\text{Multi}(\theta)$	A discrete distribution with θ as parameter

3.3 Parameter Estimation

When applying our model to a dataset, text, ratings and word vectors are all given, and we need to find the hidden parameters that can maximize the posterior likelihood. So, our goal of training is to learn the parameters that can maximize the following probability:

$$p(\boldsymbol{P}, \boldsymbol{Q}, \boldsymbol{B}_U, \boldsymbol{B}_I, \boldsymbol{\theta}, \boldsymbol{\mu}, \boldsymbol{\Sigma} | \boldsymbol{W}, \boldsymbol{R}). \tag{1}$$

Here P and Q refer to all latent vectors for items and users, B_U and B_I refer to bias terms of users and items. W refers to all the words we observe and R refers to all the ratings. μ and Σ represent all means and covariance matrices of the Gaussian distributions of all topics. The hyperparameters are omitted in the formula. As there is no closed form solution for our problem, we use Gibbs-EM algorithm [28] for parameter estimation. For each iteration, we alternate between Gibbs sampling and gradient descent. More specifically, in each iteration, we first perform Gibbs sampling based on parameters learned in the last iteration, which will be fixed in the sampling stage. Then based on the sampled hidden variables, we optimize our objective function using gradient descent.

E-step: We fix the parameters θ, μ and Σ and collect samples of the hidden variables Z to approximate the distribution $P(Z|W, \theta, \mu, \Sigma)$. The distribution of the hidden labels for Gibbs Sampling is:

$$P(z_{ij} = t) \propto \theta_{it} \cdot \mathcal{N}(v_{w_{ij}}|\mu_t, \Sigma_t). \tag{2}$$

Here, z_{ij} is the topic assignment of the word at the jth position of text of item i and w_{ij} denotes the corresponding word.

M-step: With the collected samples of Z, we need to find the values of P, Q, B_U, B_I, θ, μ and Σ that maximize the following objective function:

$$\mathcal{L} = \sum_{Z \in \mathcal{S}} \log P(Z, W, R, P, Q, B_U, B_I, \theta, \mu, \Sigma | \alpha, \sigma_U, \sigma_I, \sigma_R, \Gamma_U, \Gamma_I), \tag{3}$$

where $\mathcal{S}$ is the set of samples collected in the E-step.

It is noted that θ for any document is constrained to be a multinomial distribution. To transform this constrained optimization problem to an unconstraint one, we use a set of auxiliary variables λ_{it} to replace θ_{it} with $\frac{\exp(\lambda_{it})}{\sum_{t'} \exp(\lambda_{it'})}$. We use gradient descent to find the optimal value of P, Q, B_U, B_I, θ. μ and Σ can be updated using the following equations:

$$\mu_{ti} = \frac{1}{N_{t.}} \sum_{w=1}^{V} N_{tw} v_{wi} \qquad \Sigma_t^{ii} = \frac{1}{N_{t.}} \sum_{w=1}^{V} N_{tw} (v_{wi} - \mu_{ti})^2. \tag{4}$$

where N_{tv} is the number of times word type w is assigned to topic t, $N_{t.}$ is the number of times all word types are assigned to topic t and Σ_t^{ii} is the element at row i, column i of matrix Σ_t.

After all parameters in the model are learned, we use $\hat{r}_{ui} = p_u^{\mathsf{T}} q_i + b_u + b_i$ to predict the rating of item i by user u. In our implementation, we perform 600 runs of Gibbs EM. Because Gibbs sampling is time consuming, in each run we only perform one iteration of Gibbs sampling and collect that one sample. We then have 60 iterations of gradient descent. The gradient descent algorithm we use is L-BFGS, which is efficient for large scale data set [22]. We downloaded word vectors from the homepage of word2vector[2] and use them as our word embedding vectors.

---

[2] https://code.google.com/p/word2vec/

4   Experiment

Our model can be applied to many recommendation tasks on social media where user-generated content plays an important role. To test it, we pick two representative social media platforms for experiments. The first is Meetup[3], an event-based online social network. Meetup allows users to create interest groups and organize events. A commonly studied recommendation task on Meetup is how to recommend an interest group to a user. The second is Amazon's product review platform. We use user-generated product reviews as additional textual information to help product recommendation. The content in these two platforms are also representative. In the online social network website we use, content contains tags given by users. Because there is not a controlled vocabulary of tags and the number of tags assigned to each item can be small, the data is very sparse. In online review website, users can write their reviews in free form. So the content is relatively rich but the diversity is still high.

For each dataset, we use 10% of the data as the development set and another 10% of the data as the testing set. The remaining 80% of the data is used for training. We tune all models according to the development set and test them on the testing set. As our model does not update word embedding vectors. Those words with no pre-trained vectors are of no use to CET. So we just delete them all. The average percentage of words with embedding vectors is 54.7% over all datasets. To show the effectiveness of our model, we choose several appropriate state-of-the-art recommendation techniques for comparison. Besides showing their performance, we also do statistical significance test of results using Wilcoxon signed-rank test.

4.1   Group Recommendation in Meetup

The first experiment is conducted on a Meetup dataset [16]. Meetup is an online event-based social network. In this website, users can build or join groups and each group can organize and publish offline events for people to participate in. Users and groups can use tags to label themselves to show their interests. The text we use is tags associated with groups. The dataset we use is a random sample from the data used in [16]. There are 2225 users, 6950 groups, 8015 user-group membership pairs and each group has 7.06 tags on average. This data is very sparse as only 0.04% of its user-group matrix entries contain values. For this dataset, we only have the information about which groups a user has joined. For the groups the user has not joined, there can be different reasons. The user may not like the group or the user may be unaware of the group at all. This type of negative examples is called implicit feedback. Because of this, we choose two models that work on implicit feedback as our baselines as follows.

---

[3] http://www.meetup.com

CTR: Collaborative Topic Regression [29] is a model designed for scientific article recommendation with implicit feedback. It assumes that each article's latent factor is a deviation from its topic distribution.

OCF: One-class Collaborative Filtering [23] extends traditional matrix factorization to model implicit feedback. In our experiments, we use the re-weighting technique proposed in this paper.

Quantitative Study. We use MPR (Mean Percentage Ranking) [9] as the evaluation metric. For each user-group pair in our testing data, we randomly select 1,000 "negative" groups and mix them with the "positive" group. We rank all these 1,001 groups based the predicted rating from the target user. Then, we calculate MPR as follows:

$$\text{MPR} = \frac{1}{N} \sum_{i=1}^{N} \frac{R_i}{M}, \tag{5}$$

where R_i is the position of the adopted group in testing pair i and M is the number of ranked groups, which is 1001 in our experiment, and N is the number of pairs in testing data. For a testing instance i, Percentage Ranking (PR) is defined as $\frac{R_i}{M}$, which will be used in the next subsection.

The MPR for CET, CTR and OCF are shown in Figure 1. OCF performs the worst for this dataset. This is because the dataset is too sparse and it is very hard to learn useful item latent vectors purely based on user membership information. By utilizing tag information, CTR can obtain a much better MPR value. CET can even outperform CTR by using word embedding vectors as it utilize the semantic meaning of words. Statistical test shows that CET's performance is significantly better than CTR and OCF at 5% level. It proves that compared with the baselines, our model can learn latent factors much more effectively.

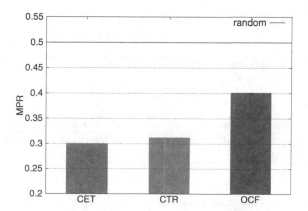

Fig. 1. Mean Average Ranking (MAR) for CET, CTR and OCF on Meetup data.

Qualitative Study. To qualitatively understand how our model outperforms CTR, we display some sample users in Table 2. The representative tags of groups they have joined as indicated in the training data, the tags of "positive" groups in the testing data (i.e. groups that should be recommended) and the corresponding Percentage Ranking (PR) by CET and CTR are also shown together. For user 1408, the tags of groups he has joined tell us that he is interested in exercises and outdoor activities. A recommendation method should rank groups related to this topic higher than others. The group with tags "aerobics" and "running" shows up in our test instances. Our CET model ranks it higher than 98% of the negative examples while CTR only ranks it higher than 42% of the negative examples. The reason is that tags used in social media is very diverse, and groups with similar properties may share no words at all. Traditional way of using lexical similarity to compute textual similarity cannot work very well in this case. So it becomes hard for them to recommend groups based on tags. However, by leveraging words' vector representation, CET can tackle this problem better. The second and third cases also prove this. It is interesting that CET is also able to recommend groups that is conceptually related but have different properties. In the fourth row, we can see that user 399 is interested in Buddhism, and he has also joined a group about vegetarian, which appears in our test dataset. Buddhism is about religion while vegetarian is about food preference. It is impossible to connect them based only on lexical similarity. However, we know that many Buddhists are also vegetarians. So these two words are semantically related and it is reasonable to recommend a vegetarian group to a person interested in Buddhism. While CTR fails to do this, our CET model successfully recommends this group based on using semantic similarity between words.

Table 2. Sampled users, the representative tags of groups they join, the tags of group we should recommend and the percentage ranking of CET and CTR.

User ID	Tags of groups they have joined	Tags of groups we should recommend	PR by CET	PR by CTR
1408	fitnees friends music meditation hiking yoga	aerobics running	0.020	0.577
1247	cooking nutrition movies fitness	volleyball	0.135	0.663
835	photo weightloss fitness	theater art museum	0.001	0.528
399	photoshop alternative mediation buddhism	vegetarian nutrition	0.042	0.563

We also show the top words of the topics learned by CET and CTR in Table 3. As we can see, topics learned by CET look much neater. We can find some noisy words in topics learned by CTR. For example, dance and Japanese are in the same topic and hiking and dogs are also in the same topic. Previous work has shown that LDA, which is used to model text in CTR, is not able to learn topics well when documents are very short [32]. The average number of tags for each meetup group is only 7, so it is really hard for LDA to learn good topics. However, by using the embedding vectors, which carry semantic meanings of words, CET can cluster word much better and learn neater and more meaningful topics.

Table 3. Top words of sampled topics learned from Meetup data by CET and CTR.

CET	dance dancing salsa tango salsa-dancing latin-dance flamenco dance-lessons ballet latin-dancing
	hiking excursionismo-hiking kayaking camping outdoors snowshoeing skiing backpacking walkers paddling
	dogs puppy cats pets chihuahua pug yorkie sheltie dachshund dog-lovers
	language culture spanish-culture english french-culture language-and-culture languages japanese-language german-culture european-culture
	movies films movie film movie-nights arthouse movies-dinner movies-and-dinner cinema-and-films dinner-and-a-movie
CTR	dance wellness group-fitness-training japanese dance-lessons cloud-computing english-conversation python democrat korean
	hiking outdoor-recreation startup-ventures javascript creative-writing new-york-city dogs singles-who-love-to-travel activities css
	business-networking weightloss stress foodie crosscultural socializing-dogs dog-lovers london liberty anime
	social language theater bike beer backpackers business-and-social-networking museum rockclimbing men
	fitness movies movie-nights exercise-nutrition business film snowboard cinema-and-films movies-dinner mountain-biking

4.2 Product Recommendation in Online Review Website

For the second experiment, we use data from Amazon, which is composed of 9 datasets used in [19]. We have users' explicit ratings at scale 1-5 of items and their reviews. Similar to [19], we use the aggregated reviews of an item as the associated text of it. Users and items with fewer than 3 reviews are filtered out. Statistics of this type of dataset are shown in Table 4.

We choose two state-of-the-art techniques that model both explicit ratings and text information as our baselines.

HFT: Hidden Factors as Hidden Topics [19] is a model that directly ties each dimension of hidden factors in matrix factorization of ratings to one hidden topic in review text by using an exponential transformation function.

Table 4. Dataset statistics, which show number of users, number of items, number of reviews, total number of word types, average number of tokens per review in each column.

dataset	#users	#items	#reviews	#word types	#tokens/review
office	691	313	4034	12652	46.33
patio	748	344	6814	8691	32.7
software	314	235	2468	14317	83.03
beauty	4281	1817	33290	22208	33.91
sports	8039	5545	91294	37645	30.23
tools	4935	3346	38998	68390	55.14
toys	3479	2776	25951	51224	50.07
games	9919	6124	88684	301829	115.83
health	4529	2460	35123	39674	40.36

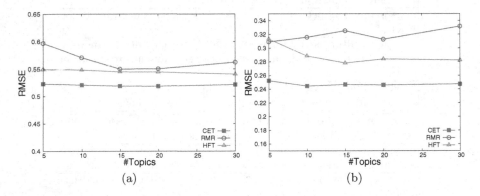

Fig. 2. RMSE over topic numbers on two datasets. The left one is *office* dataset, the right one is *patio* dataset.

RMR: Ratings Meet Reviews [15] is a model similar to HFT except the way they link ratings with reviews. It assumes that each user has one Gaussian rating distribution on each topic, which characterizes how the user is interested in this topic.

Table 5. RMSE of CET, HFT and RMR. For each dataset, the best result is in bold font. † indicates that CET significantly outperforms RMR at 1% level. ‡ indicates that CET significantly outperforms both RMR and HFT at 1% level.

	office	patio	software	beauty	sports	tools	toys	video	health
CET	**0.521**†	**0.252**‡	**0.725**‡	**0.371**‡	**0.215**‡	**0.746**‡	**0.967**‡	1.183	**0.483**‡
RMR	0.597	0.309	0.767	0.484	0.351	0.802	1.013	**1.138**	0.595
HFT	0.552	0.283	0.776	0.444	0.262	0.813	1.146	1.172	0.548

Table 6. Top words of sampled topics learned by CET, HFT and RMR.

Topics learned by CET					
work	tape	paper	product	pages	daughter
office	file	binder	products	templates	old
job	files	printed	price	interface	home
desk	tapes	printing	buy	page	son
working	folder	printer	purchase	multiview	mother
phone	folders	print	buying	text	father
cabinet	video	binders	brand	functionality	niece
offices	taped	pencil	purchasing	webpage	grandmother
works	filing	sheets	brands	app	granddaughter
telephone	clips	ink	pricing	template	dad
Topics learned by HFT					
desk	folders	binder	pen	cards	black
keyboard	tabs	binders	markers	paper	color
mouse	folder	rings	fine	card	folders
pad	file	pages	pens	print	look
hp12c	reinforced	open	colors	business	good
feet	tab	one	ink	avery	colors
rest	manila	pockets	write	printer	great
wrist	use	ring	sharpie	printed	nice
holder	smead	front	use	quality	side
platform	box	plastic	highlighters	make	well
Topic learned by RMR					
desk	folders	binder	markers	cards	folders
keyboard	files	binders	colors	paper	black
mouse	hanging	rings	pens	card	color
pad	using	open	ink	avery	look
rest	still	pockets	pen	print	file
wrist	drawer	pages	sharpie	business	good
holder	pendaflex	ring	highlighters	printer	great
feet	bottom	front	great	printed	nice
platform	product	avery	write	make	one
tray	capacity	cover	marker	professional	colors

Quantitative Study. For Amazon review dataset, we use RMSE (Root Mean Squared Error) [19] as the evaluation metric, which is defined as:

$$\text{RMSE} = \sqrt{\frac{1}{N} \sum_{i=1}^{N} (\hat{r}_i - r_i)^2}, \tag{6}$$

where r_i is the true rating for the ith testing instance and $\hat{r}_i$ is the prediction. The results over all 9 datasets are shown in Table 5. We can see that CET significantly outperforms RMR and HFT on most datasets. It means our model can effectively learn users' interests by modeling rating and text information.

To have a closer look at how the performance of all three models change over different number of topics, we pick two datasets *office* and *patio* and show the results in Figure 2. We can see that CET outperforms both baselines when using different numbers of topics. Its performance is also more stable over topic numbers compared with the other two.

Qualitative Study. In this subsection, we show the top words of some sampled topics learned by CET, RMR and HFT in Table 6. All topics are from the *office* domain and the number of topics is set to 30 for all models. As we can see, CET can learn meaningful topics like office, file, paper, purchase and so on as well as HFT and RMR. By taking a closer look at the top words of these topics, we can find that the top words of CET are cleaner. Most of the top words are about the same topic and there is less noise in these words. However, there exist some noisy words in the top word list of HFT and RMR, many of them are general words like "one", "use", "well", etc. By using word vector to represent words, words can be clustered better compared with models like HFT and RMR. It is interesting that CET also discover a topic, family members, which cannot be learned by RMR and HFT. This may be a topic worth mining for recommendation as it probably reflects who the product is bought for. However, CET is not perfect and it fails to discover the topic about pens.

5   Conclusions and Future Work

In this work, we have proposed a recommendation model for social media based on users' ratings, text and word embedding vectors. Compared with existing work, our model is able to find the similarity between two pieces of text based on their semantic similarity rather than simply lexical similarity. This makes it more effective for recommendation problems in social media. Extensive experiments on two recommendation problems in social media show that this model can outperform state-of-the-art methods. A closer look at topics also tells us that by using the semantic meanings reflected in embedding vectors, our model can learn cleaner topics. When documents have as few as 7 words on average, our model can still learn meaningful topics and get good recommendation results.

We have shown that using vectors learned from neural network based model can improve both recommendation and topic discovery in social media. It would be interesting to try vectors learned by other word embedding models such as topical word embedding [17], multi-prototype word embedding [21] and so on. Besides, it is a promising direction to model text in other ways beyond bag of words. Models which take order into consideration, like Recursive Neural Network [27] and Convolutional Neural Networks [10] are worth trying.

References

1. Baltrunas, L., Makcinskas, T., Ricci, F.: Group recommendations with rank aggregation and collaborative filtering. In: Proceedings of the Fourth ACM Conference on Recommender Systems, pp. 119–126 (2010)

2. Bao, Y., Fang, H., Zhang, J.: TopicMF: Simultaneously exploiting ratings and reviews for recommendation. In: Proceedings of the Twenty-Eighth AAAI Conference on Artificial Intelligence, pp. 2–8 (2014)

3. Blei, D.M., Ng, A.Y., Jordan, M.I.: Latent dirichlet allocation. Journal of machine learning research **3**, 993–1022 (2003)

4. Chen, G., Chen, L.: Recommendation based on contextual opinions. In: Dimitrova, V., Kuflik, T., Chin, D., Ricci, F., Dolog, P., Houben, G.-J. (eds.) UMAP 2014. LNCS, vol. 8538, pp. 61–73. Springer, Heidelberg (2014)

5. Clinchant, S., Perronnin, F.: Aggregating continuous word embeddings for information retrieval. In: Proceedings of the Workshop on Continuous Vector Space Models and their Compositionality, pp. 100–109 (2013)

6. Das, R., Zaheer, M., Dyer, C.: Gaussian LDA for topic models with word embeddings. In: Proceedings of the 53nd Annual Meeting of the Association for Computational Linguistics (2015)

7. Gorla, J., Lathia, N., Robertson, S., Wang, J.: Probabilistic group recommendation via information matching. In: Proceedings of the 22nd International Conference on World Wide Web, pp. 495–504 (2013)

8. Guo, G., Zhang, J., Yorke-Smith, N.: Trustsvd: collaborative filtering with both the explicit and implicit influence of user trust and of item ratings. In: Proceedings of the Twenty-Ninth AAAI Conference on Artificial Intelligence, pp. 123–129 (2015)

9. Hu, Y., Koren, Y., Volinsky, C.: Collaborative filtering for implicit feedback datasets. In: Proceedings of the 2008 Eighth IEEE International Conference on Data Mining, pp. 263–272 (2008)

10. Kalchbrenner, N., Grefenstette, E., Blunsom, P.: A convolutional neural network for modelling sentences. In: Proceedings of the 52nd Annual Meeting of the Association for Computational Linguistics, pp. 1749–1751 (2014)

11. KgebLck, M., Mogren, O., Tahmasebi, N., Dubhashi, D.: Extractive summarization using continuous vector space models. In: Proceedings of the 2nd Workshop on Continuous Vector Space Models and their Compositionality, pp. 31–39 (2014)

12. Koren, Y., Bell, R., Volinsky, C.: Matrix factorization techniques for recommender systems. Computer **42**, 30–37 (2009)

13. Krestel, R., Fankhauser, P., Nejdl, W.: Latent dirichlet allocation for tag recommendation. In: Proceedings of the Third ACM Conference on Recommender Systems, pp. 61–68 (2009)

14. Kywe, S.M., Lim, E.-P., Zhu, F.: A survey of recommender systems in twitter. In: Aberer, K., Flache, A., Jager, W., Liu, L., Tang, J., Guéret, C. (eds.) SocInfo 2012. LNCS, vol. 7710, pp. 420–433. Springer, Heidelberg (2012)

15. Ling, G., Lyu, M.R., King, I.: Ratings meet reviews, a combined approach to recommend. In: Proceedings of the 8th ACM Conference on Recommender Systems, pp. 105–112 (2014)

16. Liu, X., He, Q., Tian, Y., Lee, W.C., McPherson, J., Han, J.: Event-based social networks: linking the online and offline social worlds. In: Proceedings of the 18th ACM SIGKDD International Conference on Knowledge Discovery and Data Mining, pp. 1032–1040 (2012)

17. Liu, Y., Liu, Z., Chua, T., Sun, M.: Topical word embeddings. In: Proceedings of the Twenty-Ninth AAAI Conference on Artificial Intelligence, pp. 2418–2424 (2015)

18. Ma, H., Yang, H., Lyu, M.R., King, I.: Sorec: social recommendation using probabilistic matrix factorization. In: Proceedings of the 17th ACM Conference on Information and Knowledge Management, pp. 931–940 (2008)

19. McAuley, J., Leskovec, J.: Hidden factors and hidden topics: Understanding rating dimensions with review text. In: Proceedings of the 7th ACM Conference on Recommender Systems, pp. 165–172 (2013)

20. Mikolov, T., Sutskever, I., Chen, K., Corrado, G.S., Dean, J.: Distributed representations of words and phrases and their compositionality. Advances in Neural Information Processing Systems **26**, 3111–3119 (2013)

21. Neelakantan, A., Shankar, J., Passos, A., McCallum, A.: Efficient non-parametric estimation of multiple embeddings per word in vector space. In: Proceedings of the 2014 Conference on Empirical Methods in Natural Language Processing, pp. 1059–1069 (2014)

22. Nocedal, J.: Updating QUASI-Newton matrices with limited storage. Mathematics of Computation **35**(151), 773–782 (1980)

23. Pan, R., Zhou, Y., Cao, B., Liu, N.N., Lukose, R., Scholz, M., Yang, Q.: One-class collaborative filtering. In: Proceedings of the 2008 Eighth IEEE International Conference on Data Mining, pp. 502–511 (2008)

24. Pennacchiotti, M., Gurumurthy, S.: Investigating topic models for social media user recommendation. In: Proceedings of the 20th International Conference Companion on World Wide Web, pp. 101–102 (2011)

25. Pennington, J., Socher, R., Manning, C.: Glove: Global Vectors for Word Representation. In: Proceedings of the 2014 Conference on Empirical Methods in Natural Language Processing (EMNLP), pp. 1532–1543 (2014)

26. Shi, Y., Larson, M., Hanjalic, A.: Collaborative filtering beyond the user-item matrix: A survey of the state of the art and future challenges. ACM Computation Survey **47**(1), 3:1–3:45 (2014)

27. Socher, R., Perelygin, A., Wu, J., Chuang, J., Manning, C.D., Ng, A.Y., Potts, C.: Recursive deep models for semantic compositionality over a sentiment treebank. In: Proceedings of the 2013 Conference on Empirical Methods in Natural Language Processing, pp. 1631–1642 (2013)

28. Wallach, H.M.: Topic modeling: Beyond bag-of-words. In: Proceedings of the 23rd International Conference on Machine Learning, pp. 977–984 (2006)

29. Wang, C., Blei, D.M.: Collaborative topic modeling for recommending scientific articles. In: Proceedings of the 17th ACM SIGKDD International Conference on Knowledge Discovery and Data Mining, pp. 448–456 (2011)

30. Wang, F., Pan, W., Chen, L.: Recommendation for new users with partial preferences by integrating product reviews with static specifications. In: Carberry, S., Weibelzahl, S., Micarelli, A., Semeraro, G. (eds.) UMAP 2013. LNCS, vol. 7899, pp. 281–288. Springer, Heidelberg (2013)

31. Wang, H., Wang, N., Yeung, D.: Collaborative deep learning for recommender systems. In: Proceedings of the 21th ACM SIGKDD International Conference on Knowledge Discovery and Data Mining, pp. 1235–1244 (2015)

32. Yan, X., Guo, J., Lan, Y., Cheng, X.: A biterm topic model for short texts. In: Proceedings of the 22nd International Conference on World Wide Web, pp. 1445–1456 (2013)

33. Zhang, W., Wang, J., Feng, W.: Combining latent factor model with location features for event-based group recommendation. In: Proceedings of the 19th ACM SIGKDD International Conference on Knowledge Discovery and Data Mining, pp. 910–918 (2013)

Author Index

Printed in the United States
By Bookmasters